INTERNATIONAL SERIES
OF
MONOGRAPHS ON PHYSICS

SERIES EDITORS

INTERNATIONAL SERIES OF MONOGRAPHS ON PHYSICS

Problems of Condensed Matter Physics

Quantum coherence phenomena
in electron-hole and coupled
matter-light systems

Edited by

ALEXEI L. IVANOV

School of Physics, Cardiff University

SERGEI G. TIKHODEEV

*A. M. Prokhorov General Physics Institute,
Russian Academy of Sciences, Moscow*

OXFORD

UNIVERSITY PRESS

OXFORD
UNIVERSITY PRESS

Great Clarendon Street, Oxford OX2 6DP

Oxford University Press is a department of the University of Oxford.
It furthers the University's objective of excellence in research, scholarship,
and education by publishing worldwide in

Oxford New York

Auckland Cape Town Dar es Salaam Hong Kong Karachi
Kuala Lumpur Madrid Melbourne Mexico City Nairobi
New Delhi Shanghai Taipei Toronto

With offices in

Argentina Austria Brazil Chile Czech Republic France Greece
Guatemala Hungary Italy Japan Poland Portugal Singapore
South Korea Switzerland Thailand Turkey Ukraine Vietnam

Oxford is a registered trade mark of Oxford University Press
in the UK and in certain other countries

Published in the United States
by Oxford University Press Inc., New York

British Library Cataloguing in Publication Data

Data available

Library of Congress Cataloging in Publication Data

Data available

Printed in Great Britain
on acid-free paper by
Biddles Ltd. www.biddles.co.uk

ISBN 978-0-19-923887-3 (Hbk)

1 3 5 7 9 10 8 6 4 2

Dedicated to Professor Leonid V. Keldysh on his 75th anniversary

CONTENTS

PREFACE

We are very pleased, as former research students of Professor Leonid V. Keldysh, to introduce this Festschrift dedicated to him on his 75th birthday. The brilliant contributions of Professor Keldysh, which include the Franz–Keldysh effect, an electron–hole liquid, the nonequilibrium (Keldysh) diagram technique, Bose–Einstein condensation (of excitons) and a "metal–dielectric" transition, acoustically-induced superlattices, multiphoton transitions and impact ionization in solids, etc., in many aspects influenced and formed the paradigm of modern condensed matter physics. It is also a great responsibility for us to edit the issue: many famous researchers enthusiastically agreed to contribute to the Festschrift.

The key point of the book was to collect a number of review papers in order to spot probably the hottest and most interesting topics of condensed matter physics at the present time. Not surprisingly, there are many references to the pioneering works by Leonid Keldysh. In a sense, we consider the Festschrift as a "guide-book" of modern condensed matter physics. The following topics are covered by the review articles of the book:

- Bose–Einstein condensation of excitons and the excitonic insulator
- Electron–hole liquid
- Metal–dielectric transitions
- Semiconductor and organic quantum wells, superlattices, microcavities and other nanostructures
- Disordered systems in condensed matter
- Many-body theory and the Keldysh diagram technique
- Composite fermions and the quantum Hall effect
- Spintronics and quantum computation
- Resonant acousto-optics of semiconductors
- Coherent optical phenomena in semiconductor nanostructures
- Inelastic electron tunneling spectroscopy

Thus the volume is addressed to a very broad audience of condensed matter physicists, from young researchers to experienced academics. In his article published in this book, Vitalii L. Ginzburg overviews his concept of the "Physical Minimum", a list of research problems of physics and astrophysics which are particularly important, challenging and interesting in the beginning of the twenty-first century. From the condensed matter physics side, the list includes nine problems, 2–10 (see the article). In this connection, the papers of the Festschrift directly relate to nearly all of these hot topics of modern condensed matter physics.

Professor Leonid V. Keldysh in his office, P. N. Lebedev Physical Institute, Russian Academy of Sciences, Moscow, 2006.

Apart from the first recollection paper by Elias Burstein, the second article on the "Physical Minimum" by Vitalii Ginzburg, and the last two review papers by the Editors, the articles are alphabetically ordered following the name of the first author.

The paper by V. M. Agranovich deals with the physics of conceptually new hybrid organic–inorganic semiconductor nanostructures. The resonant coupling between the Frenkel and Wannier–Mott exciton states in two different materials of the nanostructure can lead to a fast and efficient noncontact and nonradiative energy transfer. Furthermore, the exciton-mediated nonlinearities are strongly enhanced in these artificial structures. The author also discusses a new family of light-emitting devices based on hybrid organic–inorganic semiconductor nanostructures.

The use of surface acoustic waves (SAWs) in an electron-spin based quantum state processor is reviewed in the paper by C. H. W. Barnes and M. Pepper. The authors discuss a SAW-induced electric current of discrete single or controllable numbers of electrons which can interact and become entangled. Variable g-factor materials are proposed as a means of applying electrical control to single-qubit rotation and as a form of optical input and readout. The processing scheme can be used as a component in a quantum communications network.

D. M. Basko, I. L. Aleiner and B. L. Altshuler review their recent work on the nonzero-temperature metal–insulator transition which occurs in a system of weakly interacting electrons in the presence of a disorder potential strong enough to localize all the single-particle states. The authors show that for the short-range weakly interacting electrons, the many-body localization transition is well-defined to all orders of perturbation theory.

In the paper by E. Burstein, the author overviews electric-field-induced Raman scattering of the light field which resonates with the interband transition in a semiconductor. The role played by the Franz–Keldysh effect in surface space-charge electric-field-induced Raman scattering is particularly emphasized. A strong enhancement of the efficiency of "forbidden" exciton-mediated Raman scattering by applying a static electric field, the effect predicted and observed by the author and his colleagues, is also discussed.

L. V. Butov reviews his recent achievements in experimental realization of a cold but still dense gas of long-lived indirect excitons in GaAs-based coupled quantum well structures. The unusual dynamics (photoluminescence jump) and pattern formation (inner and external rings, as well as fragmentation of the external ring in circular-shaped spots), which were detected in time-resolved and spatially-resolved photoluminescence associated with indirect excitons, are described in detail.

The fractional quantum Hall effect and composite fermions in a two-dimensional electron system are analyzed in the paper by I. V. Kukushkin, J. H. Smet and K. von Klitzing. The underlying picture of a composite fermion, a quasiparticle consisting of one electron and two magnetic flux quanta, is detailed and qualitatively illustrated. The authors also discuss detection of the composite fermions in cryogenic transport and optical experiments.

In connection with cavity quantum electrodynamics, the optical properties of a few-mode semiconductor microcavity with embedded quantum dots are reviewed in the article by V. D. Kulakovskii and A. Forchel. The authors discuss both the weak (Purcell effect) and strong (polariton effect) coupling limits of the "microcavity single-mode light – quantum dot" interaction. The experimental realization of the strong coupling limit is illustrated for a high-quality (high-Q) GaAs-based pillar microcavity.

M. Kuwata-Gonokami overviews mid-infrared pump–probe spectroscopy which is applied to visualize picosecond dynamics of excitons and electron–hole ensembles in bulk CuCl and Cu_2O. For CuCl, the author describes his experimental results on the Mott transition from the excitonic phase to an electron–hole plasma. The observed spatial segregation of the two phases is also discussed. For Cu_2O, the measurements of the infrared Lyman series associated with orthoexcitons are reviewed, and an effective generation of paraexcitons through scattering of cold orthoexcitons is proposed.

In the review paper by P. B. Littlewood, a collective coherent ground state in "exciton-like" many-body systems (excitons, microcavity polaritons, quantum Hall bilayers, dimer spin systems, ultracold atomic fermi gases, etc.) is discussed in terms of Bose–Einstein condensation. Both cases, the high-density limit (BCS limit) and low-density limit (BEC limit), are described in a unified way. Finally, the author outlines the decoherence effects which are relevant to a thermodynamically open system such as short-lived microcavity polaritons.

Inelastic light scattering by low-lying excitations of two-dimensional quantum Hall fluids in semiconductor nanostructures at very low temperatures and large magnetic fields is reviewed in the contribution by V. Pellegrini and A. Pinczuk. The authors show that the optical methods yield unique access to the elementary excitations of a many-electron system in the quantum Hall regime and discuss their measurements of charge and spin excitations for various values of the filling factor ν.

L. P. Pitaevskii outlines the Lifshitz equation, which describes the interaction potential of an atom with the surface of a bulk dielectric medium, and shows how to obtain this equation in a much simpler and more straightforward way comparing to the original derivation by E. M. Lifshitz. The key methodological point of the proposed approach is to neglect the retardation effects and evaluate the Green function of the longitudinal light field. Finally, the author recollects his first meeting with Leonid Keldysh which took place in 1958.

In the paper by E. I. Rashba, the author reviews the fundamentals of semiconductor spintronics and recent developments in this field. In order to manipulate and control the electron spin, such methods as electrical spin orientation, optical and electrical spin injection are discussed. Furthermore, the author outlines the spin interference phenomena, transport in media with spin–orbit coupling, and the spin Hall effect.

Metal–insulator transitions in disorder-free crystals are described in the paper by T. M. Rice. The author reviews excitonic insulators, electron–hole liquids and metal–insulator transitions due to band crossing. The realization of an exciton

(P se–Einstein) condensate in the high-density limit is illustrated by experiments on a quantum Hall bilayer. In the second part of the paper, the Mott transition from a metal to a Coulomb localized insulator is discussed in terms of the one-band Hubbard model.

The chapter written by N. N. Sibeldin reviews an electron–hole liquid. This very interesting phase, observed mainly in bulk indirect semiconductors, Ge and Si, was thoroughly investigated in the late 1960s–mid-1980s. The first discovery of the electron–hole liquid with the subsequent study of its many unusual properties were initiated by L. V. Keldysh.

V. B. Timofeev overviews the attempts to realize Bose–Einstein condensation of excitons in bulk semiconductors and semiconductor nanostructures, with emphasis on indirect excitons in GaAs/AlGaAs coupled double quantum wells. The author details his recent experiments with indirect excitons confined in a circular electric-field-induced in-plane trap. The observed spatially-ordered patterns in the photoluminescence signal and large coherence length are interpreted in terms of Bose–Einstein condensation of indirect excitons.

The review paper by R. Zimmermann deals with Bose–Einstein condensation of excitons. After a historical survey of the search for Bose–Einstein condensation of excitons, the author concentrates on long-lived indirect excitons in coupled quantum wells: Bose–Einstein condensation of noninteracting bosons in a two-dimensional parabolic trap is outlined, and the ground-state wavefunction of indirect excitons is discussed in detail. Finally, the author presents his resent theoretical results on the concentration-dependent blue-shift and broadening of the photoluminescence line associated with indirect excitons.

A. L. Ivanov reviews the recently proposed concept of resonant acousto-optics, when interaction between the light and acoustic waves is mediated and strongly enhanced by the polarization field associated with either excitons (i.e., visible spectral band) or TO-phonons (THz band). For resonant acousto-optics, both dominant interactions, the polarization-light coupling (polariton effect) and the interaction of the polarization wave with the acoustically-induced grating, are treated nonperturbatively (strong coupling regime) and on an equal basis.

S. G. Tikhodeev and H. Ueba overview the recent results on inelastic electron tunneling spectroscopy of a single absorbed molecule. The authors analyze an adsorbate-induced resonance coupled to the molecular vibration. A theoretical description of inelastic electron tunneling is given in terms of the nonequilibrium Keldysh diagram technique.

On behalf of all contributors to the volume, we sincerely wish Professor Leonid Keldysh new research results, good health, optimism and vitality!

We are most grateful to Sir Roger Elliott for his kind support and promotion of this project through Oxford University Press, and to Elias Burstein and Dima Khmelnitskii for their helpful advice and comments during the work on the Festschrift. We thank Nikolay Gippius, who is a former student of Professor Keldysh, for providing us with the photograph of Leonid Keldysh. Finally, we would also like to thank our colleagues and research students, Anton Akimov, Cele Creatore and Leonidas Mouchliadis, for their invaluable technical help in preparing the book, and Mikhail Skorikov for translation of the papers by V. L. Ginzburg and N. N. Sibeldin from Russian into English.

Alexei L Ivanov, Cardiff University, United Kingdom
Sergei G Tikhodeev, General Physics Institute, Russia

1

RECOLLECTIONS

Elias Burstein
Mary Amanda Wood Professor of Physics, Emeritus
Department of Physics and Astronomy, University of Pennsylvania,
Philadelphia, PA 19104, USA

Prologue

In 1956, I was elected Secretary-Treasurer of the Division of Solid State Physics of
the American Physical Society (renamed Division of Condensed Matter Physics
in the 1980s). I contacted John Bardeen, who was completing his term as Chairman of the Division, to ask what else I might do, as Secretary-Treasurer, in
addition to organizing the program of contributed and invited papers for the
annual March Meetings of the Division. He suggested that I might organize conferences under the auspices of the Division of Solid State Physics that would not
otherwise occur. During the years that followed, I took his advice and initiated
a number of conferences both during the term of my office and continuing in the
years that followed.[1]

As a member of the research staff at the US Naval Research Laboratory
(from 1945 to 1958), I was not able to attend the International Conferences on
the Physics of Semiconductors that took place in other countries and therefore
did not attend the Conferences that were held in Reading (1951), Amsterdam
(1954) and Garmisch Partenkirchen (1956). I decided to take the initiative in
proposing that the 1958 Conference should take place in the US, which I would
be able to attend. I brought together an Organizing Committee which made the
decision to hold the Conference in Rochester and to invite John Bardeen to be
its Honorary Chairman. Malcom Hebb (GE Knolls Research Laboratory) was
chosen as Chairman of the Executive Committee, David Dexter (University of
Rochester) served as Secretary of the Conference, and I served as a member
of the Executive and Program Committees. For me, one of the key features of
the Conference, apart from the exciting scientific program, was the presence of a
very small number of scientists from the USSR. It included, A. Joffe, E. F. Gross,
S. G. Kalashnikov and V. S. Vavilov, who were meeting their counterparts in the
US for the first time. I remember well Gross' remark to Ben Lax and me, as he

---

[1] At a dinner during the 1967 Conference on Electron Tunneling in Solids held in Denmark,
which Leo Esaki and I organized, Bob Schrieffer, a close friend and colleague in the Physics
Department at the University of Pennsylvania, remarked that I was in reality a "living creation
operator of conferences".

wrote our names on a blackboard, that before he came to the Conference Lax and Burstein were just words, and now for the first time they were real persons. I felt the same way about meeting Gross at the Conference.

It was over four decades ago when L. V. Keldysh and I met

In the early 1960s the research of my group at the University of Pennsylvania was focused on infrared and Raman spectroscopy of crystals, on the effect of an applied electric field on infrared absorption and Raman scattering by optical phonons, and on electron tunneling phenomena in superconductors. I also continued my collaboration with former colleagues at the Naval Research Laboratory on cyclotron resonance and interband magneto-optics in semiconductors. We were keenly aware of L. V. Keldysh's theoretical work on "photon assisted interband tunneling" and its counterpart, tunneling assisted photon induced interband transitions. It occurred to me that it would be most desirable to have L. V. Keldysh, who had not attended the Rochester Conference in 1958, visit our group without using attendance at an International Conference as the basis for getting approval from the US State Department for his visit. In 1961, I send a letter of invitation to Keldysh to visit the University of Pennsylvania, and contacted the State Department to obtain approval for his visit. Shortly after making this contact I received a phone call from an individual in the State Department who informed me that L. V. Keldysh was a woman. I assured him that the scientist whom I wished to invite to visit the University of Pennsylvania was definitely a man. I learned later that the individual he was referring to was actually Keldysh's mother, Ludmila V. Keldysh, a distinguished mathematician in the USSR. Permission for his visit was granted and, as was the custom, the State Department listed the places in the US that he would not be allowed to visit.

Keldysh arrived in Philadelphia together with V. A. Chuenkov, a physics colleague and a close friend. My first meeting with them took place fortuitously in an elevator of the hotel in which they were staying. My wife and three young daughters were with me and it was a charming moment when my daughters each said "how do you do" and "pleased to meet you", and shook hands graciously with Keldysh and Chuenkov.

During the early part of his visit I asked Keldysh whether he would like to attend a symphony concert, visit a museum or do some sight seeing in the Delaware Valley. He informed me without hesitation that what he would like to do is meet J. Robert Oppenheimer at the Institute for Advanced Study in Princeton, which for him was the Mecca of Physics. I contacted Oppenheimer's office and had no difficulty making arrangements for Keldysh to see Oppenheimer, whom I myself had never met. The visit with Oppenheimer was, clearly, a highlight of Keldysh's visit to the US.

After 1961, the next time we saw each other was in 1968 when I attended the International Conference on the Physics of Semiconductors in Moscow. I had been to a Conference in Paris and was on the same plane to Moscow with

Leonid Keldysh, Elias Burstein and Vasilii Chuenkov (from left to right) in Philadelphia in 1961.

William (Bill) Paul, Professor of Physics at Harvard University. Bill Paul was an "invited honored guest" of B. M. Wul, head of an experimental group at the P. N. Lebedev Physical Institute, whose paper in *Nature* on barium titanate was well known world wide. When we landed at the airport in Moscow, Bill Paul was greeted by Tatiana Galkina, a member of Wul's group, and personally escorted past customs and driven to the hotel where attendees at the Conference were staying. By virtue of being on the same flight from Paris on a Soviet plane with Paul, I was treated as a "semi-honored guest", and also escorted through customs and taken by car to the hotel. I had first met Wul in Paris in 1963 at a meeting which included Pierre Aigrain and representatives from other countries to discuss my proposal for a new journal, *Solid State Communications*, to start publication in 1963. Wul urged me to have a non-American as Editor-in-Chief of the Journal, which was why Pierre Aigran was initially its Editor-in-Chief, and I was initially Secretary of the Board of Editors with overall responsibility of the Editorial Operations of the Journal.

I visited Wul and his group at the P. N. Lebedev Physical Institute, and had interesting discussions with him about solid state physics, and his thoughts about the role played by *Solid State Communications*. I also met Victor Bagaev (husband of Tatiana Galkina), a colleague and friend of Keldysh, who was doing very interesting experimental work on excitons and electron–hole droplets[2].

---

[2] V. S. Bagaev, L. V. Keldysh, J. E. Pokrovsky and M. Voos were joint recipients of the 1975 Hewlett Packard Europhysics Prize for their work on "the condensation of excitons".

I had a good visit with Keldysh and also with members of the Moscow State University, but no longer remember the details of what we discussed. I do remember attending an exciting soccer game with Keldysh and Bagaev involving teams from two of the major cities in the USSR.

Epilogue

When Keldysh and I met in Trieste in May 2004 at the time of the ICTP Research Conference on Advancing Frontiers of Optical and Quantum Effects in Condensed Matter, he reminded me that since his visit to Philadelphia in 1961 we had been meeting each other every seven years. Although I can remember most of our meetings, I am still trying to recall some of the occasions.

I had my 88th Birthday in September of 2005. In Japan there is great respect for the elderly and the 88th birthday is highly celebrated – the Japanese characters for 80 + 8 ("BeiJu") signify longevity and rice, a symbol of longevity. It would certainly be great to get together with Leonid Keldysh in 2011, seven years since our last meeting. Whether or not this will be, I take this opportunity to send him best wishes for the years ahead with the fervent hope that he will reach his BeiJu birthday and beyond with continued health, creativity and achievement.

2

ONCE AGAIN ON THE "PHYSICAL MINIMUM"

V. L. Ginzburg

P. N. Lebedev Physical Institute, Russian Academy of Sciences, Moscow, Russia

The present volume issued in honor of Leonid Veniaminovich Keldysh (or LV, as I will write below) on his 75th birthday consists mainly of scientific papers devoted to physics. This fully agrees with the well-established practice of publishing such anniversary volumes (Festschrifts). At the same time, this practice doesn't rule out the possibility of including articles written *à propos*, i.e., papers on occasion and devoted to any theme.

The editors were kind enough to ask me, among others, for a paper to be included in this collection. Unfortunately, currently I have no original or review article on physics or astrophysics that has not been published yet. This isn't surprising, as I'm already 90 and in poor health. At the same time, I would like to participate in this volume for obvious reasons – just think that I remember LV from 1942 or 1943. The story is that during World War II many of the institutes of the USSR Academy of Sciences were evacuated to Kazan. Both I and my family and LV's parents with him were there. We lived in the same building, the hostel of the Kazan University handed over to the Academy. We were acquainted with LV's parents, although not too closely, and my memory keeps just one image. I entered their room and saw a serious-looking boy with a small knapsack on his shoulders: it seems he was going somewhere for provisions with his mother. Next I remember LV only in 1954, when he came to me at home (apparently, it wasn't easy to visit me at the Lebedev Institute because of "secrecy"). He graduated from the Physical Faculty of the Moscow University and had to enter the PhD postgraduate courses (*aspirantura*) at our Institute. On this occasion he had to fill in a long questionnaire with biographical data, etc. so that I could submit this to the personnel department (I was deputy head of the Theory Department at the Lebedev Institute). LV was admitted, and I became his supervisor. He worked very successfully, graduating from aspirantura in 1957; since then he has been a member of the I. E. Tamm Theoretical Physics Division of the Lebedev Physical Institute, which is the current name for the former Theory Department founded by I. E. Tamm in 1934. As an official supervisor, I certainly had quite a lot of contact with LV, but he worked totally on his own and we haven't any common papers. Actually, at the end of the 1950s, we did write together one paper devoted to the theory of superconductivity, but it hasn't been published;

evidently, it became of no interest since the well-known paper of Bardeen, Cooper and Schrieffer (1957) had appeared then.

As I've said, my participation in LV's work was insignificant and hardly of much importance for him; I'm not in a position to judge about this issue. I think, if he benefited in any way from my involvement, then it would only be on a small scale and, so to say, indirectly, as a member of the Department.

LV's work was very successful, and in 1976 he was already elected a full member of the Academy of Sciences, while in 1988 he replaced me as the head of the Theory Department. The tale of our later intercourses is of no public interest and, obviously, the above story by itself would not justify my participation in this volume.

Thus, I proceed to the subject I decided to take on here, namely, to some remarks about the "Physical Minimum"; I used this term for designating an area of my activities that I pursued for many years – a "project", as it is customary to say now.

I began teaching in 1945 at the University of Nizhni Novgorod (then Gor'kii) (it was an additional job – I traveled from Moscow and back). This teaching lasted for quite a number of years, as I was connected with Nizhni Novgorod owing to my family affairs.[3] Since 1968 I started teaching at the Moscow Institute of Physics and Technology; here, I organized the chair "Problems of Physics and Astrophysics", which is still doing well. Now, observation of the practices of even good universities – the two mentioned above certainly belong to this category – had led me to the conclusion that, as a rule, the scientific horizon of graduating students is rather narrow. Of course, they know well that field of physics in which they did their diploma thesis, but don't know anything in a great number of fields that are topical and intensively studied at the moment. This fact is quite understandable, as physics has spread so much that the formula "Everything about something and something about everything" is no longer valid; instead, one cannot but remember the advice of Koz'ma Prutkov: "Nobody can comprehend incomprehensible". So, what can be done?

The essence of the "Physical Minimum" project is rather simple: it consists in compiling a list of problems in physics and astrophysics that are currently most important and urgent. It is obvious, first, that such a "list" varies with time, as far as the solution to some problems appears and others may become less important. It is obvious, second, that what is to be placed in such a "list" is unavoidably a subjective matter. Thus, it is desirable to have lists compiled by different authors – though, I believe that, at any given time, such "lists" would be mainly identical.

Third, issues not included in such a list certainly cannot be regarded as "unimportant" or "uninteresting". Properly speaking, any problem of real physical content, whose formulation, of course, shouldn't contradict the established

---

[3]It would be completely inappropriate to expand on these matters here. The interested reader can refer to my "Nobel" autobiography, see Ginzburg (2006).

laws of Nature, may hardly be proclaimed unimportant or uninteresting. What I imply is just no more than a subjective selection.

Fourth, what I'm talking about is not a kind of program (I have come across such an interpretation of my "list"), but rather a somewhat trivial enumeration of issues on which any physicist should have an idea in order to be regarded as a highly educated specialist. Of course, this requirement is just complementary, and any broad erudition ("... something about everything") cannot replace the knowledge necessary to work in one's own field. Clearly, the emphasis isn't on how somebody may be "regarded"; the point is that broad general erudition in physics (like any other specialty, of course) is both a pledge of successful work and a source of joy for a person who loves science. Broad erudition comes, I would say, automatically, if one attends good and diverse seminars and reads systematically popular scientific magazines for physicists (and scientists in the areas close to physics) like *Physics World* (Britain) and *Physics Today* (USA), as well as more specialized and, thus, necessarily more narrowly-oriented journals like *Reviews of Modern Physics* and *Uspekhi Fizicheskikh Nauk*.[4] Unfortunately, under the current state of affairs – especially here in Russia – not every university and even not every city may boast the existence of good broad-profile seminars and availability of necessary journals.

So, I've arrived at describing the goal of the "Physical Minimum" project, however clear it may be from the above discussion. This goal consists in helping young people, and others as well, to master this necessary minimum of information on the modern physics (see above).

The first step in this direction is compiling the "list" itself. The next step is, so to speak, filling the "list" with some content. This can be accomplished through a diverse course of lectures devoted to every topic in the "list". The volume of such a course may be from several (say, three) to many lectures. In the latter case, a single lecturer, apart from some exceptions, cannot cover the entire existing material, as it goes without saying that he should know the material on a much deeper and broader scale than the audience is to learn.

It is reasonable to implement such a course as a large article or a book.

This is exactly the line of action I adhered to by myself. The first of such articles was published in 1971 (Ginzburg, 1971). Later, an extended version of this paper appeared as part of my book (Ginzburg, 1995), several editions of which were published over the years with the addition of new material. The latest edition, cited above, was published in English, again with some additions (Ginzburg, 2001b). Next, the "Physical Minimum" was the subject of the first article in my other book (Ginzburg, 2004), which was also translated into English (Ginzburg, 2005). Finally, I spoke on the "Physical Minimum" in my Nobel Lecture given in Stockholm on December 8, 2003 (Ginzburg, 2003). This part of the lecture is titled " 'Physical Minimum' – What problems of physics and

---

[4]It is worth mentioning that, in addition to specialized review articles, we try to publish reports on "hot" scientific news in *Uspekhi*. This practice, however, cannot provide a substitute for such magazines as *Physics World*.

astrophysics seem now to be especially important and interesting in the beginning of the twenty-first century?".

My Nobel Lecture had a somewhat unusual (or nonstandard) format due to the reasons explained in the lecture itself. It was not necessary at all to include the issue of the "Physical Minimum" in the lecture, but I did it for two reasons. First, Nobel Lectures are open events and they are attended by young people, to whom the issue of the "Physical Minimum" is, in any case, more interesting than the issues of the theory of superconductivity and superfluidity. Second, and most important, I wished to use actually my last possibility to bring attention to the "Physical Minimum". For, I should confess, previously I expected that the "Physical Minimum" would not be left nearly unnoticed, which evidently took place for reasons I can only guess about. I don't know, and won't know, the opinion of my colleagues on that score, so there is nobody to argue with. Nevertheless, I remain with my opinion that the issue of the "Physical Minimum" deserves some degree of attention. Thus, I included it in the lecture (Ginzburg, 2003) and decided to dwell upon it in this paper.

What is my intention now? First, I will present the whole "list" under consideration in its latest version (Ginzburg, 2005; Ginzburg, 2003). Next, I will make several comments related to the content of the list.

The following is the list of problems that seem to be most important and interesting at the beginning of the twenty-first century:

1. Controlled nuclear fusion.
2. High-temperature and room-temperature superconductivity (HTSC and RTSC).
3. Metallic hydrogen. Other exotic substances.
4. Two-dimensional electron liquids (anomalous Hall effect and other effects).
5. Some questions of solid-state physics (heterostructures in semiconductors, quantum wells and dots, metal–dielectric transitions, charge and spin density waves, mesoscopics).
6. Second-order and related phase transitions. Some examples of such transitions. Cooling (in particular, laser cooling) to superlow temperatures. Bose–Einstein condensation in gases.
7. Surface physics. Clusters.
8. Liquid crystals. Ferroelectrics. Ferrotoroics.
9. Fullerenes. Nanotubes.
10. The behavior of matter in superstrong magnetic fields.
11. Nonlinear physics. Turbulence. Solitons. Chaos. Strange attractors.
12. X-ray lasers, gamma-ray lasers, superhigh-power lasers.
13. Superheavy elements. Exotic nuclei.
14. Mass spectrum. Quarks and gluons. Quantum chromodynamics. Quark–gluon plasma.
15. Unified theory of weak and electromagnetic interactions. $W^{\pm}$ and Z^0 bosons. Leptons.

16. Standard model. Grand unification. Superunification. Proton decay. Neutrino mass. Magnetic monopoles.
17. Fundamental length. Particle interaction at high and superhigh energies. Colliders.
18. Nonconservation of CP-invariance.
19. Nonlinear phenomena in vacuum and in superstrong electromagnetic fields. Phase transitions in a vacuum.
20. Strings. M-theory.
21. Experimental verification of the general theory of relativity.
22. Gravitational waves and their detection.
23. The cosmological problem. Inflation. Λ-term and "quintessence". Relationship between cosmology and high-energy physics.
24. Neutron stars and pulsars. Supernova stars.
25. Black holes. Cosmic strings.
26. Quasars and galactic nuclei. Formation of galaxies.
27. The problem of dark matter (hidden mass) and its detection.
28. The origin of superhigh-energy cosmic rays.
29. Gamma-bursts. Hypernovae.
30. Neutrino physics and astronomy. Neutrino oscillations.

This "list" is already several years old. Not without satisfaction I note that, compiling such a list now (December 2006), I would introduce – or, if you like, I am introducing – just a few changes. Specifically, in item 3 I'm adding the question of that wonderful two-dimensional film, graphene (Nete *et al.*, 2006). In items 4 and 5 (or in a separate item) I would add the question of quantum phase transitions and several problems of optics (Glauber, 2006; Hall, 2006; Hansch, 2006; Agranovich and Gartstein, 2006). I'm leaving behind quantum computers and discussions on related problems; I'm not at all acquainted with these matters. Another point that should be mentioned is spintronics, a branch of electronics where, apart from the electron charge, electron spin plays an important role. Probably, the issue of polariton and magnon Bose–Einstein condensation should be noted, too (Kasprzak *et al.*, 2006; Demokritov *et al.*, 2006). In item 12, free-electron lasers need to be mentioned. On item 13, I may note that elements with $Z = 116$ and $Z = 118$ have already been created, but, apparently, nothing new in relation to the "island of stability" around $Z = 114$ was added. Of course, work goes on in the areas covered by items 14–19, but I know nothing of principal novelty there. Incidentally, mention of magnetic monopoles in item 16 may be dropped; apparently, these monopoles do not exist in Nature. On the other hand, a lot has been written on magnetic monopoles in the past, and it isn't bad to know what this stuff is about – if only for historic reasons.

I would add one more item, the one about "new" extra dimensions existing apart from the four (three coordinates and time) we know. Although this problem is far from being novel, it resounded really loudly only in recent years. Regarding astrophysics, mention of cosmic strings has to be crossed out from the "list". They weren't actually discovered, and there is no ground to include in the minimum any purely hypothetic objects.

One can see from the above discussion that the content of the "list" is rather conservative, which, however, is not an indication of the conservatism of physics itself. Quite the reverse, it develops at as rapid a pace as previously in many areas "disguised" somewhat under titles of these or those items of the "list". New achievements are to be reflected in lectures or articles devoted to the "minimum".

It's a different matter that in certain cases a lot depends on giant facilities taking many years to be constructed. In such areas, development proceeds in leaps and bounds. For example, the final decision on the construction of the Large Hadron Collider (LHC) was taken in 1994, and its coming into service is due in 2007–2008. And thus, scientists specializing in high-energy physics are awaiting results from the LHC, literally chattering their teeth in hunger for many years. By the way, the cost of the LHC, which is a multinational project, will be about ten billion dollars. Colliding proton beams at the LHC will make it possible to study processes at energies as high as $14\,\text{TeV} = 1.4 \times 10^{13}\,\text{eV}$. For a long time until now (i.e., before the LHC), the most advanced accelerator was the collider operated at Fermilab in the USA. There, energies of 2 TeV for counterpropagating proton–antiproton beams can be attained. So to speak, it is in such order-of-magnitude jumps (from 2 to 14 TeV) that progress in high-energy physics occurs. To my knowledge, no real project for building a "next-generation" proton accelerator exists. The next large collider will be the International Linear Collider (ILC), where counterpropagating beams of electrons and positrons of energies 0.5–1 TeV each will be smashed together. The construction of the ILC will commence not earlier than 2009 and will cost five to seven billion dollars.

Very much is anticipated from the LHC. There is no possibility to enlarge here on this issue (see articles in *Physics World* and *Physics Today*). I restrict myself only to mentioning the search for Higgs bosons and new supersymmetry-type particles. These are sort of expected results. The most interesting outcome would be if the expectations do not realize. Especially, if Higgs bosons don't exist, that would be literally a catastrophe for current high-energy particle physics. Still, this is very unlikely, while supersymmetric particles may well be just creations of the imagination.

Unfortunately, I couldn't suggest qualified comments on many of the items in the "list" now, since this requires competence in the latest data and latest studies. But, in many of the areas listed, new accomplishments were not numerous.

The main event related to problem No. 1, controlled nuclear fusion, is the end of a years-long dispute on the construction site for ITER (a tokamak-type international experimental thermonuclear reactor). The decision has been made, ITER will be built in France. This will take up to several (about ten) years and it will cost around ten billion dollars. This is an immense amount of work, and an immense amount of money, but hardly anyone doubts the necessity of building this facility. At the same time, as far as I know, researchers do not wish to drop efforts aimed at the development of nuclear-fusion reactors (experimental installations) based on schemes other than the "tokamak". I mention so-called "probkotrons" that utilize magnetic traps of linear configuration. Building of facilities (sometimes huge ones) for "inertial" fusion continues as well; in this scheme, reaction occurs upon rapid compression of a target containing deuterium and tritium nuclei (d+t). Let's take this opportunity to illustrate what I mean by "Physical Minimum" for the case of, say, controlled thermonuclear fusion. One needs to be acquainted with the formulation of the problem and the information on reactions d+d and d+t, etc. Next, understanding of the tokamak arrangement and, in particular, of the ITER design, is required. Finally, one needs to know about "inertial" fusion and, say, about "probkotrons".

Item 2 of the list represents, actually, my narrow specialty, and I'd written a lot on this topic. Here, I just refer to article 3 in my book (Ginzburg, 2006), in particular, in the English edition. The main goal in this area is the creation of superconductors whose critical temperature T_c exceeds room temperature (or, better, $T_c \sim 100°C$). Can this goal be attained? Nobody may guarantee; I believe that it is realizable. What may be the solutions? Different ideas were put forward. In my opinion, the most attractive way at the moment is related to the creation of new materials possessing appropriate phonon spectra and, simultaneously, characterized by strong electron–phonon coupling (Pickett, 2006). Such a statement could have been expressed years ago. However, there existed – and, actually, exist until now – doubts whether such substances can be stable. At the same time, the discovery in 2001 of a MgB_2 superconductor with $T_c \simeq 40\,\mathrm{K}$, as well as a number of other examples and considerations, make the possibility of obtaining such materials more realistic (Pickett, 2006).

In this context, one remark seems pertinent. Our advances, in solid state physics in particular, lead us to forget, so to say, that we are still unable to create any conceivable substance at our wish. The calculational capabilities are limited, but, even if according to calculations some stable or even metastable substance may exist, this doesn't mean at all that the route to synthesizing it is clear. I feel myself totally incompetent in this field, where many people were working and work now. However, the scale of these studies is apparently insufficient, and the situation has to be changed.

On item 3 of the "list" I know nothing new, apart from some calculation results (Maksimov and Savrasov, 2001). They indicate that, at a pressure of 20 Mbar, metallic hydrogen is a superconductor with $T_c \simeq 600\,\mathrm{K}$. It's a poor consolation, as (speaking of sufficiently low temperatures) metallic hydrogen still hasn't been obtained and realistic ways to attaining this goal remain unclear.

I've already mentioned above the two-dimensional electron gas (item 4). Actually, it is more appropriate to imply two-dimensional systems in general. In this context, a notable event, probably the most interesting discovery in recent years, occurred in 2004, when so-called graphene (Nete *et al.*, 2006) was obtained and studied. Graphene is simply two-dimensional carbon, or, if you like, a single graphite plane (although a proper statement now would be that graphite is a stack of graphene layers). Conduction electrons in graphene behave like particles of zero mass and, thus, are similar in many respects to relativistic particles – only with their velocity being 300 times smaller than the speed of light in free space. A surprising and, certainly, very interesting fact is the simplicity of the fabrication of graphene layers. Promising applications of graphene are foreseen.

In relation to item 6, one should mention the discussion of polariton Bose–Einstein condensation (Kasprzak *et al.*, 2006). Polaritons are nothing else than normal electromagnetic waves in a medium, in particular, in a crystal (Agranovich and Ginzburg, 1979). In other words, these are the same photons but propagating in a medium. It is hardly reasonable to speak of Bose–Einstein condensation of photons in a vacuum, when their behavior is fully described by quantum electrodynamics. A photon in a medium represents a new particle, and, in my opinion, the concept of Bose–Einstein condensation is relevant in this case. This problem may be of importance, taking into account that, in superconductivity theories, polaritons are considered as possible replacement for phonons; see article 3 in my book (Ginzburg, 2006).

As I've already noted, a renovated list should include (e.g., in item 6) advances in optics. The corresponding references were given above (Glauber, 2006; Hall, 2006; Hansch, 2006; Agranovich and Gartstein, 2006). The latter review article is devoted to optical phenomena in media with negative group velocity; see Agranovich and Ginzburg (1979) (just to remind readers, the group velocity is $\mathbf{v}_{\mathrm{g}} = \partial \omega / \partial \mathbf{k}$, where $\mathbf{k}$ and ω are the wavevector and the frequency of the wave, respectively; in conventional isotropic media, $\mathbf{v}_{\mathrm{g}}$ is oriented along $\mathbf{k}$, while in media with negative group velocity $\mathbf{v}_{\mathrm{g}}$ is oriented along $-\mathbf{k}$).

Another issue already mentioned, in my opinion deserving to be included in the renewed "list", is the problem of extra dimensions, i.e., the possibility that the world we live in has more than four dimensions. This issue is widely discussed today, especially in relation to string theory (item 20). It is for spaces with 10 and 11 dimensions that the latter is being developed. By and large, I don't know how to deal with the new dimensions. But the idea of "compactification" is as follows. If, for instance, a particle in a five-dimensional space moves along a curve $(x = 0, y = 0, z = 0 + \zeta, \zeta = \zeta(t))$ and the value of the fifth coordinate ζ is very small, then, to a good approximation, we simply obtain motion along the z-axis. In other words, if the values of the "new" coordinates are small, then, for a certain range of phenomena, new coordinates may be disregarded. In essence, this is what is named "compactification". In some of the theories, the typical scale of the "new length" is on the order of the so-called gravitational, or Planck length $l_{\mathrm{g}} = \sqrt{G\hbar/c^3} = 1.6 \times 10^{-33}$ cm (here, $G = 6.67 \times 10^{-8}$ cm^3/(g × s^2) is the gravitational constant, $c = 3 \times 10^{10}$ cm/s is the speed of light and $\hbar = 1.055 \times 10^{-27}$ erg × s is the quantum constant). Note that the smallest length-scale that can be probed using modern accelerators is $\sim 10^{-17}$–10^{-18} cm, and, thus, length-scales of the order of $l_{\mathrm{g}} \sim 10^{-33}$ cm may be of importance only in the vicinity of a cosmologic singularity and other singularities, say, in black holes.

Some comments on all items in the "list" may be found in the respective articles of my books (Ginzburg 1995, 2001*b*, 2004, 2005) and references cited therein. Here, I would only enlarge on the fundamental issues of "dark matter" and "dark energy", which figure in items 23 and 27 of the "list", dark energy being called "quintessence" in item 23.

Just think what a staggering fact is the observation that baryonic matter in the Universe accounts to only 4% of all matter, with the share of dark matter being nearly 20% and dark energy, 75%. Dark matter, as it is commonly believed, consists of some still unknown particles that interact with the baryonic matter only through gravitation forces. These particles are being searched for intensively but, despite considerable efforts, they have still not been found. Then, maybe, "dark matter" is something completely different, some "new physics"? Such hypotheses were put forward, and, currently, the situation is totally unclear. The more so is the situation with the "dark energy".

First of all, it may be said that "dark energy" represents a kind of antigravitation source, i.e., repulsion between massive bodies instead of attraction due to conventional (Newtonian) gravitation. In the case of conventional gravitation, a material ball initially being at rest will collapse if it isn't rotating; only a rotating ball may be in a steady state, which we know by the example of the solar system. The simplest model for the Universe is a ball that does not rotate, and, thus, depending on the initial conditions, it will either collapse or expand. By the way, such a description may be obtained, to a first approximation, in the framework of Newton's theory of gravitation (which wasn't noticed at first). However, such a model was actually proposed for the first time on the basis of the equations of general relativity (GR) by A. A. Friedmann in 1922 and 1924. Meanwhile, such a nonstationary model ("expanding-Universe model") had not been considered as late as at the beginning of the last century; the Universe was believed stationary. Einstein had finished the GR theory in 1915 (Einstein, 1915) and written a large review article in 1916 (Einstein, 1916); it is often cited as the publication that completed the creation of the GR theory. Quite naturally, he wished to apply GR to cosmology, but he used the stationary model that was the only one considered at the time. And he saw that no required steady-state solution of the GR equations exists. It appeared, however, that the GR equations can be generalized by the addition of one more term, called the Λ-term. The case can be clarified most easily if we write down the GR equations for the metric tensor g_{ik} (for details, see any textbook, such as Landau and Lifshitz (1982) *Theory of Fields*):

$$R_{ik} - \frac{1}{2} R g_{ik} - \Lambda g_{ik} = \frac{8\pi G}{c^4} T_{ik}\,, \tag{2.1}$$

where T_{ik} is the energy–momentum tensor of matter, and R_{ik} and R are functions of g_{ik}. Physically, the Λ-term (i.e., the term Λg_{ik}) for $\Lambda > 0$ means repulsion. If the Λ-term is moved to the right-hand side of equation (2.1), then this term will be equivalent to the introduction of an additional energy–momentum tensor of the vacuum:

$$T_{ik}^{\mathrm{v}} = \frac{c^4 \Lambda}{8\pi G}\, g_{ik}\,. \tag{2.2}$$

The equation of state of this "vacuum matter", assuming $g_{00} = -1$ and $g_{\alpha\alpha} = 1$, can be written as

$$\varepsilon_{\rm v} = -p_{\rm v} = \frac{c^4 \Lambda}{8\pi G}, \qquad (2.3)$$

which, for positive energy density $\varepsilon_{\rm v} > 0$, corresponds to negative pressure $p_{\rm v} = -\varepsilon_{\rm v}$. Next, in GR, the "effective gravitational mass" per unit volume equals $(\varepsilon + 3p)/c^2$, i.e., the pressure possesses weight. Thus, if in equation (2.3) $\Lambda > 0$, a unit volume of a vacuum possesses gravitational mass

$$m_{\rm v} = -\frac{c^4 \Lambda}{4\pi G}. \qquad (2.4)$$

Curiously, after Friedmann's work, Einstein considered the introduction of the Λ-term as almost a mistake, and, in any case, as an unsatisfactory move. Others, e.g., Pauli, agreed with him, and Landau literally wouldn't hear about the Λ-term. For myself, I always considered the introduction of the Λ-term as a beautiful and interesting hypothesis, not because of any deep ideas but just by the reason that this is the only possibility for generalizing Einstein's GR theory without resorting to more complicated equations.

And now, astronomic observations of recent years have demonstrated that the expansion of the Universe proceeds as if the Λ-term really exists. More accurately, it is only possible to conclude from the experimental data that the vacuum possesses energy and momentum density – maybe, the relationship $\varepsilon_{\rm v} = -p_{\rm v}$ is not satisfied exactly, but, nevertheless, $\varepsilon_{\rm v} + 3p_{\rm v}$ is negative (the so-called quintessence). To the best of my knowledge, the latest data indicate that $\varepsilon_{\rm v} = -p_{\rm v}$, i.e., the dark energy is described exactly by Einstein's Λ-term.

All this represents a huge achievement, but may we confine ourselves just to assertion that the Λ-term does exist? I believe, like many others, that this isn't the case: new ideas will appear and deeper understanding will be required.

Unfortunately, providing explanations for many of the items in the "list" even on the above scale would be inappropriate for this article. I only wish to make one more remark regarding string theory (item 20). This is, so to say, the latest fashion in theoretical particle physics and high-energy physics. It is based on a hypothesis that primordial matter is composed of stringlets with the size on the order of $l \sim l_{\rm g} \sim 10^{-33}$ cm. Many very strong theorists work in this area, and it is stated that string theory, which is already about 30 years old, has already gone through two revolutions. But I should confess that I don't know what, after all, string theory gives to physics, how it is related to the real world, and what it predicts. I know from the literature that many other people encounter similar questions.

I think that, in the context of the "Physical Minimum", it is sufficient to give a fairly limited account of string theory, but, most likely, even that should be done by a specialist.

The size of this article has grown considerably beyond my expectations, but still I don't know whether I managed to convince the reader of the usefulness of the "Physical Minimum". Is it a right way, how is it efficient, especially with respect to the young? I cannot answer, but, frankly speaking, I'm surprised that, among the great number of physicists, no one has compiled his own "list" and comment on it.

Meanwhile, however arguable may be the question of the "Physical Minimum" and its relevance, it seems beyond doubt that care about the advancement of the qualification of young people and broadening of their outlook is necessary anyway. Making use of the "Physical Minimum" is one of possible steps. Another measure, especially important and urgent for Russia, is related to organizing and taking advantage of a high-quality popular scientific journal like *Physics World*. Such journals shouldn't be confused with conventional popular scientific magazines. Fairly good periodicals of the latter type, like *Nauka i zhiz'n* (*Science and Life*) and *Priroda* (*Nature*) do exist in our country. Publications in a "popular scientific journal for physicists" should consist of papers and notes providing the latest information on the situation in physics and neighboring disciplines. For example, not later than the end of November 2006 I received the November issue of *Physics World* with the information on the above-cited papers from *Nature* (Kasprzak *et al.*, 2006; Demokritov *et al.*, 2006) and a large article about graphene, which is written by researchers who work actively in this field (Nete *et al.*, 2006). In general, such a journal (let's call it arbitrarily *Physics and Astronomy Today*) should publish both articles and notes of the editorial staff and papers and interviews with actively working physicists and astronomers. Thus, a reader of such a journal will be up-to-date with the latest developments and receive information on original publications in *Nature* and in specialized journals. The existence of such a Russian-language journal is absolutely necessary, as subscription to foreign literature is rather limited, and, after all, it's impossible to order 10 000 copies of *Physics World*.[5] My suggestion to organize the publication of such a widely accessible journal under the title *Physics and Astronomy Today* was put forward in 2002 and 2003 and published in the media (Ginzburg, 2001*a*). Surely, I also addressed this proposal to the Division of Physical Science (DPS) of the Russian Academy of Sciences. It was discussed at the DPS Bureau and, seemingly, got approval, but no more came of it. There appeared no enthusiasts who would undertake the enterprise of publishing the journal; or rather nobody sought them. Meanwhile, the DPS currently counts 70 full members and 111 corresponding members, to say nothing of quite a number of physicists in other divisions of the Academy. Almost all of them have aides and employees, and it wouldn't be difficult to duly organize functioning of *Physics and Astronomy Today*, and it isn't costly either. However, authorities don't bother about these

---

[5] As far as I know, the circulation of *Physics World* is 35 000 copies. Of them, 28 000 are distributed within Britain. Apparently, all members of the British Institute of Physics receive the magazine. I think that, in our country, the circulation of *Physics and Astronomy Today* should be no less than the above figures. Unfortunately, we do not have an active Physical Society. Hopefully, the task of organizing such a journal may stimulate the activity of this Russian Physical Society.

matters.

Here, I would allow myself one general comment. We hear moans about under-development of science in Russia, and, certainly, reproaches about poor funding are justified, both with respect to equipment and low salaries. However, success-ful work in science (and not only in science) requires, apart from "daily bread" (in the form of money, sufficient, in particular, for decent salaries), some moral factor, so to speak. I mean enthusiasm, devotion to the cause, desire to reach a success. And, in my opinion, it's not alright with this moral factor in Russia nowadays, and I do have what to compare with – only at the Lebedev Physical Institute I have worked since 1940. The reasons for this apathy and lack of ini-tiative – fortunately, far from being all-embracing – are not clear for me, it's an intricate social phenomenon.

Finally I wish to complete this article with an appeal, especially targeted at young scientific workers: It is a great privilege to be occupied in science, this driving force of the modern civilization, which has fallen to your lot. So, appreciate your fortune and don't spare your efforts – and not only for your own studies, but for public work in the interests of the entire scientific community.

References

Agranovich, V. M. and Gartstein, Yu. N. (2006). *Usp. Fiz. Nauk.*, **176**, 1051 [*Phys. Usp.* **49**, 1029].

Agranovich, V. M. and Ginzburg, V. L. (1979). *Kristallooptika s prostranstven-noi dispersiey i teoriya eksitonov* (Russian title). Nauka, Moscow. [En-glish translation: (1984) *Crystal Optics with Spatial Dispersion, and Excitons*, Springer, Berlin].

Bardeen, J., Cooper, L. N. and Schrieffer, J. R. (1957). *Phys. Rev.* **108**, 1175.

Demokritov, S. O., Demidov, V. E., Dzyapko, O., Melkov, G. A., Serga, A. A., Hillebrands, B. and Slavin, A. N. (2006). *Nature* **443**, 430.

Einstein, A. (1915). *Sitzungsber. preuss. Akad. Wiss.* **47**, 831.

Einstein, A. (1916). *Ann. d. Phys.* **49**, 284.

Friedmann, A. (1922). *Zeits. für Physik* **10**, 377.

Friedmann, A. (1924). *Zeits. für Physik* **21**, 326.

Ginzburg, V. L. (1971). *Usp. Fiz. Nauk* **103**, 87 [*Sov. Phys. Usp.* **14**, 21].

Ginzburg, V. L. (1995). *O fizike i astrofizike* (Russian title) [*On physics and astrophysics*]. Byuro Kvantum, Moscow.

Ginzburg, V. L. (2001*a*). *Izvestia*, January 11, 2001; also, *Literaturnaya Gazeta*, June 9, 2003. This article is also accessible on the site www.ufn.ru, section "Tribuna", entry 57.

Ginzburg, V. L. (2001*b*). *The Physics of a Lifetime. Reflections on the Problems and Personalities of* 20th *Century Physics*. Springer, Berlin.

Ginzburg, V. L. (2003). The Nobel Lecture. See in the first part of (Ginzburg, 2006); (2004) *Usp. Fiz. Nauk* **174**, 1240; and also (2004) *Rev. Mod. Phys.* **76**, 981.

Ginzburg, V. L. (2003,2004). *O nauke, o sebe i o drugikh* (Russian title). Fizmatlit, Moscow.

Ginzburg, V. L. (2005). *About Science, Myself, and Others.* IOP Publ., Bristol.

Ginzburg, V. L. (2006). *On superconductivity and superfluidity. Autobiography.* [in Russian]. Fizmatlit, Moscow. [An English translation of this book is expected to appear from Springer].

Glauber, R. J. (2006) *Rev. Mod. Phys.* **78**, 1267.

Hall, J. L. (2006) *Rev. Mod. Phys.* **78**, 1279.

Hansch, T. W. (2006) *Rev. Mod. Phys.* **78**, 1297.

Landau L. D. and Lifshitz E. M. (1982). *Classical Theory of Fields* . Pergamon, New York.

Kasprzak, J., Richard, M., Kundermann, S., Baas, A., Jeambrun, P., Keeling, J. M. J., Marchetti, F. M., Szymańska, M. H., André, R., Staehli, J. L., Savona, V., Littlewood, P. B., Deveaud, B. and Le Si Dang. (2006). *Nature* **443**, 409.

Maksimov, E. G. and Savrasov, D. Yu. (2001). *Solid State Commun.* **119**, 569.

Nete, A. C., Guinea, F., and Peres, N. G. (2006). *Physics World* **19** (11), 33.

Pickett, N. E. (2006). *Journal of Superconductivity* **19** 291.

3

HYBRID ORGANIC–INORGANIC NANOSTRUCTURES AND LIGHT–MATTER INTERACTION

V. M. Agranovich
NanoTech Institute, The University of Texas at Dallas, Richardson, TX
75083-0688, USA
Institute for Spectroscopy, Russian Academy of Sciences, Troitsk, Moscow
district, 142092 Russia

Abstract

The combination of organic and inorganic materials in a single nanostructure may lead to novel devices which benefit from the advantages of both classes of materials. Following this idea, we discuss the properties of electronic excitations in nanostructures based on a combination of organic materials and inorganic semiconductors, having, respectively, Frenkel and Wannier–Mott (W-M) excitons with nearly equal energies. The resonant coupling between Frenkel and W-M excitons in quantum wells (QWs) may lead to many striking novel effects, among them: (i) strong enhancement of the resonance all-optical nonlinearity in the strong coupling regime and (ii) fast and efficient noncontact and nonradiative excitation energy transfer from inorganic quantum wells to organic or inorganic materials (with a broad spectrum and strong relaxation) in the weak coupling regime. The latter effect permits us to put forward a new concept for light-emitting devices: the electrical pumping of excitations in the semiconductor quantum well can efficiently turn on the organic material luminescence. We briefly mention analogous processes in a microcavity configuration where giant polariton Rabi splitting drastically changes the kinetics of luminescence and the conditions for polariton condensation.

3.1 Introduction: Hybrid organic–inorganic nanostructures

The need for systems with better optoelectronic properties to be used in applications has moved researchers to materials science in order to develop novel compounds and novel structures. The progress in the field is impressive due to the use of innovative growth techniques and the realization of systems in low-dimensional confined geometries. We have now many newly developed inorganic and organic structures with very unusual properties. Particularly, as a result of the fast development of organic materials science, the tremendous advances in organic-based electronic materials applications have recently been reported, from organic lasers to molecular-scale transistors. The advances have generated a vital and growing interest in organic materials research, and could potentially revolutionize future electronic applications.

In these brief notes we will discuss the optoelectronic properties of hybrid organic–inorganic nanostructures containing semiconductor and organic components with nearly equal energies of excitons. We will demonstrate that such structures, with properly selected materials of organic and inorganic components, can occupy a suitable place in the development of organics based material science. The structures are interesting from the point of view of basic science as well as for applications in optoelectronics because in one and the same hybrid structure we can combine ingeniously the high conductivity of the semiconductor component (important for electrical pumping) with the strong light–matter interaction of the organic component. In this way, the desirable properties of both the organic and inorganic material unite and in many cases can overcome the basic limitations of each.

Quoting Oulton *et al.* (2003), "In the future, the combination of both materials systems will allow the rules that apply to matter and light to be further stretched, potentially seeding a new paradigm in optoelectronic devices". This sentiment, substantiated by the results already obtained, evidently calls for an intense research effort.

We will describe only the most important qualitative features of hybrid structures and also some recent experimental observations of the predicted effects. It is important to have in mind that, in consideration of resonant interaction of systems, the dissipation of excited states plays a very important role. At the resonance of two states the splitting of these states in the energy spectrum can be formed, but this takes place only if a dissipative width of resonating states is small in comparison with the splitting value. This limiting case (a strong coupling regime) has been considered in the beginning of our studies. However, for organic materials the comparatively large dissipative width of excited electronic states is typical. Therefore the weak coupling can be expected for many hybrid organic–inorganic resonant structures. The consideration of both limiting cases is interesting because the physics behind them as well as the possible applications are very different. The hybrid structures in the limit of strong coupling between Frenkel and W-M excitons have been investigated in different geometrical configurations such as QWs (Agranovich *et al.*, 1994; La Rocca *et al.*, 1995), quantum wires (Yudson *et al.*, 1995) and quantum dots (Engelmann *et al.*, 1998). The resonant interaction between Frenkel excitons in an organic layer and W-M excitons in a semiconductor QW was also considered in a microcavity configuration (Agranovich *et al.*, 1997b; Benisti *et al.*, 1998), where organic and inorganic materials are separated. The hybrid structures in the weak coupling regime have been investigated in (Agranovich *et al.*, 1997a; Basko *et al.*, 1999; Agranovich and Basko, 1999; Basko *et al.*, 2000) In these papers the electronic excitation energy transfer from an inorganic QW to an organic overlayer has been considered for different configurations; it was shown that in many cases this process is very fast, much faster than radiative or nonradiative processes in the considered inorganic QWs. Having in mind that inorganic QWs can easily be pumped electrically, this observation gave the possibility for formulation of a new concept of light-emitting devices (Agranovich *et al.*, 2001). Before considering the above mentioned results – more information can be found in (Agranovich *et al.*, 1998) and in (Agranovich and Bassani, 2003) – we have to recall some peculiarities of excitons in organic and inorganic materials.

3.1.1 *Excitons in inorganic and organic materials*

It is known that the electronic excitations known as excitons play a fundamental role in the optical properties of dielectric solids. They correspond to a bound state of one electron and one hole and can be created by light or can appear as a result of relaxation processes of free electrons and holes, which, for example, can be injected electrically. There are three models conventionally used to classify excitons: the small-radius Frenkel and charge transfer exciton models and the large-radius W-M exciton model. In principle, there is also a possibility to have strongly enhanced W-M excitons in layered inorganic–organic structures (Ishihara *et al.*, 1990; Muljarov *et al.*, 1995; Ishihara, 1995), because of the effect of Coulomb interaction enhancement (Rytova, 1967; Keldysh, 1979) when a W-M exciton in a narrow inorganic QW with large dielectric constant is surrounded by an organic material with a smaller dielectric permittivity. Below we will discuss only the structures with Frenkel and W-M excitons.

The internal structure of W-M excitons can be represented by hydrogen-like wavefunctions. Such a representation results from the two-particle, Coulombic electron–hole states in a crystalline periodic potential. The mean electron–hole distance for this type of exciton is typically large (in comparison with the lattice constant). On the other hand, the Frenkel exciton is represented as an electronic state of a crystal in which electrons and holes are placed on the same molecule. We can say that Frenkel excitons in organic crystals have radii a_F which is comparable to the lattice constant, $a_F \sim a \sim 5\,\text{Å}$. In contrast, weakly bound W-M excitons in semiconductor QWs have large Bohr radii ($a_B \sim 100\,\text{Å}$ in III-V materials and $a_B \sim 30\,\text{Å}$ in II-VI ones; in both cases $a_B \gg a$). The oscillator strength of a Frenkel exciton is close to the molecular oscillator strength F and, due to the strong overlapping of electron and hole wavefunctions in a molecule, can sometimes be very large (order of unity), whereas the oscillator strength f of a W-M exciton is usually much weaker: in a QW $f \sim a^3\, a_B^{-2}\, L^{-1}\, F$, where L is the QW width ($a_B > L > a$). In high-quality semiconductors as well as in organic crystalline materials, the optical properties near and below the bandgap are dominated by the exciton transitions and the same situation takes place also for organic and inorganic QWs (or wires or dots). The excitonic optical nonlinearities in semiconductor QWs can be large because the ideal bosonic approximation for W-M excitons breaks down as soon as they start to overlap with each other, i.e., when their 2D density n becomes comparable to the saturation density $n_S \sim 1/(\pi a_B^2)$ (n_S is, typically, $10^{12}\,\text{cm}^{-2}$). Then, due to phase space filling, exchange and collisional broadening, the exciton resonance is bleached.

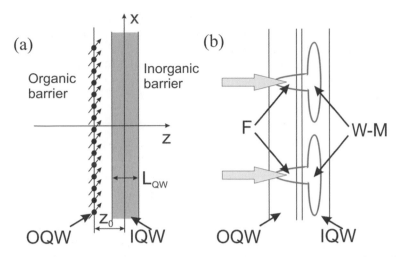

FIG. 3.1. (a) The configuration under study: organic and inorganic QWs (OQW and IQW). (b) Frenkel (F) and W-M excitons in the nanostructure. Light pumping (grey arrows) through Frenkel excitons and exciton–exciton interaction through W-M excitons cause a strong enhancement of the resonant $\chi^{(3)}$ nonlinearity.

The generic figure of merit for all optical nonlinearities scales like $I_P^{-1}(\Delta\chi/\chi)$, where $\Delta\chi$ is the nonlinear susceptibility change in the presence of pump of intensity I_P. As $\Delta\chi/\chi \sim n/n_S \sim n a_B^2$, but also $n \propto f\,I_P \propto a_B^{-2}\,I_P$, such a figure of merit in semiconductors can be large but nearly independent of the exciton Bohr radius (Green *et al.*, 1990). This conclusion did not leave any hope that the transition to organic materials could be useful for the creation of devices with enhanced optical nonlinearity.

3.2 Strong coupling regime

3.2.1 *Hybrid 2D Frenkel–Wannier–Mott excitons at the interface of organic and inorganic quantum wells*

The situation changes drastically in hybrid nanostructures in which Frenkel and W-M excitons are in resonance with each other and coupled through their dipole–dipole interaction at the interface (Agranovich *et al.*, 1994). Qualitatively, this effect can be understood taking into account that in semiconductor and organic nanostructures, interacting at the interface, new eigenstates can be realized by appropriate coherent linear combinations of large-radius exciton states in the inorganic material and small-radius exciton states in the organic one (the so-called strong coupling regime). These hybrid electronic excitations are characterized by a radius dominated by their 2D W-M component and by an oscillator strength dominated by their Frenkel component. One of the most natural choices to implement this idea is a layered structure with an interface between a covalent

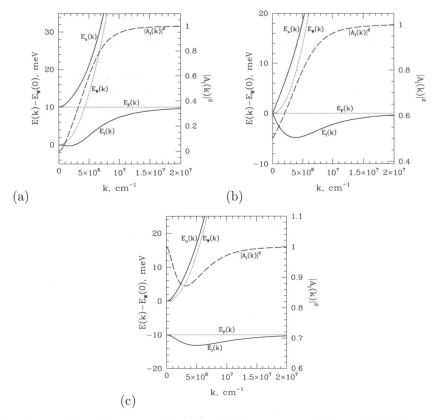

FIG. 3.2.   The dispersion $E_{\mathrm{u,l}}(k)$ of the upper and lower hybrid exciton
branches (solid lines) and that of the unperturbed Frenkel and Wannier
excitons (dotted line). The "weight" of the Frenkel exciton component in
the lower branch $|A_1(k)|^2$ is shown by the dashed line. The interaction pa-
rameters are: the electric transition dipole between the conduction and valence band $d^{\mathrm{vc}} = 12$ Debye, the transition dipole moment in organic molecule
$d^{\mathrm{F}} = 5$ Debye, $a_{\mathrm{B}} = 2.5$ nm, $a_{\mathrm{F}} = 0.5$ nm, $L_{\mathrm{W}} = 1$ nm, $z_0 = 1$ nm, effective
mass of Wannier exciton $m_{\mathrm{W}} = 0.7\,m_0$, dielectric constants in QW and in
organics $\varepsilon_\infty = 6$ and $\tilde{\varepsilon}_\infty = 4$, respectively. The detuning $\delta = E_{\mathrm{F}}(0) - E_{\mathrm{W}}$
is 10, 0 and -10 meV in panels (a), (b) and (c), respectively. From (Agra-
novich *et al.*, 1994). Adapted figure with permission from Elsevier Ltd: [V. Agranovich,
R. Atanasov and F. Bassani, *Solid State Commun.* **92**, 295 (1994)], copyright (1994).

semiconductor and a crystalline molecular semiconductor, see Fig. 3.1(a). The
hybrid excitons can have a large oscillator strength F and, at the same time,
a small saturation density n_{S}. Generally speaking, we can say that the hybrid
exciton interacts with light through its Frenkel component and interacts with

another hybrid exciton due to the large-radius W-M component (Fig. 3.1b).

In this way, we can expect that the desirable properties of both the inorganic and organic material combine to overcome the basic limitation mentioned above for the figure of merit of the exciton resonance nonlinearities. Figure 3.2 demonstrates the dispersion of the new hybrid exciton states calculated for an inorganic QW representative of II-VI semiconductor (e.g., ZnSe/ZnCdSe) QWs: $\varepsilon = \varepsilon_\infty = 6$, $d^{vc}/a_B \approx 0.1e$ (which corresponds to $d^{vc} \simeq 12$ Debye and a Bohr radius of 2.5 nm), the exciton mass $m_W = 0.7\,m_0$ and the well width $L_W = 1$ nm. For the organic part of the structure, typical parameters relevant to such a media are taken: $\tilde{\varepsilon} = \tilde{\varepsilon}_\infty = 4$, the transition dipole for the molecules in the monolayer $d^F = 5$ Debye and $z_0 = 1$ nm. The dispersion curves of the upper and lower branches, $E_{u,l}(k)$, are plotted in Fig. 3.2 for three values of the detuning $\delta = 10$ meV, $\delta = 0$ and $\delta = -10$ meV. The characteristic interaction energy of Frenkel and W-M excitons, which determines the value of splitting at resonance as it follows from calculations (Agranovich et al., 1994), depends on the wavevector of interacting Frenkel and W-M excitons, and for the considered parameters is about 10 meV. This means that, as can be estimated from the uncertainty principle, the characteristic energy transfer time between excitons in the structure is of the order of 0.1 ps, i.e., much shorter than the radiative lifetime of Frenkel as well as W-M excitons. Qualitatively this result can be understood taking into account that the exciton–photon interaction energy which determines the exciton radiative lifetime is roughly a product of the exciton transition dipole moment and photon electric field amplitude. In contrast to a photon in a microcavity, a photon in free space is distributed over three-dimensional (3D) space; its electric field at the place where the exciton dipole moment is located is much smaller than the electric field of the nearby second exciton transition dipole moment. We will see that this observation of short energy transfer time is also important for the hybrid structures with weak coupling of W-M excitons with an organic overlayer.

The above theory of a hybrid state allows us also to estimate the $\chi^{(3)}$ resonant optical nonlinearity of the hybrid structure. It was demonstrated analogously, that the combined organic–inorganic nanostructures are able to overcome the above-mentioned basic limitation for the figure of merit of the exciton resonance nonlinearities. The calculation of M_{hybr}, the figure of merit of the resonant $\chi^{(3)}$ nonlinearity of hybrid nanostructures made by La Rocca et al. (1995), showed that its ratio $\eta = M_{\text{hybr}}/M_{\text{semQW}}$ to the figure of merit of the semiconductor component alone, M_{semQW}, is equal to $\eta \simeq [a_B/a_F]^2$, where a_B and a_F are the Bohr radii of the W-M and Frenkel excitons, respectively. Thus, the resonant nonlinearity in a hybrid structure in the strong coupling regime and for a given pump intensity can be about two orders of magnitude larger than that in a typical semiconductor QW. It is not a simple task to produce a hybrid strongly coupled structure because it needs a crystalline organic component with nearly perfect interfaces with the barrier layer or directly with the semiconductor QW. Nevertheless, attempts to create the described organic–inorganic structures are in progress and an example of such an effort is demonstrated in (Braun *et al.*, 1999).

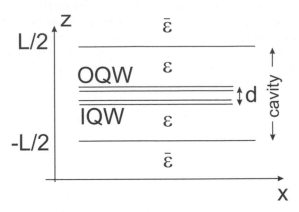

FIG. 3.3.   Schematic of a microcavity embedded in organic and inorganic QWs.

3.2.2   *Hybridization of Frenkel and W-M excitons in a 2D microcavity*

It is known that polaritons in 3D cubic crystals have a gap in the dispersion curve in the vicinity of the exciton resonance equal to so-called excitonic transverse-longitudinal splitting. This splitting arises due to the interaction of 3D transverse photons and 3D transverse excitons (or optical phonons) and can be observed if this splitting is large in comparison with the dissipative width of the exciton resonance. Using modern language accepted in the nanoscience literature, we can say that a strong exciton–photon coupling takes place in this case. The new states (polaritons) are superposition of excitonic and photonic states and are hybrid in this sense. It is clear that an analogous effect should exist also for interacting 2D light and 2D excitations. Indeed, such a splitting (the effect of strong coupling) has been found for 2D surface polaritons interacting with dipole active excitations in thin films on the surface supporting surface waves (Agranovich and Malshukov, 1974). The magnitude of splitting was $\sim \sqrt{d/\lambda}$ where d is the film thickness and λ is the surface polariton wavelength. Later the splitting in the spectrum of surface waves was observed in the infrared, visible and ultraviolet spectral ranges; for a review see (Agranovich and Leskova, 1988) and (Vinogradov and Leskova, 1990). More recently, a strong coupling between surface plasmons and excitons in a layer of organic cyanine dye J-aggregates deposited on a silver film was demonstrated (Bellessa *et al.*, 2004). At room temperature the resonant splitting of Frenkel excitons and surface plasmons was found to be as much as 180 meV. The work of Bellessa *et al.* (2004) is actually the second demonstration of a strong coupling between the Frenkel exciton and surface plasmon, after Pockrand *et al.* (1982) (with a dye Langmuir–Blodgett bimonolayer on silver). However, in the latter reference, the observed splitting was found to be much smaller.

The strong coupling regime for 2D light has also been observed by Weisbuch *et al.* (1992) in the spectrum of 2D cavity polaritons in a microcavity containing a semiconductor QW with the excitonic resonance. The corresponding splitting, now called the Rabi splitting, is usually of the order of 5 meV in a microcavity with inorganic semiconductor QWs. In a microcavity containing spatially separated organic and inorganic QWs (see Fig. 3.3) we can expect a resonant interaction between different exciton states through a virtual cavity photon exchange. Similarly to a microcavity with a single QW, the interaction between organic and inorganic QWs can be strong. This takes place if the cutoff frequency of the cavity photon is close to the exciton resonances. In this case the hybrid states are superpositions of three states: two exciton states and one cavity photon state. In this case new hybrid Frenkel-W-M exciton + cavity photon states can be tailored to engineer the fluorescence efficiency and relaxation processes in a microcavity (Agranovich *et al.*, 1997b; Benisti *et al.*, 1998).

The strong coupling of semiconductor W-M excitons with cavity photons is a mechanism working only in a small fraction of k (momentum) space close to the light line, with k much smaller than that of the thermalized excitons. Relaxation of excitons to these small k is a slow and inefficient process ($\gtrsim 1$ ns) mediated by weak acoustic phonons (Sumi, 1976; Stanley *et al.*, 1996). This bottleneck precludes the use of strong coupling to speed up the spontaneous emission whose dynamics remains dominated by uncoupled excitons. Relaxation processes in hybrid organic–inorganic microcavities capitalize on the strong interaction with the organic's phonons, both for intrabranch and interbranch relaxation, and the relaxation process time-constant can be estimated to be of the order of ~ 100 ps. Up to now only strong optical coupling of two different Frenkel excitons from two separated organic layers in a microcavity has been observed (Lidzey *et al.*, 2001). As could be expected (Agranovich *et al.*, 1997b, Benisti *et al.*, 1998), three cavity polariton branches were found in this paper, with the middle branch containing a significant component of the cavity photon and both of the exciton states.

3.2.3 *Giant Rabi splitting in organic microcavities*

The prediction of a giant Rabi splitting in a microcavity containing organic material with a large exciton transition oscillator strength has been an important by-product of the paper by Agranovich *et al.* (1997b). According to the estimation made in this paper, the Rabi splitting in an organic microcavity with appropriate selection of organic material can be about two orders of magnitude larger than that in a microcavity with the inorganic semiconductor QW. This prediction was very soon confirmed in the experiments performed by Bradley and his collaborators (Lidzey *et al.*, 1998, 1999; Tartakovskii *et al.*, 2001; Schouwink *et al.*, 2002; Takada *et al.*, 2003): a Rabi splitting as large as 200–500 meV has been observed in organic microcavities. An excellent discussion of the similarities and differences between the inorganic and organic semiconductor microcavities operating within the strong coupling regime and also of the prospects for inorganic/organic hybrid materials for optoelectronic devices can be found in

(Oulton *et al.*, 2003). This field of investigation is now growing quickly, see for example (Holmes and Forrest, 2004, 2005). In these papers the authors examine the influence of singlet–triplet intersystem crossing and excited state molecular relaxation on strong exciton–photon coupling in optical microcavities filled with small molecular weight organic materials. As intersystem crossing and molecular relaxation quench the strong coupling, the interplay between strong coupling and relaxation processes offers a unique opportunity to probe directly the fundamental ultrafast excitonic phenomena and can be used for estimation of their relative transition rates.

In all these cases a strong coupling regime has been achieved in microcavities with amorphous organics (porphyrin dyes, J-aggregated cyanine dyes and σ-conjugated silane polymers). In such microcavities in the region of exciton resonances the co-existence of coherent and localized excitations in the excitation spectrum has been predicted (Agranovich *et al.*, 2003; Agranovich and La Rocca, 2005; Michetti and La Rocca, 2005). This rather complicated structure of the cavity exciton–polariton spectrum gives rise to a very unusual cavity polariton dynamics. The first investigations of such dynamics have been performed by Song *et al.* (2004).

As far as we know, the strong coupling regime in microcavities with crystalline organics has not been observed yet. Nevertheless, the cavity polariton spectra in crystalline organics and the rates of their relaxation processes have been studied theoretically (Litinskaya *et al.*, 2004a, 2004b). The polariton nonlinear dynamics in such a microcavity could also be interesting (Zoubi and La Rocca, 2005a, 2005b). Having in mind that the kinematic interaction between Frenkel excitons has a different nature in comparison with such interaction between W-M excitons in inorganic microcavities, we can expect that collective properties of polaritons in organic microcavities will have many peculiarities in their dynamics and condensation.

3.3 Weak coupling regime in hybrid nanostructures

3.3.1 *The Förster energy transfer*

In order to discuss the weak coupling between nanostructures, it is instructive to recall the nonradiative mechanism of Förster resonant energy transfer from an excited molecule (a donor) to some another molecule (an acceptor) which can be in the ground or excited state. The probability of such transfer is determined by the Coulomb nonretarded (instantaneous) dipole–dipole interaction between the molecules. It is proportional to R_{F}^{6}/R^{6}, where R_{F} is the Förster radius and R is the distance between molecules. For organic materials the Förster radius is usually about several nanometers and depends strongly on the overlapping of the donor fluorescence spectrum with the acceptor absorption spectrum. In this formula any footprints of dissipation are absent, because actually this formula is correct only in the limit of very strong dissipation in the acceptor. Thus, using the Förster mechanism to describe the energy transfer between molecules, we assume that dissipation in the acceptor is much faster than the energy transfer in the opposite direction, from acceptor to donor; for more details see (Agranovich and Galanin, 1982).

The same physical picture can be applied to hybrid nanostructures in the weak coupling regime. In this case the dissipation of the excited state in the acceptor (organic or inorganic) component of the nanostructure is very strong, but the energy of resonant Coulomb coupling between the semiconductor nanostructure and the organic material is not strong enough to produce a stable coherent superposition of excited states of organic and inorganic components of the nanostructure. The energy transfer probability on the distance between components of the nanostructure strongly depends on the geometry of nanostructures and differs, for example, in QWs and quantum dots. Below we describe the results of calculations for the energy transfer from a semiconductor QW to an organic overlayer in the planar geometry only.

3.3.2 *Förster energy transfer in a planar geometry*

The calculations of the energy transfer rate in this case can be simplified, because the nonradiative (Förster-like) energy transfer from the QW to organics is nothing but the Joule losses in the organic material. They are produced by the penetration of the electric field generated by the semiconductor exciton polarization into the organics which has to be calculated in the framework of the microscopical exciton theory. Thus, to consider the energy transfer from a semiconductor QW to an overlayer, we can use a semiclassical approach and characterize the overlayer (it can be organic or something else) as a medium with a dielectric function $\varepsilon_{ij}(\omega)$. Then, as the result of quantum-mechanical calculations, it is necessary to find $\mathbf{P}(\omega, \mathbf{r}, t)$, the density of the exciton transition dipole moment in the semiconductor microcavity. In the considered case it is not possible to use the approximation of a point transition dipole because the distance between the semiconductor QW and the overlayer is of the order of the thickness of the QW. Having in mind that $\rho = -\mathrm{div}\,\mathbf{P}$ is the coupled charge density in a QW and using the Poisson equation with appropriate boundary conditions we can determine the distribution of potential and electric field $\mathbf{E}(x, y, z)$ in the overlayer induced by the exciton polarization. The knowledge of this field and also of the overlayer material dielectric function gives us the possibility to calculate the Joule losses of the electromagnetic field created by the exciton polarization in the overlayer by using the well-known formula from the electrodynamics of continuous media (Landau and Lifshitz, 1958). For the rate of energy dissipation of a monochromatic electromagnetic field we have:

$$\mho = \frac{1}{2\pi\hbar} \int \mathrm{Im}\varepsilon_{ij}(\mathbf{r}, \omega_{\mathrm{exc}}) E_i E_j d^3\mathbf{r}\,,$$

and the corresponding exciton energy $\hbar\omega_{\mathrm{exc}}$ transfer time from the semiconductor QW to the overlayer can now be found from the relation

$$\tau = \hbar\omega_{\mathrm{exc}}/\mho\,.$$

Although in such an estimation an infinitely thick layer of organic molecules

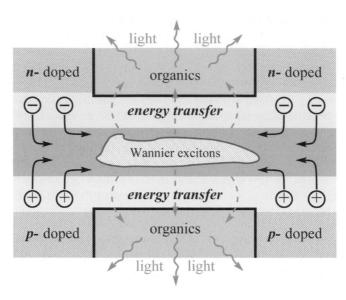

FIG. 3.4. Schematic of the planar structure under discussion.

was assumed, the obtained results are correct with rather high accuracy even for ∼10 nm-thick layers because the exciton polarization induced electric field penetrates into the organic layer only on the thickness of the semiconductor QW and the barrier between the QW and organic layer.

Such a calculation procedure, proposed originally in (Agranovich*et al.*, 1997a) as well as in (Agranovich *et al.*, 1994), is now used in many other papers. For the case considered here, i.e., weak coupling, it corresponds exactly to the results of consecutive microscopic quantum mechanical calculations in the framework of Fermi's Golden Rule and can be applied also for the derivation of the energy transfer in the weak coupling regime for the case of other types of nanostructures, quantum wires, quantum dots and so on. Since the organic material is described here only by its dielectric function, the same theory can be applied if another material with fast relaxation and a broad spectrum in the region of the W-M exciton transition is used in place of the organic material as acceptor. For instance, it was applied recently to study the regimes of energy transfer from epitaxial QW to a proximal monolayer of semiconductor nanocrystals (Kos *et al.*, 2005).

The numerical estimations in (Agranovich *et al.*, 1997a; Basko *et al.*, 1999; Agranovich and Basko, 1999; Basko *et al.*, 2000) have been performed for electron–hole (e-h) excitations in a semiconductor QW including free and localized excitons and unbound e-h pairs in typical QWs of II-VI and III-V semiconductors and typical organic materials (see the schematic in Fig. 3.4). These calculations demonstrate that nonradiative energy transfer from the semiconductor QW to the organic overlayer in combination with strong dissipation of excited states in organic materials occurs on time-scales of several tens of picoseconds for II-VI semiconductors and several hundreds picoseconds for III-V semiconductors. In both cases it is significantly shorter than the semiconductor excitation lifetime in the absence of such a transfer. The authors of similar calculations of the energy transfer from epitaxial QW to a proximal monolayer of semiconductor nanocrystals (Kos *et al.*, 2005) indicate that with a careful design of the geometrical and electric parameters of the system, the Förster transfer can be used as an efficient "noncontact" pumping mechanism in nanocrystal QD-based light-emitting devices.

For the studies of the energy transfer dynamics from a semiconductor QW to organics it is very important to know the origin of exciton luminescence in the semiconductor QW at electrical or optical nonresonant pumping as well as the characteristic building-up time of the exciton line as a function of excitation density and temperature. In more general terms it is important to know the dynamics which determine the concentrations of free carriers as well as of free and localized W-M excitons. An interesting discussion of these problems can be found in (Deveaud *et al.*, 2005) where the time-resolved photoluminescence experiments in a very high-quality InGaAs QW sample have been carried out and the exciton versus free carrier luminescence has been investigated in the time domain.

3.3.3 *Non-contact pumping of light emitters via nonradiative energy transfer: A new concept for light-emitting devices*

Over the past few years, considerable progress has been made in the development of organic light-emitting devices (LEDs) in the visible range using polymers as well as small molecular compounds. In all such organics-based devices a fundamental role is played by electroluminescence, the generation of light by electric excitation. Already in early studies (Pope *et al.*, 1963) it was established that the process responsible for the electroluminescence requires an injection of electrons from one electrode and holes from the other, the transport of one or both charges, the capture of oppositely charged carriers on the same molecule (or recombination center), and the radiative decay of the resulting excited electron–hole state. For inorganic semiconductors, where the recombining electron–hole state can be a W-M exciton, all the above mentioned processes are investigated and well documented (Weisbuch and Vinter, 1991). In organics, in contrast to inorganic semiconductors, researchers have met many problems in the use of electroluminescence for the creation of devices conceptually similar to what was done with the use of semiconductor materials. The main reasons are the small mobility of carriers and strong chemical interaction between organics and metals which prevents the injection of charges into organics. Due to the small mobility of charge carriers in organic materials the current is principally bulk-limited, through the built-up of a space charge.

To avoid the above mentioned problems for light-emitting devices, Agranovich *et al.*, (1997a, 2001) proposed using hybrid organic–inorganic structures

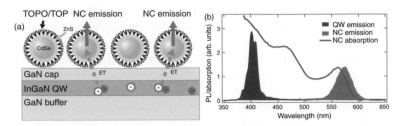

FIG. 3.5.   Schematics of the hybrid QW/nanocrystal structure and illustration of its optical properties. (a) The structure consists of an In-GaN/GaN QW heterostructure with a monolayer of trioctylphosphine oxide/trioctylphosphine (TOPO/TOP)-capped CdSe/ZnS core/shell nanocrystals on top of it. Electron–hole pairs from the QW can transfer resonantly and nonradiatively into nanocrystals. The nanocrystals excited by energy transfer emit light with a wavelength determined by the nanocrystal size. (b) The emission of the QW (black) overlaps spectrally with the absorption of the nanocrystals (line). For CdSe nanocrystals of 1.9 nm radius, the emission wavelength is around 575 nm (gray). From (Achermann *et al.*, 2004). Adapted figure with permission from Macmillan Publishers Ltd: [M. Achermann, M. A. Petruska, S. Kos, D. L. Smith, D. D. Koleske and V. I. Klimov, *Nature* **429**, 642 (2004)], copyright (2004).

to combine the comparatively good transport properties of semiconductors (e.g., electrically pumped) and good light-emitting properties of organic substances. If an organic material with a broad absorption band in the optical range overlapping with the semiconductor exciton resonance is placed near the semiconductor, the resonant dipole–dipole interaction between the two substances in combination with the fast dephasing, common for many organic substances (e.g., due to the scattering by phonons), will result, as we discussed above, in the efficient nonradiative, noncontact transfer of the semiconductor excitation energy to the organics. Of course, in this case it is important to use organic materials with a high-luminescence quantum yield. Due to the energy transfer from the semiconductor QW to the organics, the nonradiative processes with a characteristic time longer than that of energy transfer will be suppressed. As a result, we can expect that the quantum efficiency of the considered layer structure will be increased in comparison with the quantum efficiency of the semiconductor QW in the absence of the energy transfer to the organics, provided that the luminescence quantum yield of the emissive organic layer is high enough.

3.3.4   *First experiments*

Although in theoretical papers (Agranovich *et al.*, 1997a, 2001) organic materials were used as a model acceptor, the first experimental demonstration of the LEDs of the new concept was performed by Klimov's group (Achermann *et al.*, 2004) in an investigation of the energy transfer from a QW (donor) to a monolayer of semiconductor nanocrystals (acceptor). The latter, however, had a broad absorption spectrum like most organic materials. The structure studied in these experiments is depicted schematically in Fig. 3.5(a). It consists of an InGaN QW, on top of which a close-packed monolayer of highly monodisperse CdSe/ZnS core/shell nanocrystals was assembled using the Langmuir–Blodgett technique. The nanocrystals consist of a CdSe core (radius 1.9 nm) overcoated with a shell of ZnS (0.6 nm thickness), followed by an external layer of organic molecules, trioctylphosphine (TOP) and trioctylphosphine oxide (TOPO). These nanocrystals show efficient emission centered near 575 nm and a structured absorption spectrum with the lowest 1S absorption maximum at 560 nm (Fig. 3.5b). QW samples were grown on sapphire substrates by metal–organic chemical-vapor deposition. They consisted of a 20 nm GaN nucleation layer, a 3 nm GaN bottom barrier, and a 3 nm InGaN QW. The concentration of In in the QWs was 5–10%, which corresponds to an emission wavelength of ~ 400 nm (Fig. 3.5b). This wavelength is in the range of strong nanocrystal absorption, which provides strong coupling of QW excitations to the absorption dipole of nanocrystals, and should allow efficient energy transfer. The emission color of semiconductor nanocrystals can be modified by changing their size as a result of quantum-confinement effects. Such spectral tunability, together with large photoluminescence quantum yields and high photostability, makes nanocrystals attractive for the use in a variety of light-emitting technologies. It is difficult to achieve the electrical pumping in the case of nanocrystals largely due to the presence of an insulating organic capping layer on its surface.

An approach with indirect injection of electron–hole excitations into nanocrystals (see also Fig. 3.6) by this noncontact nonradiative Förster-like energy transfer from a proximal QW, which can in principle be pumped either electrically or optically, allows us to solve the problem of nanocrystal pumping. The results obtained by Klimov's group indicate that this energy transfer is fast enough to compete with the electron–hole recombination in the QW, causing greater than 50 % energy transfer efficiencies in the tested structures. The measured energy-transfer rates (Achermann *et al.*, 2004) are sufficiently large to provide pumping even in the stimulated emission regime, indicating the feasibility of nanocrystal-based optical amplifiers and lasers based on this approach; see also (Klimov and Achermann, 2005).

The shortcoming of a nanocrystal monolayer is in its small thickness. Actually the electric field created by the QW exciton polarization spreads out on a length much larger than the thickness of the monolayer of nanocrystals and can pump a larger volume of emissive material. In the case of an organic overlayer this shortcoming can easily be eliminated. Using organic materials to create LEDs of the new type will enrich this field of investigations. There is no doubt that organic LEDs of the new type will be brighter, cheaper and much more multicolored.

The first demonstration of the hybrid structure with an organic overlayer which we discussed in our papers (Agranovich *et al.*, 1997, 1999, 2001) has been recently done by Donal Bradley and his collaborators (Heliotis *et al.*, 2006). In

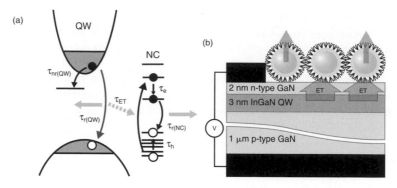

FIG. 3.6. Carrier relaxation and energy-transfer processes in the hybrid QW/nanocrystal structure and a schematic of an electrically driven energy-transfer device. (a) The QWNC energy transfer competes with radiative and nonradiative decay processes in the QW. High-energy excitations created in the nanocrystals through energy transfer, relax rapidly to the nanocrystal band edge, preventing the back transfer. Subscript *e* stands for electron and *h* for hole. (b) An electrically powered hybrid QW/nanocrystal device that can be used to realize the energy transfer color-converter in the regime of electrical injection. It comprises an InGaN QW sandwiched between p-type bottom and thin n-type top GaN barriers with metal contacts attached. From (Achermann *et al.*, 2004). Adapted figure with permission from Macmillan Publishers Ltd: [M. Achermann, M. A. Petruska, S. Kos, D. L. Smith, D. D. Koleske and V. I. Klimov, *Nature* **429**, 642 (2004)], copyright (2004).

this paper the fabrication and study of a hybrid inorganic/organic semiconductor structure has been reported, employing an organic polyfluorene thin film with the opportunity to place it in sufficiently close proximity to the underlying InGaN QW. As we mentioned, such an architecture has the potential to take advantage of the complementary properties of the two classes of semiconductors that it contains and thus might lead to high-performance devices with desirable electrical and optical characteristics. A schematic of the fabricated structures is shown in Fig. 3.7. The UV light-emitting InGaN QW (2 nm thick) is spaced from the blue-light-emitting poly(9,9-dioctylfluorene-co-9,9-di(4-methoxy)phenylfluorene) film (5 nm thick) by GaN cap layers of variable thickness.

Energetic alignment (as illustrated in the energy-level diagram of Fig. 3.7) is thus needed in order to maximize the resonant coupling between the inorganic and organic excitations. Radiative decay of the latter should then result in light emission with the characteristic spectrum of the organic layer. Figure 3.7 shows the corresponding InGaN QW photoluminescence (PL) emission and the absorption and PL spectra of the organic film. The QW has a narrow PL band that peaks at 385 nm, coincident with the organic layer absorption maximum. The organic layer, in turn, emits blue light with a broad structured PL band

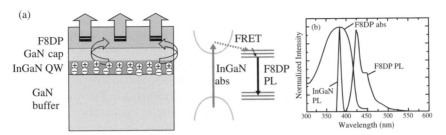

FIG. 3.7. (a) Schematic of the hybrid inorganic/organic semiconductor heterostructures and a simplified energy-level diagram illustrating the proposed
energy-transfer scheme: the excitations initially generated by absorption
(ABS) in the QW layer (InGaN) can resonantly Förster transfer (FRET)
their energy to the singlet excited states of the poly(9,9-dioctylfluorene-
co-9,9-di(4-methoxy)phenylfluorene) (F8DP) polymer layer. Following internal conversion (vibrational relaxation), emission should then occur via radiative decay to the polymer ground state (photoluminescence, F8DP PL). (b)
Absorption spectrum (labelled F8DP ABS) of F8DP and photoluminescence
(labelled PL) emission spectra of the InGaN QW and the F8DP thin film.
From (Heliotis *et al.*, 2006). Adapted figure with permission from: [G. Heliotis, G. Itskos, R. Murray, M. D. Dawson, I. M. Watson, and D. D. C. Bradley, *Adv. Mater.* **18**,
334 (2006)], copyright Wiley-VCH Verlag GmbH & Co. KGaA.

(vibronic features at 425, 450 and 480 nm). For these experiments, three hybrid
heterostructures were fabricated with different GaN cap thicknesses (namely, 15,
4 and 2.5 nm) between the QW and the organics. The variation in GaN cap-layer
thickness allowed them to tune the strength of the dipole–dipole interaction between the QW and organics and, hence, to look for the expected improvement in
organics emission efficiency (relative to the radiative transfer) under nonradiative Förster-like energy transfer from the QW to the organics at small distances
from the QW and the organics. In their concluding remarks the authors (Heliotis
et al., 2006) stress that by adjusting the separation between the inorganic and
organic layers they found an intensity enhancement of organic layer radiation
by 20 times, at least in the case of purely radiative transfer from the QW to
organics. Such structures, as stressed by the authors, are able to take advantage
of the complementary properties of organic and inorganic semiconductors, which
may lead to devices with highly efficient emission across the entire visible band
of the electromagnetic spectrum.

The first practical implementation of a high-efficient nonradiative energy
transfer in an electrically pumped hybrid LED has been performed recently
(Achermann *et al.*, 2006).

Acknowledgements

The results of the investigation of hybrid structures described in these notes were completed by the author during his stays in Italy, France and Germany in tight and fruitful collaborations with Franco Bassani and Giuseppe La Rocca (Scuola Normale Superiore, Pisa), Henri Benisti and Claude Weisbuch (Ecole Polytechnique, Paris), Peter Reineker (University of Ulm), Vladimir Yudson (Institute of Spectroscopy, Troitsk) and with PhD students Denis Basko and Marina Litinskaya. The author is sincerely thankful to all of them. He acknowledges also the partial support through grants from Russian Foundation of Basic Research, from Program "Physics of Nanostructures" and from Russian Ministry of Science and Technology.

References

Achermann, M., Petrushka, M. A., Kos, S., Smith, D. L., Koleske, D. D. and Klimov, V. I. (2004). *Nature* **429**, 642.

Achermann, M., Petrushka, M. A., Koleske, D. D., Crawford, M. H. and Klimov, V. I. (2006). *Nano. Lett.*. **6**, 1396.

Agranovich, V. M. and Galanin, M. D. (1982). *Electronic Excitation Energy Transfer in Condensed Matter*. North-Holland, Amsterdam.

Agranovich, V. M. and Leskova, T. A. (1988). *Progress in Surface Science* **29**, 169.

Agranovich, V. M. and Malshukov, A. G. (1974). *Opt. Commun.* **11**, 169.

Agranovich, V. M., Atanasov, R. and Bassani, F. (1994). *Solid State Commun.* **92**, 295.

Agranovich, V. M., La Rocca, G. C. and Bassani, F. (1997a). *Pisma Zh. Eksp. Teor. Fiz.* **66**, 714 [*JETP Lett.* **66**, 748].

Agranovich, V. M., Benisty, H. and Weisbuch, C. (1997b). *Solid State Commun.* **102**, 631.

Agranovich, V. M., Basko, D. M., La Rocca, G. C. and Bassani, F. (1998). *J. Phys.: Condens. Matter* **10**, 9369.

Agranovich, V. M. and Basko, D. M. (1999). *Pisma Zh. Eksp. Teor. Fiz.* **69**, 232 [*JETP Lett.* **69**, 250].

Agranovich, V. M., Basko, D. M., La Rocca, G. C. and Bassani, F. (2001). *Synth. Met.* **116**, 349.

Agranovich, V. M. and Bassani, F. (2003). *Electronic Excitations in Organic Based Nanostructures*. Elsevier, Amsterdam.

Agranovich, V. M., Litinskaya, M. L. and Lidzey, D. G. (2003). *Phys. Rev. B* **67**, 85311.

Agranovich, V. M. and La Rocca, G. C. (2005). *Solid State Commun.* **135**, 544.

Basko, D. M., La Rocca, G. C., Bassani, F. and Agranovich, V. M. (1999). *Eur. Phys. J.* B **8**, 353.

Basko, D. M., Agranovich, V. M., Bassani, F. and La Rocca, G. C. (2000). *Eur. Phys. J.* B **13**, 653.

Bellessa, J., Bonnand, C., Plenet, J. C. and Mugnier, J. (2004). *Phys. Rev. Lett.* **93**, 036404.

Benisti, H., Weisbuch, C. and Agranovich, V. M. (1998). *Physica* E **2**, 909.

Braun, M., Tuffentsammer, W., Wachtel, H. and Wolf, H. C. (1999). *Chem. Phys. Lett.* **303**, 157.

Deveaud, B., Kappei, L., Berney, J., Morier-Genoud, F., Portella-Oberli, M. T., Szczytko, J. and Piermarocchi, C. (2005). *Chem. Phys.* **318**, 104.

Engelmann, A., Yudson, V. I. and Reineker, P. (1998). *Phys. Rev.* B **57**, 1784.

Green, B. I., Orenstein, J., and Schmitt-Rink, S. (1990). *Science* **247**, 679.

Heliotis, G., Itskos, G., Murray, R., Dawson, M. D., Watson, I. M. and Bradley, D. D. C. (2006). *Adv. Mater.* **18**, 334.

Holmes, R. J. and Forrest, S. R. (2004). *Phys. Rev. Lett.* **93**, 186404.

Holmes, R. J. and Forrest, S. R. (2005). *Phys. Rev.* B **71**, 235203.

Ishihara, T., Takahashi, Jun and Goto, T. (1990). *Phys. Rev.* B **42**, 11099.

Ishihara, T. (1995). *Optical Properties of Low-Dimensional Materials* (ed by T. Ogawa and Y. Kanemitsu). World Scientific, Singapore, p. 287.

Keldysh, L. V. (1979). *Pisma Zh. Eksp. Teor. Fiz.* **29**, 716. [*JETP Lett.* **29**, 658].

Klimov, V. I. and Achermann, M. (2005). *Non-contact pumping of light emitters via non-radiative energy transfer*. U.S. Pat. Appl. Publ., CODEN: USXXCO US 2005253152 A1 20051117.

Kos, S., Achermann, M., Klimov, V. I. and Smith, D. L. (2005). *Phys. Rev.* B **71**, 205309.

La Rocca, G. C., Bassani, F and Agranovich, V. M. (1995). *Nuovo Cimento* D **17**, 1555.

Landau, L. D. and Lifshitz, E. M. (1958). *Electrodynamics of Continuous Media*. Elsevier.

Lidzey, D. G., Bradley, D. D. C., Skolnik, M. S., Virgili, T., Walker, S. and Whittaker, D. M. (1998). *Nature* **395**, 53.

Lidzey, D. G., Bradley, D. D. C., Virgili, T., Armitage, A. and Skolnik, M. S. (1999). *Phys. Rev. Lett.* **82**, 3316.

Lidzey, D. G., Bradley, D. D. C., Skolnik, M. S. and Walker, S. (2001). *Synth. Met.* **124**, 37.

Litinskaya, M., Reineker, P. and Agranovich, V. M. (2004a). *J. Luminescence* **110**, 364.

Litinskaya, M., Reineker, P. and Agranovich, V. M. (2004b). *Phys. stat. sol.* (c) **1**, 646.

Michetti, P. and La Rocca, G. C. (2005). *Phys. Rev.* B **71**, 115320.

Muljarov, E. A., Tikhodeev, S. G., Gippius, N. A. and Ishihara T. (1995). *Phys. Rev.* B **51**, 14370.

Oulton, R. F., Takada, N., Koe, J., Stavrinou, P. N. and Bradley, D. D. C. (2003). *Semicond. Sci. Technol.* **18**, 419.

Pockrand, I., Brillante, A. and Mobius, D. (1982). *J. Chem. Phys.* **77**, 6289.

Pope, M., Kalmann, H. and Magnante, P. (1963). *J. Chem. Phys.* **38**, 2042.

Rytova, N. S. (1967). *Vestn. Mosk. Univ.* **3**, 30 (in Russian).

Song, J. H., He, Y., Nurmikko, A. V., Tischler, J. and Bulovic, V. (2004). *Phys. Rev. B.* **69**, 235330.

Schouwink, P., Lupton, J. M., von Belepsch, H., Daehne, L. and Mahr, R. F. (2002). *Phys. Rev. B* **66**, 081203(R).

Stanley, R. P. , Houdre, R., Weisbuch, C., Oesterle, U. and Ilegems, M. (1996). *Phys. Rev. B* **53**, 10995.

Sumi, H. (1976). *J. Phys. Soc. Jpn.* **41**, 526.

Takada, N., Kamuta, T. and Bradley, D. D. C. (2003). *Appl. Phys. Lett.* **82**, 1812.

Tartakovskii, A. I., Emam-Ismail, M., Lidzey, D. G., Skolnik, M. S., Bradley, D. D. C., Walker, S. and Agranovich, V. M. (2001). *Phys. Rev. B*, **63**, 121302.

Vinogradov, E. A. and Leskova, T. A. (1990). *Phys. Rep.* **194**, 271.

Weisbuch, C. and Vinter, B. (1991). *Quantum Semiconductor Structures. Fundamentals and Applications*. Academic Press, Boston.

Weisbuch, C., Nishioka, M., Inshikawa, A. and Arakawa, Y. (1992) *Phys. Rev. Lett.* **69**, 3314.

Yudson, V. I., Reineker, P. and Agranovich, V. M. (1995). *Phys. Rev. B* **52**, R5543.

Zoubi, H. and La Rocca, G. C. (2005a). *Phys. Rev. B* **71**, 235316; *ibid* **72**, 125306 (2005).

Zoubi, H. and La Rocca, G. C. (2005b). *Phys. Lett. A* **339**, 171.

4

THE ACOUSTIC-WAVE DRIVEN QUANTUM PROCESSOR

C. H. W. Barnes and M. Pepper
Cavendish Laboratory, University of Cambridge, J J Thomson Ave, Cambridge
CB3 OHE, United Kingdom

Abstract

This paper discusses the role of surface acoustic waves in an electron-spin based quantum state processor. The waves can be used to generate an acousto-electric current of discrete single or controllable numbers of electrons which can interact and become entangled. Variable g-factor materials are proposed as a means of applying electrical control to single-qubit rotation and as a form of optical input and readout. This relies on the coherent creation and recombination of entangled excitons and is introduced as a means of allowing the processor to be used as a component in a quantum communications network.

4.1 Introduction

Keldysh (1962) first proposed that a surface acoustic wave (SAW) will modify the electron energy levels in a semiconductor by imposing an additional potential, so giving rise to bandgaps in addition to those created by the crystal field. The scattering of conduction electrons from a SAW has been used for some time to explore electron dynamics (Paalanen *et al.*, 1992). Application of a SAW of sufficient amplitude will replace the scattering of electrons, from the potential wave, by a dragging motion which then creates an acousto-electric current. This situation will emerge when the SAW amplitude exceeds the Fermi energy and corresponds to the theoretically considered quantum pump (Niu, 1990; Thouless, 1983). An experimental demonstration of this was achieved in the context of a possible current standard which would allow a high-accuracy measurement of the electron charge (Shilton *et al.*, 1996). In this and subsequent work, the number of electrons transported by a SAW in the 2D electron gas of a GaAs-AlGaAs heterostructure was controlled by the action of surface gates as the acousto-electric current passed through a one-dimensional electrostatically defined channel. The number of electrons can be controlled to better than 1 % for the case of single-electron transport, with decreasing accuracy as the number increases. At this level of accuracy a simple extension of these devices has been proposed as a means to construct a quantum processor. Here we discuss this device and some modifications that will allow it to be used as a nonlinear optical element.

An N-qubit quantum processor is capable of storing and manipulating vectors with 2^N elements; this potentially vast capacity for computation and manipulation has received a great deal of interest and many research groups are now actively engaged in an attempt to identify N-particle systems that not only satisfy the necessary DiVincenzo criteria (DiVincenzo, 2000), but which will also minimize the time and cost of production of even a single quantum processor. We consider two refinements to the previously proposed SAW-based quantum computer (Barnes *et al.*, 2000) that will make it faster to operate, easier to control, easier to program and easier to measure the final state. The modifications we suggest also make it more attractive for use as a component in a quantum-communication network, because both the input and the output can be in the form of entangled photons.

The original design for a SAW quantum processor consists of a GaAs/AlGaAs heterostructure containing a set of N depleted parallel channels defined either by surface etching (Ebbecke *et al.*, 2000) or surface gates (Thornton *et al.*, 1986). The magnetic field along each of the channels is modulated by a pattern of single-domain magnetic surface gates of different orientations, lengths and field strengths. Occasional electrostatically controlled breaks in the barriers between channels allow virtual tunneling of electrons between channels. N electrons are captured in each cycle of the SAW, so that each SAW minimum has one electron trapped in each of the parallel channels. Each set of N electrons in each SAW minimum is a single quantum processor. Each of these quantum processors is put into the same initial state - through the physics of the SAW electron capture process (Barnes *et al.*, 2000; Robinson and Barnes, 2001). The state of each processor is manipulated in exactly the same way as it passes through single-qubit gates, produced by the magnetic gates, and the two-qubit gates, defined by the electrostatic tunnel barriers. Measurement of each qubit occurs at the end of each channel and the average over a large number of measurements of the same qubit in successive processors is the output of the processor device. This measurement is similar to the ensemble measurement made in nuclear-magnetic-resonance (NMR) quantum computation where a large number of identical systems are prepared, manipulated and then measured. The most important difference between a SAW quantum processor and an NMR quantum processor is that, owing to its low-temperature operation, the initial states are likely to be nearly identical and highly pure. The SAW-based scheme is also more amenable to scaling up to large numbers of qubits. The modifications that are proposed and discussed in this paper are: the use of variable g-factor materials for single-qubit manipulation and the use of optical input and readout.

4.2   Variable g-factor materials

Many solid-state approaches to produce a quantum processor use electron spin as the qubit. In particular, this is the case for the original SAW quantum processor (Barnes *et al.*, 2000). Within its architecture, single-qubit operations are performed by placing splits, single-domain magnetic gates across a channel, or perpendicular to it, at the location where the rotation is required. Since only rotation about two perpendicular axes is required to move a spin to an arbitrary point on its Bloch sphere, this combination is sufficient to produce universal single-qubit gates. The main problems with such gates are: fairly large magnetic fields are required if the length of each magnetic gate is not to be longer than the spin dephasing length (typical gate lengths in GaAs, for 1 T surface fields, exceed 1 μm); once the magnetic gates have been patterned, there is no easy way to change their magnetic field strength, especially if there is a distributed pattern of similar magnetic gates nearby; each single-domain magnet would produce unwanted stray fields affecting other channels and the orientation of domains in other magnets; and owing to the nature of their fabrication, evaporation through a mask, it is very difficult to determine which field strength will be achieved for a magnetic gate produced using a given mask. A degree of control over single-qubit rotation can be achieved if electromagnets are used in place of permanent magnets, but in order to achieve sufficient field strengths, very high currents (1 A) in very short pulses (1 ms) are required.

The use of variable g-factor materials solves all of these problems. The original proposal suggesting the use of such materials (Vrijen *et al.*, 2000) consists of a SiGe heterostructure containing a single quantum well that is split into two parts with different g-factors. This is possible because there are two types of SiGe with different Ge content that have the same bandgap, band offset and are lattice matched, but have different g-factors. The doping in their proposed heterostructure is arranged so that the two-dimensional subband wavefunction sits entirely in one half of the quantum well. A continuous-wave off-resonance global radio frequency (r.f.) signal and a constant external magnetic field are applied to the device. When a single-qubit operation is required, a potential is applied to a surface gate above the qubit so that the electron is pulled to the other side of the quantum well where it comes onto resonance with the external r.f. field. σ_x and σ_y rotations are achieved by carefully timing the gate-potential pulses with the precession of the electron spin in the external field. A very similar scheme is possible in III-IV materials. There are two basic options: to create a double quantum well structure in which the upper and lower wells have different Aluminium or Indium content, or to create a single quantum well with a graded barrier of Aluminium. Aluminium content tends to reduce the g factor towards and through zero. Indium content tends to increase the g factor and can do so by roughly an order of magnitude.

In order to apply these variable g-factor ideas to the SAW quantum processor, one must either use the SiGe heterostructures and use flip-chip technology to bind them to a piezoelectric substrate such as quartz (Vrijen and Yablonovitch, 2001) or use one of the two III-IV options. σ_x and σ_y rotations are achieved by placing the surface gates distances apart that correlate with the precession of the electron spins, in a constant external magnetic field, with their position in the device as they are moved through by the SAW.

4.3   Two-qubit gate

The two-qubit operation for electron spins in a SAW quantum processor is based on the exchange interaction. It occurs when two electrons are brought into tunnel contact by reducing a barrier separating them. For a SAW quantum processor device, the barrier is defined electrostatically by a short metallic Schottky gate placed over a gap in the barrier between channels. These gates are typically an order of magnitude faster than the single-qubit magnetic gates. The necessary gate length for root-of-swap operations can be tens of nanometers and the effect of each gate can be made very local since the tunneling probability is exponentially dependent on the surface gate voltage. This implies that their operation is unlikely to affect the operation of the single qubit gates. However, on the contrary, the potential applied to a single-qubit gate could change an adjacent tunnel barrier. Therefore, in order to define a set of single and two-qubit gates using variable g-factor materials, a process of calibration will need to be carried out to define any quantum algorithm.

4.4   Optical readout

From the GaAs optical selection rules, a spin $+1/2$ electron will recombine with a spin $+3/2$ hole to produce a left-handed circularly polarized photon ($m_j = 3D - 1$) (D'yakonov and Perel, 1970; Pierce *et al.*, 1975; Vrijen and Yablonovitch, 2001). Similarly, a spin $-1/2$ electron will recombine with a spin $-3/2$ hole to produce a right-handed circularly polarized photon.

Optical readout of SAW electron-spin qubits can therefore be achieved by bringing the SAW electrons into contact with holes at the end of each channel and measuring the polarization of the output light. This can be done in two ways: placing a hole gas at the end of each channel (Foden *et al.*, 2000; North *et al.*, 1999; Vijendran *et al.*, 1999) or, at the input stage, creating excitons so that each electron carries a hole along with it (Sogawa *et al.* 2001).

For either case, as each electron enters the recombination region, it will be decohered by elastic collisions with holes so that its state will be spin-up with probability $|\alpha|^2$ and spin-down with probability $|\beta|^2$. The two different hole spin types occur with known probabilities in the relevant magnetic field range and therefore the numbers of left and right circularly polarized photons emitted will be in direct relation to these probabilities. The output light from each channel is passed through a beam splitter and perpendicular polarizing elements to measure the ratio $|\alpha|^2 : |\beta|^2$.

This optical readout method does not transfer the quantum state of the entangled electrons to the emitted photon field. For quantum communication though, accurate quantum information transfer is clearly essential. Vrijen and Yablonovitch (2001) have suggested two methods to achieve coherent quantum information transfer. They both require placing each hole gas in a high-finesse cavity and straining the GaAs or applying an external magnetic field to remove the valence-band degeneracy. A suitable high-finesse cavity can be produced during molecular beam epitaxy growth. Alternating layers of material with different dielectric constants are grown in bands above and below the hole gas. These layers then act as Bragg reflectors. The introduction of strain is possible through introducing layers of material with different lattice constants. Imagoglu *et al.* (1999) have suggested an alternate method for preserving entanglement that involves the use of conduction-band-hole Raman transitions induced by classical laser fields.

4.5 Optical input

When a single photon hits the GaAs-AlGaAs heterostructure, it is likely that it will create an exciton. If the photon is of the appropriate polarization and the heterostructure has the degeneracy of its hole bands lifted, the quantum state of the incident photon will be transferred to the electron part of the exciton wavefunction and the electron and hole parts of the exciton wavefunction will be separable (Vrijen and Yablonovitch, 2001). If there is a SAW present, then both the electron and the hole will be swept along, but kept physically separate. The electron will move to a SAW minimum and the hole to a maximum. The electron can then be guided electrostatically into a narrow channel where its spin can be manipulated and coherently re-emitted by the methods described above. With this method of coherent optical input, the recombination can be with the original hole created with the electron or another, since by design, the electron and hole quantum information are separable.

The extension of this idea to N-photons is straightforward, but does not scale well to large N. It would be necessary to arrange for each photon to arrive at the device in a different optical fibre so that the electron from each exciton created can be forced into a different channel. If delays occur, it is possible that different electrons may be captured in different SAW minima. This could be compensated for by using quantum dots to capture and delay electrons until all the N electrons have been captured or an error registered. Evidence that quantum dots can capture electrons from SAWs has recently been demonstrated (Fletcher *et al.*, 2003) and the presence or absence of a single electron in a quantum dot can be detected using a noninvasive voltage probe (Field *et al.*, 1993).

4.6 The optical quantum-state processor

A SAW-based N-qubit optical quantum-state processor would require N input optical fibres connected to GaAs regions where, by using a suitable cavity, the probability for the creation of a single exciton for each incident photon would be above some predefined error tolerance threshold. These collection regions would connect directly to N depleted parallel channels. A regular pattern of surface gates defining both single- and two-qubit operations would have potentials applied to them to carry out a specific quantum algorithm on the electron part of the N captured excitons. At the end of the device suitable screening gates, cavities and optical fibres would be attached so that recombination would create an output set of entangled photons.

A single computation would be performed by sending a set of N entangled photons along the attached optical fibres. They would create a set of N excitons, with an entangled electron part to the full many-particle wavefunction, but with the electron and hole parts separable (Vrijen and Yablonovitch, 2001). These electrons would be swept through the N channels by a SAW wave. Delay methods using quantum dots (Fletcher *et al.*, 2003; Field *et al.*, 1993) would be performed, if necessary, to make sure that all electrons were in the same SAW minimum. Single-qubit operations would be performed by electron-spin-resonance in regions where gates were placed in the center of channels. Two-qubit operations would be performed where adjacent electrons were allowed into tunnel contact. At the end of the device, recombination would cause the N photons to be re-emitted in a different state of entanglement. Amongst many other uses, such a device could be used to entangle pairs of photons, perform error correction on entangled sets of photons or to perform an arbitrary quantum-processing step as part of a network of such devices.

A simple example of the use of this type of quantum processor is in creating pairs of entangled photons using a single depleted channel with one input and one output region and an external magnetic field to remove the hole degeneracy (Barnes, 2003). Two photons would be fired at the device creating two excitons. With some probability, the electron parts of these excitons would be caught in a single SAW minimum and swept into a single depleted channel. Non-invasive techniques (Fletcher *et al.*, 2003; Field *et al.*, 1993) could be used to verify that this had happened. Over the length of the channel, the two electrons would interact strongly and decoherence would cause them to fall into the ground spin singlet-state. At the end of the channel, a Y-branch splitter would be used to separate the two electrons. They would then pass into the recombination region, where they would create two photons in a polarization singlet-state.

4.7   Summary

The suggestion by Keldysh (1962) that new physical behavior could be explored by modifying electron transport with a SAW has led to a number of new concepts including a scheme of quantum computation. In order to accomplish this we have considered single-qubit gates based on variable g-factor materials allied with electron-spin-resonance techniques. They have the advantage over the originally proposed magnetic split gates in that they can be tuned simply by changing the potential on a surface Schottky gate. Optical input and output were also described and it was shown how a SAW processor could be used as a nonlinear element, modifying the state of entanglement of a set of N photons, in an optical network.

Acknowledgments

This work was supported by EPSRC.

References

Barnes, C. H. W., Shilton, J. M. and Robinson A. M. (2000). *Phys. Rev.* B **62**, 8410.

Barnes, C. H. W. (2003). *Philos. Trans. R. Soc. London, Ser.* A **361**, 1487.

DiVincenzo, D. P. (2000). *Fortschritte der Physik-Progress of Physics* **48**, 771.

D'yakonov, M. I. and Perel, V. I. (1971). *Zh. Eksp. Teor. Fiz.* **60**, 1954 [*Soviet Phys. JETP* **33**, 1053].

Ebbecke, J., Bastian, G., Blocker, M., Pierz, K. and Ahlers, F. J. (2000). *Appl. Phys. Lett.* **77**, 2601.

Field, M., Smith, C. G., Pepper, M., Ritchie, D. A., Frost, J. E. F., Jones, G. A. C. and Hasko, D. G. (1993). *Phys. Rev. Lett.* **70**, 1311.

Fletcher, N. E., Ebbecke, J., Janssen, T. J. B. M., Ahlers, F. J., Pepper, M., Beere, H. E. and Ritchie, D. A. (2003). *Phys. Rev.* B **68**, 245310.

Foden, C. L., Talyanskii, V. I., Milburn, G. J., Leadbeater, M. L. and Pepper, M. (2000). *Phys. Rev.* A **62**, 011803.

Imamoglu, A., Awschalom, D. D., Burkard, G., DiVincenzo, D. P., Loss, D., Sherwin, M. and Small, A. (1999). *Phys. Rev. Lett.* **83**, 4204.

Keldysh, L. V. (1962). *Fiz. Tverd. Tela* **4**, 2265 [*Sov. Phys. Solid State* **4**, 1658].

Niu, Q. (1990). *Phys. Rev. Lett.* **64**, 1812.

North, A. J., Burroughes, J. H., Burke, T., Shields, A. J., Norman, C. E. and Pepper, M. (1999). *IEEE J. Quantum Electron.* **35**, 352.

Paalanen, M. A., Willet, R. L., Littlewood, P. B., Ruel, R. R., West, K. W., Pfeiffer, L. N. and Bishop, D. J. (1992). *Phys. Rev.* B **45**, 11342.

Pierce, D. T., Meier, F. and Zurcher, P. (1975). *Phys. Lett.* A **51**, 465.

Robinson, A. M. and Barnes, C. H. W. (2001). *Phys. Rev.* B **63**, 165418.

Shilton, J., Talyanski, V. I., Pepper, M., Ritchie, D. A., Frost, J. E. F., Ford, C. J. B., Smith, C. G. and Jones, G. A. C. (1996). *J. Phys.: Condens. Matter* **8**, L531.

Sogawa, T., Santos, P. V., Zhang, S. K., Eshlaghi, S., Wieck, A. D. and Ploog, K. H. (2001). *Phys. Rev. Lett.* **87**, 276601.

Thornton, T. J., Pepper, M., Ahmed, H., Andrews, D. and Davies, G. J. (1986). *Phys. Rev. Lett.* **56**, 1198.

Thouless, D. J. (1983). *Phys. Rev.* B **27**, 6083.

Vijendran, S., Sazio, P. J. A., Beere, H. E., Jones, G. A. C., Ritchie, D. A. and Norman, C. E. J. (1999). *J. Vac. Sci. Techn.* B **17**, 3226.

Vrijen, R., Yablonovitch, E., Wang, K. L., Jiang, H. W., Balandin, A. and Roychowdhury, V. (2000). *Phys. Rev.* A **62**, 12306.

Vrijen, R. and Yablonovitch, E. (2001). *Physica* E **10**, 569.

5

ON THE PROBLEM OF MANY-BODY LOCALIZATION

D. M. Basko, I. L. Aleiner and B. L. Altshuler
Physics Department, Columbia University, New York, NY 10027, USA

Abstract
We review recent progress in the study of transport properties of interacting electrons subject to a disordered potential which is strong enough
to localize *all* single-particle states. This review may also serve as a guide
to the recent paper by the authors [D. M. Basko, I. L. Aleiner, and B. L.
Altshuler, *Annals of Physics*, **321**, 1126 (2006)]. Here we skip most of the
technical details and make an attempt to discuss the physical grounds of
the finite-temperature metal–insulator transition described in the above-
mentioned paper.

5.1 Introduction

Transport properties of conducting materials at low temperature T are determined by the interplay between the interaction of the itinerant electrons with
each other and the quenched disorder which creates a random potential acting on
these electrons. In the absence of the electron–electron interaction the most dramatic phenomenon is Anderson localization (Anderson, 1958) – the dc electrical
conductivity σ can be qualitatively different depending on whether one-particle
wavefunctions of the electrons are localized or not. In the latter case $\sigma(T)$ has a
finite zero-temperature limit, while in the former case $\sigma(T)$ vanishes when $T \to 0$.
Therefore, Anderson localization of electronic states leads to the metal–insulator
transition at zero temperature.

When discussing zero-temperature conductivity $\sigma(0)$, we need to consider
only electronic states close to the Fermi level. The conductivity becomes finite
at any finite temperature, provided that extended states exist somewhere near
the Fermi level. It is commonly accepted now that localized and extended states
in a random potential cannot be mixed in the one-electron spectrum and thus
this spectrum in a general case is a combination of bands of extended states and
bands of localized states. The border between a localized and an extended band is
called the mobility edge. If the Fermi level is located inside a localized band and
inelastic scattering of the electrons is completely absent, the conductivity should
follow an Arrhenius law $\sigma(T) \propto \exp(-E_c/T)$, where E_c is the distance from
the Fermi level to the closest mobility edge. Another common belief following
from the scaling theory of Anderson localization (Thouless, 1977; Abrahams *et
al.*, 1979) is that in low dimensionality d, namely for $d = 1, 2$ all states are

localized in an arbitrarily small disorder, while for free electrons (no periodic potential) $E_c > 0$ is finite for $d = 3$.[6] This means that without inelastic processes $\sigma_{d=1,2}(T) = 0$, while for $\sigma_{d=3}(T)$ one should expect the Arrhenius law. Note that for electrons in a crystal within a given conduction band the latter conclusion is not always correct – strong enough disorder can localize the whole band.

As soon as inelastic processes are included, the situation becomes more complicated. In particular, electron–phonon interaction leads to the mechanism of conductivity known as hopping conductivity (Fritzche, 1955; Mott, 1968a; Shklovskii and Efros, 1984) – with the assistance of phonons, electrons hop between the localized states without being activated above the mobility edge. As a result, $\sigma(T)$ turns out to be finite (although small) at arbitrarily low T even when all one-electron states are localized.

Can interaction between electrons play the same role and cause the hopping conductivity? This question has been discussed in the literature for a long time (Fleishman and Anderson, 1980; Shahbazyan and Raikh, 1996; Kozub *et al.*, 2000; Nattermann *et al.*, 2003; Gornyi *et al.*, 2004) and no definite conclusion has been achieved. The problem is that although the electric noise exists inside the material with a finite ac conductivity[7] the "photons" in contrast with phonons become localized together with electrons.

In recent work (Basko *et al.*, 2006) we have demonstrated that the electron–electron interaction alone cannot cause finite conductivity even when the temperature is finite, but small enough. In the absence of phonons and extended one-electron states, the conductivity of a system of interacting electrons vanishes exactly below some critical temperature T_c. At the same time, at high temperatures $T > T_c$ the conductivity $\sigma(T)$ is finite. This means that at $T = T_c$ the system of interacting electrons subject to a random potential undergoes a genuine phase transition that manifests itself by the emerging of a finite conductivity!

This transition can be thought of as many-body localization – it applies to many-body eigenstates of the whole system. This localization occurs not in real space, but rather in Fock space. This fact does not affect the validity of the concept of mobility edge. In fact, the existence of the "metallic" state at $T > T_c$ implies that the many-body states with energies $\mathcal{E}$ above $\mathcal{E}_c$ are extended. One can estimate the difference between $\mathcal{E}_c$ and the energy of the many-body ground state $\mathcal{E}_0$ as $\mathcal{E}_c - \mathcal{E}_0 \sim T\mathcal{N}(T)$, where $\mathcal{N}(T)$ is the total number of one-particle states in the energy strip of width T. Note that the existence of the extended many-body states above the mobility edge does not contradict the fact that below T_c there is no conductivity – in contrast with the case of one-particle localization there is no Arrhenius regime since $\mathcal{E}_c - \mathcal{E}_0$ turns out to be proportional to the

---

[6]For $d = 1$ this statement was proved rigorously both for one-channel (Gertsenshtein and Vasil'ev, 1959; Berezinskii, 1973) and multichannel (Efetov and Larkin, 1983; Dorokhov, 1983) disordered wires.

[7]In this paper we mostly focus on dc conductivity. As to ac conductivity, it never vanishes, because at any frequency the density of resonant pairs of states is finite.

volume of the system, i.e., is macroscopically large (see Section 5.4 below for more details).

In order to avoid possible misunderstanding we would like to emphasize that we focus only on the inelastic collisions between the electrons, i.e., on creation or annihilation of *real* electron–hole pairs. There are other effects of electron–electron interactions which can be understood as renormalization of the one-particle random potential by the interaction. Being temperature-dependent, this renormalization leads to a number of interesting effects, such as the interaction corrections to the density of states and conductivity in disordered metals (Altshuler and Aronov, 1985). On the insulating side of the one-particle localization transition similar effects cause the well-known Coulomb gap (Shklovskii and Efros, 1984) which reduces hopping conductivity. On the other hand, this is just a correction to the time-independent random potential. As such, it may shift the position of the many-body metal to insulator transition, i.e., renormalize T_c, but is unable to destabilize the insulating or metallic phases. From now on we will simply neglect all elastic (Hartree–Fock) effects and concentrate on the *real* inelastic electron–electron collisions.

Localization of the many-body states in Fock space has been discussed by Altshuler *et al.* (1997) for the case of zero-dimensional systems with finite, although large, number of electrons. In this paper the authors proposed an approximate mapping of the Hamiltonian of a metallic grain with large Thouless conductance g and moderate interaction between the electrons to the one-particle Hamiltonian on a lattice with the topology of the Cayley tree and an on-site disorder. The latter problem has an exact solution (Abou-Chacra *et al.*, 1973; Efetov, 1987) that exhibits the localization transition. In terms of interaction electrons this transition means that one-particle excitation states below a certain energy are quite close to some exact many-body excitations. As to the one-particle excitations with energies higher than the critical one, their wavefunctions can be viewed as a linear combination of a large number of the many-body eigenstates.

For an infinite system ($d > 0$) the situation is more complex, and the Cayley tree approximation is hard to justify. Nevertheless, a consistent analysis of a model with weak and short-range interaction to all orders of perturbation theory enabled us to analyze the many-body localization transition and to demonstrate that both the metallic state at high temperatures and the insulating state at low temperatures are stable and survive all higher loop corrections to the locator expansion. Therefore, the existence of the transition is proved on the physical level of rigor.

It should be noted that such an insulating state that is characterized by exactly zero conductivity is quite different from all other known types of insulators. For example, a Mott insulator is believed to have finite, though exponentially small conductivity at finite temperatures.

The present text represents a shortened version of the paper by Basko, Aleiner and Altshuler (2006), hereafter referred to as the BAA paper. We omit most of the technical details (for which the reader will be referred to specific sections of the BAA paper), and stress the key ideas.

The remainder of the paper is organized as follows. In Section 5.2 we briefly review some well-known facts about electric conduction in Anderson insulators and pose the problem. Section 5.3 represents a sketch of the solution whose details are given in the BAA paper. We discuss the model for interacting localized electrons in Section 5.3.1 and the corresponding Fock space picture in Section 5.3.2. In Section 5.3.3 we show the formal way to characterize metallic and insulating phases. In Section 5.3.4 we introduce the main approximation used in the calculation (self-consistent Born approximation) and discuss its validity. The existence of the metallic state at high temperatures and its properties are discussed in Section 5.3.5. Section 5.3.6 is dedicated to the proof of existence of the insulating phase at low temperatures; the value of the transition temperature is obtained as the limit of stability of the insulating phase. In Section 5.4 we discuss the macroscopic implications of the problem, introducing the concepts of many-body localization and many-body mobility edge. Finally, in Section 5.5 we summarize the results and present an outlook of the future developments.

5.2 Background and formulation of the problem

5.2.1 *Non-interacting electrons in disorder potential*

Let us briefly review the basic concepts developed for the problem of one-electron wavefunctions in a disordered potential in d dimensions. Depending on the strength of the disorder potential, a wavefunction $\phi_\alpha(\vec{r})$ of an eigenstate α with the energy ξ_α can be either *localized* or *extended*:

$$|\phi_\alpha(\vec{r})|^2 \propto \begin{cases} \frac{1}{\zeta_{\rm loc}^d} \exp\left(-\frac{|\vec{r}-\vec{\rho}_\alpha|}{\zeta_{\rm loc}}\right), & \text{localized}; \\ \frac{1}{\Omega}, & \text{extended}. \end{cases} \quad (5.1)$$

Here $\zeta_{\rm loc}$ is the localization length which depends on the eigenenergy ξ_α, and Ω is the volume of the system. Each localized state is characterized by a point in space, $\vec{\rho}_\alpha$, where $|\phi_\alpha(\vec{r})|^2$ reaches its maximum, and an exponentially falling envelope. Extended states spread more or less uniformly over the whole volume of the system. Localized and extended states cannot coexist at the same energy, and the spectrum splits into bands of localized and extended states. The energies separating such bands are known as *mobility edges*. For free electrons in $d \geq 3$ disorder potential leads to only one mobility edge $\mathcal{E}_1$, so that

$$\begin{aligned} \xi_\alpha < \mathcal{E}_1 : & \quad \text{localized}, \\ \xi_\alpha > \mathcal{E}_1 : & \quad \text{extended}. \end{aligned} \quad (5.2)$$

If a finite mobility edge given by equation (5.2) exists and the Fermi level $\epsilon_{\rm F}$ lies in the band of localized states, the conductivity is determined by the exponentially small occupation number of the delocalized states

$$\sigma(T) \propto e^{-(\mathcal{E}_1 - \epsilon_{\rm F})/T}. \quad (5.3)$$

In this paper we are interested in transport properties of the systems where *all* single-particle states are localized and thus without many-body effects $\sigma = 0$

at any temperature. It is well established now that the mobility edge usually does not exist for one- and two-dimensional systems, and all single-particle states are indeed localized for an arbitrarily weak disorder. Such a situation can arise for a large d as well, if the bandwidth is finite and disorder is sufficiently strong.

5.2.2 *Role of inelastic processes and phonon-assisted hopping*

As long as all single-particle states are localized, transport occurs only because of inelastic processes, which transfer electrons between different localized eigenstates. At this stage we introduce the main energy scale of the problem: the typical energy spacing between states whose spatial separation does not exceed ζ_{loc}, so that there is overlap between their wavefunctions:

$$\delta_\zeta = \frac{1}{\nu \zeta_{loc}^d} \, , \tag{5.4}$$

where ν is the one-particle density of states per unit volume.

The conductivity is, roughly speaking, proportional to the rate of the transitions between different localized states, which can be called the inelastic relaxation rate Γ. Obviously, at $T = 0$ inelastic processes disappear, so regardless of the mechanism of inelastic relaxation the conductivity must vanish:

$$\lim_{T \to 0} \sigma(T) = 0 \, . \tag{5.5}$$

The question is how $\sigma(T)$ approaches zero for each particular mechanism.

When phonons are the main source of inelastic scattering, the answer is given by Mott's variable range hopping formula (Mott, 1968a)[8]

$$\sigma(T) = \sigma_0 \left(\frac{T}{\delta_\zeta} \right)^\alpha \exp \left[- \left(\frac{\delta_\zeta}{T} \right)^{1/(d+1)} \right] , \tag{5.6}$$

where σ_0 and α are constants. According to equation (5.6), $\sigma(T)$ remains finite as long as $T \neq 0$. The reason for $\sigma(T) \neq 0$ is that for any pair of localized states one can always find a phonon whose frequency exactly corresponds to their energy mismatch, at low temperature one should just wait long enough.

The same type of $\sigma(T)$-dependence would result from the coupling of electrons with *any* delocalized thermal bath whose energy spectrum is continuous down to zero energy. The specific nature of the bath at most affects the power-law prefactor. On the contrary, the stretched exponential factor is universal; it originates from the counting of electronic states available for the transition and does not depend on the specific scattering mechanism.

5.2.3 *Inelastic relaxation due to electron–electron interaction*

Now let us assume that there is no external bath coupled to electrons, but some electron–electron interaction is present. What will be the dependence of $\sigma(T)$?

---

[8]Here we do not consider the effects of the Coulomb interaction which are known to modify the power of the temperature in the exponent (Shklovskii and Efros, 1984).

In line with the discussion of the previous subsection, the question should be posed as follows: do electron–hole pairs themselves provide a suitable bath in a localized system, thus validating equation (5.6)?

One possible answer is "yes". Indeed, recall Mott formula for the low-temperature dissipative ac conductivity $\sigma(\omega)$ in a localized system (Mott, 1968b):

$$\sigma(\omega) = \sigma_1 \frac{\omega^2}{\delta_\zeta^2} \ln^{d+1} \frac{\delta_\zeta}{|\omega|} \,. \tag{5.7}$$

According to fluctuation-dissipation theorem, at finite temperature electromagnetic fluctuations of a finite spectral density should be present, and they might serve as a bath. The problem with this argument is that equation (5.7) is the *spatial average* of the conductivity over the whole (infinite) volume. For each given realization of disorder, excitations determining $\sigma(\omega)$ from equation (5.7) are localized, and although the total volume is infinite, the spectrum of electron–hole pairs is effectively discrete.

The crucial point is that as long as electron–electron interaction is local in space (here we do not consider long-range interactions), it effectively couples electronic states only within the same localization volume, where the spectrum of electronic states is effectively discrete. The following sections are dedicated to a systematic discussion of this problem which was first pointed out by Fleishman and Anderson (1980). The conclusion can be stated as follows: electron–hole excitations can cause finite conductivity only if the temperature of the system exceeds some critical value. At lower temperatures $\sigma(T)$ vanishes exactly.

5.3 Finite-temperature metal–insulator transition

5.3.1 *Matrix elements of electron–electron interaction between localized states: essential features of the model*

For simplicity we consider a system of spinless electrons and assume that electron–electron interaction is weak and short-range:[9]

$$V(\vec{r}_1 - \vec{r}_2) = \frac{\lambda}{\nu} \, \delta(\vec{r}_1 - \vec{r}_2) \,, \tag{5.8}$$

where $\lambda \ll 1$ is the dimensionless interaction constant, ν is the one-particle density of states per unit volume.

In the basis of localized single-particle eigenstates the Hamiltonian corresponding to the pair interaction potential (5.8) takes the form

$$\hat{H} = \sum_\alpha \xi_\alpha \hat{c}_\alpha^\dagger \hat{c}_\alpha + \frac{1}{2} \sum_{\alpha\beta\gamma\delta} V_{\alpha\beta\gamma\delta} \, \hat{c}_\alpha^\dagger \hat{c}_\beta^\dagger \hat{c}_\gamma \hat{c}_\delta \,. \tag{5.9}$$

---

[9]Interaction proportional to $\delta(\vec{r}_1 - \vec{r}_2)$ in the strict sense is equivalent to no interaction for spinless electrons, considered here, due to the Pauli principle. Here, by writing $\delta(\vec{r}_1 - \vec{r}_2)$ we only mean that the range is much smaller than the electron mean free path, so it is not a true δ-function.

Consider the structure of the matrix elements $V_{\alpha\beta\gamma\delta}$. Since $V(\vec{r})$ is short-range, they decrease exponentially when the spatial separation between the states increases, the characteristic scale being the localization length ζ_{loc}. In addition to this spatial suppression, the matrix elements decrease rapidly when the energy difference, say $\xi_\alpha - \xi_\gamma$, increases exceeding the level spacing δ_ζ. This occurs because the localized wavefunctions oscillate randomly, and the bigger the energy difference, the weaker are these random oscillations correlated (Altshuler and Aronov, 1985). Provided that the restrictions

$$|\vec{r}_\alpha - \vec{r}_\beta| \lesssim \zeta_{\mathrm{loc}}, \quad |\vec{r}_\alpha - \vec{r}_\gamma| \lesssim \zeta_{\mathrm{loc}}, \quad |\vec{r}_\beta - \vec{r}_\gamma| \lesssim \zeta_{\mathrm{loc}}, \quad \text{etc.,} \quad (5.10)$$

$$|\xi_\alpha - \xi_\delta|, \ |\xi_\beta - \xi_\gamma| \lesssim \delta_\zeta \quad \text{or} \quad |\xi_\alpha - \xi_\gamma|, \ |\xi_\beta - \xi_\delta| \lesssim \delta_\zeta, \quad (5.11)$$

are fulfilled, we have $|V_{\alpha\beta\gamma\delta}| \sim \lambda\delta_\zeta$.

There are several ways to model these essential properties of $V_{\alpha\beta\gamma\delta}$. In the BAA paper a specific model was adopted; essentially, the space and energy dependences of the matrix elements were replaced by simple rectangular cutoffs (see Section 3 of the BAA paper for details).

5.3.2 *Many-electron transitions and Fock space*

We wish to note that all the discussion of this subsection is not conceptually new. In fact, it is just a generalization of the arguments of Altshuler *et al.* (1997) to an infinite system.

Conventionally, an elementary inelastic process is a decay of one single-particle excitation (an electron occupying a state α) into three single-particle excitations – a hole in the state β and two electrons in the states γ and δ. Such a decay can be described differently: one can say that the Hamiltonian couples the single-particle excitation with the three-particle excitation by the matrix element $V_{\alpha\beta\gamma\delta}$. Further action of the interaction Hamiltonian produces five-particle excitations, seven-particle excitations, etc.:

$$\xi_\alpha \ \to \ \xi_\gamma + \xi_\delta - \xi_\beta \ \to \ \xi_1 + \xi_2 + \xi_3 - \xi_4 - \xi_5 \ \to \ \dots \ . \quad (5.12)$$

If on each stage the coupling is strong enough (i.e., the matrix element is of the same order or larger than the corresponding energy mismatch), the single-particle excitation indeed decays irreversibly into all possible many-body states. In other words, exact many-body eigenstates become delocalized in Fock space. If, oppositely, three-particle states contribute only a weak perturbative admixture to the one-particle state, the contribution from five-particle states is even weaker, etc., the initial electron will never decay completely. One can say that it is localized in Fock space.

Another way to visualize the inelastic relaxation is to look at the energy structure of the quasiparticle spectral function:

$$A_\alpha(\epsilon) = \sum_k \left| \langle \Psi_k | \hat{c}_\alpha^\dagger | \Psi_0 \rangle \right|^2 \delta(\epsilon + E_0 - E_k). \quad (5.13)$$

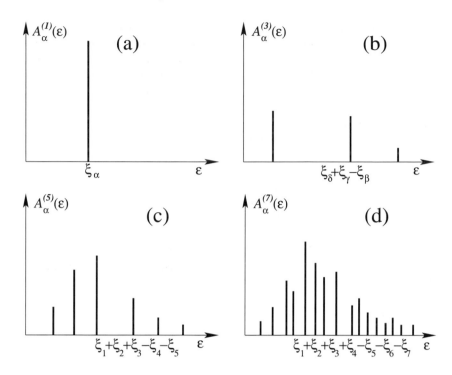

FIG. 5.1. A schematic view of the contributions to the perturbative expansion (5.14) of the spectral function $A_\alpha(\epsilon)$ from one-particle, three-particle, five-particle, and seven-particle excitations.

Here Ψ_0 and Ψ_k are many-body eigenstates (Ψ_0 is not necessarily a ground state), E_0 and E_k are the corresponding energies. Basically, $A_\alpha(\epsilon)$ shows how the single-particle excitation on top of a given eigenstate is spread over other many-body eigenstates of the system.

One can represent the spectral function in the form of expansion in powers of the interaction constant λ (see Fig. 5.1):

$$A_\alpha(\epsilon) = \sum_{n=0}^{\infty} \lambda^{2n} A_\alpha^{(2n+1)}(\epsilon). \tag{5.14}$$

$A_\alpha^{(1)}(\epsilon)$ corresponds to the bare quasiparticle peak, $\delta(\epsilon - \xi_\alpha)$. The term linear in λ represents a random Hartree–Fock shift of the energy ξ_α which is already assumed to be random. This linear term is of no interest to us, so it is not included in the expansion (5.14). The λ^2 term corresponds to the contribution of three-particle excitations. The number of these excitations is effectively finite due to the restrictions (5.10) and (5.11), so $A^{(3)}(\epsilon)$ is a collection of δ-peaks at three-particle energies $\xi_\gamma + \xi_\delta - \xi_\beta$. Again, λ^3 terms correspond only to peak shifts and are omitted. The five-particle contribution $A^{(5)}(\epsilon)$ is again a collection of

δ-peaks, however, spaced more closely than three-particle peaks. This represents the general rule: the more particles are involved, the higher is the density of the peaks. At the same time, contributions of many-particle processes are suppressed due to the smallness of λ. What is crucial to us, is the result of this competition at $n \to \infty$: will the empty spaces between peaks be filled making $A_\alpha(\epsilon)$ a continuous function, or the process will be suppressed by the powers of λ and $A_\alpha(\epsilon)$ will remain a collection of δ-peaks?

Thus, the problem of inelastic relaxation of a quasiparticle due to electron–electron interaction is related to the problem of localization or delocalization of excitations in the many-electron Fock space. The simplest model describing localization–delocalization physics is the Anderson model. Anderson (1958) considered a tight-binding model on a d-dimensional lattice. The coupling between sites is nearest-neighbor only with a fixed matrix element V. The on-site energies are assumed to be random and uncorrelated, with some typical value denoted by W.

It is thus tempting to identify the many-electron Hamiltonian (5.9) with the Anderson Hamiltonian on a certain lattice, whose sites correspond to many-particle excitations (Altshuler *et al.*, 1997). The approximate rules of correspondence then should be the following:

- $V \to \lambda \delta_\zeta$ – typical value of the coupling matrix element;
- $W \to \delta_\zeta$ – typical energy mismatch in each consecutive virtual transition, $|\xi_\alpha + \xi_\beta - \xi_\gamma - \xi_\delta| \sim \delta_\zeta$;
- finally, the coordination number $2d \to T/\delta_\zeta$. This represents the number of three-particle excitations to which a given single-particle excitation is coupled, with the energy mismatch not exceeding δ_ζ. The value T/δ_ζ is obtained as the product of the number of electrons β within the same localization volume, available for the collision with the probe electron α ($\sim T/\delta_\zeta$), and the number of ways to distribute energy allowed by the restriction (5.11), which is ~ 1.

According to Anderson (1958), the localization–delocalization transition occurs at

$$\frac{Vd}{W} \ln\left(\frac{W}{V}\right) \sim \frac{\lambda T}{\delta_\zeta} \ln\frac{1}{\lambda} \sim 1. \tag{5.15}$$

For the interacting localized electrons this would imply that below some temperature T_c the inelastic relaxation is frozen, so that the conductivity vanishes exactly; above T_c some inelastic relaxation is taking place and the conductivity is finite. This corresponds to a finite-temperature metal–insulator transition.

This analogy, however, should be used with caution. Anderson's result, in fact, can be sensitive to the structure of the lattice; possible sources of problems are listed in Section 2.2 of the BAA paper. One can develop a systematic approach to the problem, based on the diagrammatic technique for interacting electrons (Basko *et al.*, 2006). Below we outline the basic ideas of this approach to justify

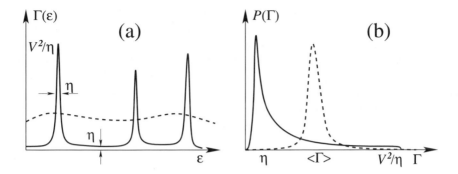

FIG. 5.2. (a) Schematic energy dependence of the quasiparticle decay rate $\Gamma(\varepsilon)$ in the metallic (dashed line) and insulating (solid line) phases for a given realization of single-particle levels $\{\xi_\alpha\}$. (b) The corresponding distribution functions $P(\Gamma)$ for $\Gamma(\epsilon)$ at a given energy ϵ in the metallic (dashed line) and insulating (solid line) phases.

the assumption that equation (5.15) indeed correctly determines the point of the metal–insulator transition.

5.3.3 *Statistics of the transition rates*

We have to admit, that all the discussion of this subsection is not conceptually new either. It rather represents a generalization of the arguments by Anderson (1958) to a many-body problem.

As discussed in previous sections, the focus of the problem is the inelastic quasiparticle relaxation, which is represented by the imaginary part of the single-particle self-energy:

$$\Gamma_\alpha(\epsilon) = \operatorname{Im} \Sigma_\alpha^A(\epsilon). \tag{5.16}$$

We stress that $\Gamma_\alpha(\epsilon)$ is a random quantity, as it depends on the positions of all single-particle levels $\{\xi_\beta\}$ and all occupation numbers $\{n_\beta\}$. One has no other choice but to perform statistical analysis of this random quantity.

How de we distinguish between metallic and insulating phases within a statistical framework? It is clear that positions of the peaks in $\Gamma(\epsilon)$ for the insulating regime wander randomly with the variation of random energies ξ_α from equation (5.9). Due to this variation, the ensemble average of the decay rate $\langle \Gamma(\epsilon) \rangle$ is the same in both phases and cannot be used for the distinction. In fact, one has to investigate the whole distribution function $P(\Gamma)$.

To understand the difference in the behavior of $P(\Gamma)$ in the two phases, it is instructive to start with the behavior of $\Gamma_\alpha(\epsilon)$ for a given realization of disorder, Fig. 5.2(a). Deep in the metallic phase, $\Gamma_\alpha(\epsilon)$ is a smooth function of energy, so its distribution $P(\Gamma)$ at a given energy is a narrow Gaussian. In the insulating phase $\Gamma_\alpha(\epsilon)$ is given by a sequence of infinitely narrow δ-peaks, so an arbitrarily chosen value of ϵ falls between the peaks with the probability 1, giving $\Gamma = 0$. If ϵ hits a δ-peak, then $\Gamma = \infty$, which happens, however, with zero probability.

One can deal with this uncertainty by introducing an infinitesimal imaginary energy shift (damping) η, and consider $\Gamma_\alpha(\epsilon + i\eta)$. Physically, it may be viewed as an infinitesimally weak coupling to a dissipative bath (e.g., phonons). Clearly, in the metallic phase it has no effect, while in the insulating phase this damping broadens δ-peaks into Lorentzians of the width η, which leads to the appearance of the tails. Now, even if the energy ϵ falls between the peaks, Γ will have a finite value proportional to η. As a result, the distribution function $P(\Gamma)$ will have the form sketched in Fig. 5.2(b). Now, having calculated the distribution function for a small but finite η, one can distinguish between the metallic and the insulating phases according to

$$\lim_{\eta \to 0} \lim_{\Omega \to \infty} P(\Gamma > 0) \begin{cases} > 0, & \text{metal}, \\ = 0, & \text{insulator}. \end{cases} \tag{5.17}$$

Here Ω is the total volume of the system. We emphasize that the order of limits in equation (5.17) cannot be interchanged, since in a finite closed system the spectrum always consists of discrete δ-peaks.

5.3.4   *Self-consistent Born approximation*

Our main object of interest is the probability distribution function of the decay rate $P(\Gamma)$. Its calculation is performed in two stages: (i) we find $\Gamma_\alpha(\epsilon)$ for a given realization of disorder, (ii) we calculate its statistics. This subsection is dedicated to the first task.

We intend to describe both metallic and insulating regimes. In the latter regime relaxation dynamics is absent, the system never reaches thermal equilibrium, and the temperature itself is ill-defined. Therefore, the only appropriate formal framework is the nonequilibrium formalism of Keldysh (1964) (see Section 4.1 of the BAA paper for details).

Our approach for the calculation of $\Gamma_\alpha(\epsilon)$ is the self-consistent Born approximation (SCBA), shown diagrammatically in Fig. 5.3, and given by

$$\Gamma_\alpha(\epsilon) = \eta + \pi \sum_{\beta,\gamma,\delta} |V_{\alpha\beta\gamma\delta}|^2 \int d\epsilon' \, d\omega \, A_\beta(\epsilon') \, A_\gamma(\epsilon' + \omega) \, A_\delta(\epsilon - \omega) \times$$

$$\times [n_\beta(1 - n_\gamma)(1 - n_\delta) + (1 - n_\beta)n_\gamma n_\delta], \tag{5.18}$$

$$A_\alpha(\epsilon) = \frac{1}{\pi} \frac{\Gamma_\alpha(\epsilon)}{(\epsilon - \xi_\alpha)^2 + \Gamma_\alpha^2(\epsilon)}. \tag{5.19}$$

Here $n_\alpha = 0, 1$ is the fermion occupation number of the single-particle state α. It is crucial to realize that one cannot replace n_α with its average equilibrium value, corresponding to the Fermi–Dirac distribution. The basis in Fock space is formed by Slater determinants corresponding to $n_\alpha = 0, 1$ only, and Γ represents the rate of a transition between two such basis states with different sets of $\{n_\alpha\}$. The temperature enters through the total energy of the state, which is determined

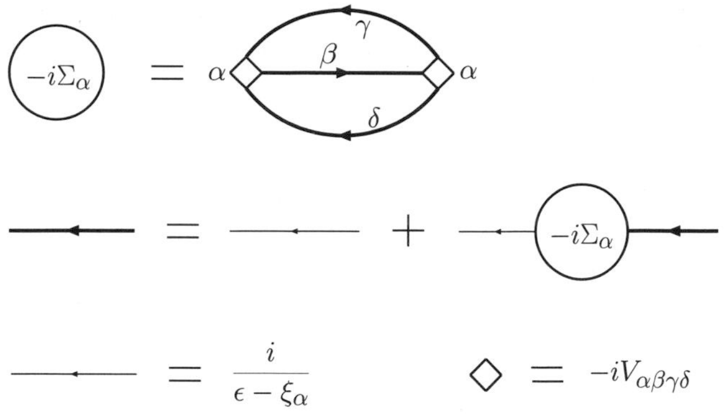

$$\left(-i\Sigma_\alpha\right) = \alpha \diamond \underset{\delta}{\overset{\gamma}{\beta}} \diamond \alpha$$

$$\longleftarrow = \longleftarrow + \longrightarrow \left(-i\Sigma_\alpha\right) \longleftarrow$$

$$\longleftarrow = \frac{i}{\epsilon - \xi_\alpha} \qquad \diamond = -iV_{\alpha\beta\gamma\delta}$$

FIG. 5.3. Diagrammatic representation of self-consistent Born approximation.

by all $\{n_\alpha\}$ and is proportional to the volume of the system (this issue will be discussed in more detail in Section 5.4 of the present paper).

Iterations of SCBA generate self-energy diagrams which have one common property: they describe decay processes where the number of particles in the final state is maximized for each given order in λ, thus maximizing the phase space available for the decay. Vice versa, each diagram satisfying this condition is taken into account by SCBA. Let us briefly discuss contributions which are neglected by equations (5.18) and (5.19); for a detailed discussion see Section 7 of the BAA paper.

- Equations (5.18) and (5.19) completely ignore the real part of the self-energy, $\operatorname{Re}\Sigma$. In most of the terms effect of $\operatorname{Re}\Sigma$ reduces to random uncorrelated corrections to already random energies, and appear to be completely negligible. In a more accurate approximation, however, some statistical correlations are present. They renormalize the numerical prefactor in the expression for the critical temperature, see Section 7.3 of the BAA paper.

- Generally, quantum mechanical probability of a transition is given by a square of the total amplitude, the latter being a sum of partial amplitudes. In fact, equations (5.18) and (5.19) correspond to replacing the square of the sum by the sum of the squares. Approximation which neglects the interference terms can be justified in the same way as in the Anderson model of a high dimensionality $d_{\mathrm{eff}} \sim T/\delta_\zeta \sim 1/\lambda \gg 1$, see Section 7.2 of the BAA paper.

- Finally, diagrams generated by SCBA correspond to taking all n-particle vertex functions in the leading approximation in λ. One can show that for $\lambda \ll 1$ vertex corrections are small, see Section 7.1 of the BAA paper. For $\lambda \sim 1$ one would have to introduce the full n-particle vertex analogously to how it is done in conventional Fermi liquid theory for $n = 2$ (Landau, 1958; Eliashberg, 1962). In this case the result for the transition temperature will be determined not by the bare interaction constant, but by the statistics of the full vertex functions. The existence and regularity of these vertex functions are basically equivalent to the assumption that in the absence of disorder the interacting system is a Fermi liquid.

5.3.5 *Metallic phase*

Starting from the self-consistent equations (5.18) and (5.19), one can straightforwardly verify (see Section 5.1 of the BAA paper) that $P(\Gamma)$ can be well approximated by Gaussian distribution with the average and dispersion

$$\langle \Gamma \rangle \sim \lambda^2 T, \quad \langle \Gamma^2 \rangle - \langle \Gamma \rangle^2 \sim \lambda^2 \delta_\zeta^2 , \tag{5.20}$$

provided that

$$\sqrt{\langle \Gamma^2 \rangle - \langle \Gamma \rangle^2} \ll \langle \Gamma \rangle \quad \Leftrightarrow \quad T \gg T^{(\text{in})} \sim \frac{\delta_\zeta}{\lambda} . \tag{5.21}$$

According to the arguments of Section 5.3.3, this is characteristic of the metallic phase. One should not think, however, that this automatically means that the system has the same transport properties as conventional metals (which would mean that the conductivity is given by the Drude formula). The system conducts in the Drude regime only when the inelastic processes completely destroy the localization, the latter manifesting itself as the weak localization correction to conductivity (Altshuler and Aronov, 1985). This occurs when the discrete levels are completely smeared:

$$\langle \Gamma \rangle \gg \delta_\zeta \quad \Leftrightarrow \quad T \gg T^{(\text{el})} \sim \frac{\delta_\zeta}{\lambda^2} . \tag{5.22}$$

It turns out that for $\lambda \ll 1$ there is a parametric range of temperatures:

$$\frac{\delta_\zeta}{\lambda} \ll T \ll \frac{\delta_\zeta}{\lambda^2} . \tag{5.23}$$

In this interval electron–electron interaction is already sufficient to cause inelastic relaxation, however, the localized nature of the single-particle wavefunctions remains important. Conduction in this regime can be viewed as an analog of hopping conduction discussed in Section 5.2.2, in the sense that electrons themselves indeed provide a good bath. However, as $T \gg \delta_\zeta$, there is no exponential factor in the temperature dependence $\sigma(T)$. This dependence can be obtained from the kinetic equation describing electron transitions between localized states and is given by a power law, $\sigma(T) \propto T^\alpha$, where α can be model-dependent. Also, in this temperature range, Wiedemann–Frantz law can be violated. For further details the reader is referred to Section 5.2 of the BAA paper. We only note that this regime is somewhat analogous to the phonon-assisted conduction discussed by Gogolin *et al.*, (1975).

5.3.6 *Insulating phase*

The self-consistent equations (5.18) and (5.19) represent a system of nonlinear integral equations whose coefficients are random due to randomness of level energies $\{\xi_\alpha\}$ and occupation numbers $\{n_\alpha\}$. Apparently, for $\eta = 0$ these equations have a solution $\Gamma_\alpha(\epsilon) = 0$, corresponding to the insulating phase. One must check, however, whether this solution is stable with respect to an infinitesimal damping η.

To perform the standard linear stability analysis of equations (5.18) and (5.19) we linearize equation (5.19) as

$$A_\alpha(\epsilon) = \delta(\epsilon - \xi_\alpha) + \frac{1}{\pi} \frac{\Gamma_\alpha(\epsilon)}{(\epsilon - \xi_\alpha)^2} + O(\Gamma^2)\,, \qquad (5.24)$$

in complete analogy with Abou-Chacra *et al.* (1973), substitute this into equation (5.18), and obtain a linear integral equation:

$$\Gamma_\alpha(\epsilon) = \eta + \sum_{\beta,\gamma,\delta} |V_{\alpha\beta\gamma\delta}|^2 \frac{2\Gamma_\gamma(\epsilon + \xi_\beta - \xi_\delta) + \Gamma_\beta(\xi_\gamma + \xi_\delta - \epsilon)}{(\epsilon + \xi_\beta - \xi_\gamma - \xi_\delta)^2}$$

$$\times \left[n_\beta(1 - n_\gamma)(1 - n_\delta) + (1 - n_\beta)n_\gamma n_\delta \right]. \qquad (5.25)$$

The solution of this equation may be sought in the form of a perturbation series

$$\Gamma_\alpha(\epsilon) = \sum_{n=0}^{\infty} \Gamma_\alpha^{(n)}(\epsilon)\,, \qquad (5.26)$$

where $\Gamma^{(n)}$ is of the order $|V|^{2n} \sim (\lambda\delta_\zeta)^{2n}$ and is obtained after n iterations of equation (5.25) starting from $\Gamma^{(0)} = \eta$. Each term in this expansion can be calculated explicitly and its statistics can be determined (details of this rather cumbersome calculation are given in Section 6 of the BAA paper). The resulting probability distribution function is controlled by a single parameter γ_n, which determines the typical scale of the random quantity $\Gamma^{(n)}$:

$$P_n(\Gamma^{(n)}) = e^{-\gamma_n/\Gamma^{(n)}} \sqrt{\frac{\gamma_n/\pi}{\left[\Gamma^{(n)}\right]^3}}\,, \qquad \gamma_n = C_1 \eta \Lambda^{2n}\,, \qquad \Lambda \equiv C_2 \frac{\lambda T}{\delta_\zeta} \ln \frac{1}{\lambda}\,. \qquad (5.27)$$

Here $C_1 \sim 1$ and $C_2 \sim 1$ are model-dependent numerical constants (equation (172) of the BAA paper contains their values for the specific model adopted there).

The behavior of this distribution function as $n \to \infty$ is qualitatively different, depending on whether Λ in equation (5.27) is smaller or larger than 1, as illustrated by Fig. 5.4. In the first case the typical scale $\gamma_n \to 0$ as $n \to \infty$ and the distribution function for the total Γ is always concentrated around $\Gamma \sim \eta$ which tends to zero as $\eta \to 0$. This perfectly matches the insulating behavior described in Section 5.3.3 and shown in Fig. 5.2(b).

On the other hand, for $\Lambda > 1$ the scale $\gamma_n \to \infty$; we emphasize that the limit $n \to \infty$ should be taken *prior* to $\eta \to 0$. This means that the distribution function $P(\Gamma)$ does not shrink to $\delta(\Gamma)$ as $\eta \to 0$; in fact, it cannot be determined from the linearized equation (5.25), the full self-consistent problem, equations (5.18) and (5.19), should be solved. The divergent linear solution signals the instability of the insulating state and the onset of the metallic state.

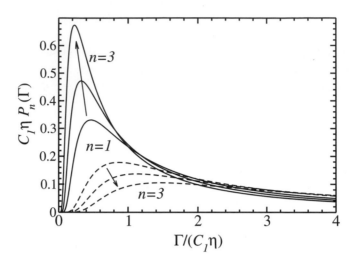

FIG. 5.4. Plot of the distribution functions $P_n(\Gamma^{(n)})$ as given by equation (5.27) for $n = 1, 2, 3$ for $T/T_c = 0.7$ (solid lines) and $T/T_c = 1.3$ (dashed lines).

These arguments enable us to identify the temperature at which $\Lambda = 1$ with the temperature of the metal–insulator transition:

$$T_c = \frac{\delta_\zeta}{C_2 \lambda \ln(1/\lambda)} . \qquad (5.28)$$

5.4 Metal–insulator transition and many-body mobility edge

In the previous section we considered the decay of a quasiparticle excitation, and found that possibility or impossibility of such decay may be viewed as delocalization or localization of this excitation in the many-electron Fock space. In this section we discuss the macroscopic implications of this picture. How can the very notion of localization be applied to many-body states?

Consider a many-body eigenstate $|\Psi_k\rangle$ of the interacting system, with the corresponding eigenenergy E_k. In the coordinate representation, the many-body wavefunction $\Psi_k\left(\{\vec{r}_j\}_{j=1}^N\right)$ depends on the coordinates of all N particles in the system. Let us create an electron–hole pair on top of $|\Psi_k\rangle$. The resulting state, which is not an eigenstate of the system, can be expanded in terms of other eigenstates:

$$\hat{c}_\alpha^\dagger \hat{c}_\beta |\Psi_k\rangle = \sum_{k'} C_{\alpha\beta}^{kk'} |\Psi_{k'}\rangle ; \quad \sum_{k'} \left|C_{\alpha\beta}^{kk'}\right|^2 = 1. \qquad (5.29)$$

It is possible that the number of terms contributing to the sum is effectively finite, i.e.,

$$\lim_{\mathcal{V} \to \infty} \left[\sum_{k'} \left| C_{\alpha\beta}^{kk'} \right|^4 \right]^{-1} < \infty. \tag{5.30}$$

This corresponds to *insulating* or *localized many-body* state; excitation cannot propagate over all states allowed by the energy conservation.

The opposite case, when expansion (5.29) contains an *infinite* number of eigenstates

$$\lim_{\mathcal{V} \to \infty} \left[\sum_{k'} \left| C_{\alpha\beta}^{kk'} \right|^4 \right]^{-1} = \infty, \tag{5.31}$$

corresponds to *metallic* or *extended many-body* state.

A developed metallic state is formed when the expansion (5.29) involves all the eigenstates with close enough energies:

$$\left| C_{\alpha\beta}^{kk'} \right|^2 \propto \text{``}\delta(E_k + \omega_{\alpha\beta} - E_{k'})\text{''}, \tag{5.32}$$

where the δ-function should be understood in the thermodynamic sense: its width, although sufficiently large to include many states, vanishes in the limit $\Omega \to \infty$. Only in this regime, which may also be called *ergodic many-body* state, can the electron–electron interaction bring the system from the initial Hartree–Fock state to the equilibrium corresponding to spanning all the states permitted by the energy conservation. In this case, the averaging over the exact many-body eigenfunction is equivalent to averaging over the microcanonical distribution, and temperature T can be defined as a usual Lagrange multiplier. It is related to E_k by the thermodynamic relation:

$$E_k - E_0 = \int_0^T C_V(T_1)\, dT_1, \tag{5.33}$$

where E_0 is the ground-state energy, and $C_V(T) \propto \Omega$ is the specific heat.

The temperature of the metal–insulator transition, found above, in fact, determines the extensive *many-body mobility edge* $\mathcal{E}_c \propto \Omega$. In other words, (i) states with energies $E_k - E_0 > \mathcal{E}_c$ are extended, inelastic relaxation is possible, and the conductivity $\sigma_k = \sigma(E_k)$ in this state is finite; (ii) states with energies $E_k - E_0 < \mathcal{E}_c$ are localized and the conductivity $\sigma(E_k) = 0$.

Let us now assume that the equilibrium occupation is given by the Gibbs distribution. One could think that it would still imply the Arrhenius law (5.3) for the conductivity. However, this is not the case for the many-body mobility threshold. In fact, in the limit $\Omega \to \infty$

$$\sigma(T) = 0; \quad T < T_c, \tag{5.34}$$

where the critical temperature is determined by equation (5.33):

$$\int_0^{T_c} dT_1 \, C_V(T_1) = \mathcal{E}_c \,. \tag{5.35}$$

Therefore, the temperature dependence of the dissipative coefficient in the system shows the singularity typical for a phase transition.

To prove the relations (5.34) and (5.35) we use the Gibbs distribution and find

$$\sigma(T) = \sum_k P_k \sigma(E_k) = \frac{\int_0^\infty dE \, e^{[S(E)-E]/T} \sigma(E)}{\int_0^\infty dE \, e^{[S(E)-E]/T}} \,,$$

where the entropy $S(E)$ is proportional to volume, and E is counted from the ground state. The integral is calculated in the saddle point or in the steepest decent approximations, exact for $\Omega \to \infty$. The saddle point $E(T)$ is given by

$$\left. \frac{dS}{dE} \right|_{E=E(T)} = \frac{1}{T} \,.$$

Taking into account $\sigma(E) = 0$ for $E < \mathcal{E}_c$ we find

$$\sigma(T) = \sigma\left[E(T)\right] \,, \quad E(T) > \mathcal{E}_c \,;$$

$$\sigma(T) \propto \exp\left(-\frac{\mathcal{E}_c - E(T)}{T}\right) \,; \quad E(T) < \mathcal{E}_c \,.$$

As both energies entering the exponential are extensive, $E(T), \mathcal{E}_c \propto \Omega$, we obtain equations (5.34) and (5.35).

To be able to establish the thermal equilibrium in such an insulating state the system should be coupled to an external bath (i.e., phonons). The presence of the finite electron–phonon interaction (as phonons are usually delocalized), smears out the transition, and $\sigma(T)$ becomes finite for any temperature. Nevertheless, if the electron–phonon interaction is weak, the phenomenon of the many-body metal–insulator transition manifests itself as a sharp crossover from phonon induced hopping at $T < T_c$ to the conductivity independent of the electron–phonon coupling at $T > T_c$.

5.5   Conclusions and perspectives

We have considered the inelastic relaxation and transport at low temperatures in disordered conductors where all single-particle states are localized, and no coupling to phonons or any other thermal bath is present. The main question is whether the electron–electron interaction alone is sufficient to cause transitions between the localized states producing thereby a finite conductivity. The answer to this question turns out to be determined by the Anderson localization–delocalization physics in the many-particle Fock space and is summarized on Fig. 5.5.

It should be emphasized that the many-body localization, which we discuss in this paper, is qualitatively different from conventional finite temperature metal

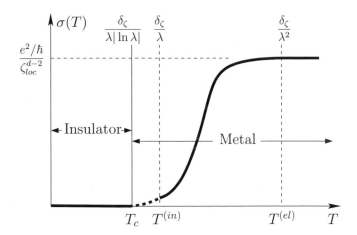

FIG. 5.5. Schematic temperature dependence of the dc conductivity $\sigma(T)$. Below the point of the many-body metal–insulator transition, $T < T_c$, no inelastic relaxation occurs and $\sigma(T) = 0$. The temperature interval $T \gg T^{(in)} > T_c$ corresponds to the developed metallic phase, where equation (5.21) is valid. At $T \gg T^{(el)}$ the high-temperature metallic perturbation theory (Altshuler and Aronov, 1985) is valid, and the conductivity is given by the Drude formula. From (Basko *et al.*, 2006). Adapted figure with permission from Elsevier Ltd: [D. M. Basko, I. L. Aleiner, B. L. Altshuler, *Ann. of Phys.* **321**, 1126 (2006)], copyright (2006).

to insulator transitions, such as formation of a band insulator due to the structural phase transition or Mott–Hubbard transition. In these two cases, at a certain temperature T^* a gap appears in the spectrum of charge excitation (Mott insulator) or all excitations (band insulator). In both cases the conductivity remains finite although exponentially small as long as $T > 0$. This is not the case for many-body localization, which implies exactly zero conductivity in the low-temperature phase.

Is the many-body localization a true thermodynamical phase transition with corresponding singularities in all equilibrium properties? This question definitely requires additional studies, however, some speculations can be put forward. The physics described in the present paper is associated with the change of the characteristics of the many-body wavefunctions. It is well known that for noninteracting systems localization–delocalization transition does not affect the average density of states, i.e., it does not manifest itself in any macroscopic thermodynamic properties. Application of the same logic to the exact many-body eigenvalues would indicate that the many-body localization transition is not followed by any singularities in the static specific heat, etc. On the other hand, at this point we cannot rule out the possibility that this conclusion is an artifact of treating the real parts of the electron self-energies with an insufficient accuracy. The most likely scenario, to our opinion, is that the insulating phase behaves like a glass (spin or structural) and demonstrates all the glassy properties (Fisher and Hertz, 1991; Bouchaud *et al.*, 1998), like the absence of ergodicity (even when some coupling with phonons is included), effects of aging, etc. Discussion of the equilibrium susceptibilities in the latter case becomes quite meaningless.

The quantitative theory built in the BAA paper assumes that the interaction is weak. On the other hand, qualitative consideration of the localization of many-body excitations does not rely on this assumption. The important ingredients are (i) localization of single-particle excitations and (ii) Fermi statistics. Consider, as an example, a Wigner crystal (Wigner, 1934). It is well known that a strong enough interaction leads to a spontaneous breaking of the translational symmetry in d-dimensional clean systems at $d \geq 2$. In a clean system Wigner crystallization is either a first-order phase transition ($d = 3$), or a Kosterlitz–Thouless transition ($d = 2$). Even weak disorder destroys both translational and orientational order (Larkin, 1970) and pins the crystal. The symmetry of this state is thus not different from the symmetry of a liquid, and the thermodynamic phase transition is commonly believed to be reduced to a crossover.

We argue that the many-body localization provides the correct scenario for the finite-temperature "melting" transition between the insulating phase, which may be called "solid", and the metallic phase, which may be called "liquid". Indeed, the conductivity of the pinned Wigner crystal is provided by the motion of defects. At low temperatures and in the absence of the external bath, all defects are localized by the one-particle Anderson mechanism. Phonon modes of the Wigner crystal are localized as well, so the system should behave as a many-body insulator. As the temperature is increased, the many-body metal–insulator transition occurs. It is not clear at present, whether it occurs before or after the crystalline order is destroyed at distances smaller than Larkin's scale. Construction of an effective theory of such a transition is a problem which deserves further investigation.

References

Abou-Chacra, R., Anderson, P. W. and Thouless, D. J. (1973). *J. Phys. C: Solid State Phys.* **6**, 1734.

Abrahams, E., Anderson, P. W., Licciardello, D. C. and Ramakrishnan, T. V. (1979). *Phys. Rev. Lett.* **42**, 673.

Altshuler, B. L. and Aronov, A. G. (1985). *Electron-Electron Interactions in Disordered Systems* (ed by A. L. Efros and M. Pollak). North-Holland, Amsterdam.

Altshuler, B. L., Gefen, Y., Kamenev, A. and Levitov, L. S. (1997). *Phys. Rev. Lett.* **78**, 2803.

Anderson, P. W (1958). *Phys. Rev.* **109**, 1492.

Basko, D. M., Aleiner, I. L. and Altshuler, B. L. (2006). *Ann. of Phys.* **321**, 1126.

Berezinskii, V. L. (1973). *Zh. Eksp. Teor. Fiz.* **65**, 1251 [*Sov. Phys. JETP* **38**, 620].

Bouchaud, J.-P., Cugliandolo, L. F., Kurchan, J. and Mezard, M. (1998). *Spin Glasses and Random Fields* (ed by A. P. Young). World Scientific, Singapore.

Dorokhov, O. N. (1983). *Zh. Eksp. Teor. Fiz.* **85**, 1040 [*Sov. Phys. JETP* **58**, 606].

Efetov, K. B. (1987). *Zh. Eksp. Teor. Fiz.* **92**, 638; *ibid.* **93**, 1125 [*Sov. Phys. JETP* **65**, 360; *ibid.* **66**, 634].

Efetov, K. B. and Larkin, A. I. (1983). *Zh. Eksp. Teor. Fiz.* **85**, 764 [*Sov. Phys. JETP* **58**, 444].

Eliashberg, G. M. (1962). *Zh. Eksp. Teor. Fiz.* **42**, 1658 [*Sov. Phys. JETP* **15**, 1151].

Fischer, K. H. and Hertz, J. A. (1991). *Spin glasses.* Cambridge University Press, Cambridge.

Fleishman, L. and Anderson, P. W. (1980). *Phys. Rev.* B **21**, 2366.

Fritzsche, H. (1955). *Phys. Rev.* **99**, 406.

Gertsenshtein, M. E. and Vasil'ev, V. B. (1959). *Teor. Veroyatn. Prim.* **4**, 424; *ibid.*, **5**, 3(E) [*Theory of Probability and its Applications* **4**, 391; *ibid.*, **5**, 340(E)].

Gogolin, A. A., Melnikov, V. I. and Rashba, É. I. (1975). *Zh. Eksp. Teor. Fiz.* **69**, 327 [*Sov. Phys. JETP* **42**, 168].

Gornyi, I. V., Mirlin, A. D. and Polyakov, D. G. (2004). Cond-mat/0407305, version 1.

Keldysh, L. V. (1964). *Zh. Eksp. Teor. Fiz.* **47**, 1945 [*Sov. Phys. JETP* **20**, 1018].

Kozub, V. I., Baranovskii, S. D. and Shlimak, I. (2000). *Solid State Commun.* **113**, 587.

Landau, L. D, (1958). *Zh. Eksp. Teor. Fiz.* **35**, 97 [*Sov. Phys. JETP* **8**, 70].

Larkin, A. I. (1970). *Zh. Eksp. Teor. Fiz.* **58**, 1466 [*Sov. Phys. JETP* **31**, 784].

Mott, N. F. (1968a). *J. of Non-Crystalline Solids* **1**, 1.

Mott, N. F. (1968b). *Phil. Mag.* **17**, 1259.

Nattermann, T., Giamarchi, T. and Le Doussal, P. (2003). *Phys. Rev. Lett.* **91**, 056603.

Shahbazyan, T. V. and Raikh, M. E. (1996). *Phys. Rev.* B **53**, 7299.

Shklovskii, B. I. and Efros, A. L. (1984). *Electronic Properties of Doped Semiconductors.* Springer-Verlag, Berlin.

Thouless, D. J. (1977). *Phys. Rev. Lett.* **39**, 1167.

Wigner, E. P. (1934). *Phys. Rev.* **46**, 1002.

6

RAMAN SCATTERING BY LO PHONONS IN SEMICONDUCTORS: THE ROLE OF THE FRANZ–KELDYSH EFFECT

Elias Burstein
Mary Amanda Wood Professor of Physics, Emeritus
Department of Physics and Astronomy, University of Pennsylvania,
Philadelphia, PA 19104, USA

Abstract
In honor of Leonid Keldysh, a valued colleague and friend for over four decades, who is celebrating his 75th birthday this year, I am pleased to contribute the following short paper on the research that was carried out by my group at the University of Pennsylvania in the 1960s and 1970s on electric-field-induced Raman scattering by optical phonons and, specifically, on the role played by the Franz–Keldysh effect (tunneling assisted interband transitions) in surface space-charge electric field-induced Raman scattering by longitudinal optical phonons in semiconductors.

6.1 Introduction

The elementary three-step processes in Raman scattering by optical phonons in semiconductors involve three "momentum-conserving" virtual electronic transitions which are accompanied, in any time order, by the annihilation of a photon of the incident radiation (i); the creation or annihilation of an optical phonon (j); and the creation of a photon of the scattered radiation (s). The first step involves a direct interband transition creating an excited electron and hole. The second step involves either an intraband scattering of the excited electron or hole by an optical phonon, or an interband scattering of the excited electron or hole to a third band. The third step involves the recombination of the excited electron and hole returning the medium to its electronic ground state. The scattering of the electron and the hole in the second step contribute coherently to the overall Raman scattering. Processes in which the second step involves an intraband excitation are termed two-band process and those in which the second step involves an interband excitation of the excited electron (or hole) to a third band are termed three-band processes (Figs. 6.1 and 6.2).

In Raman scattering by LO phonons involving two-band processes, the matrix elements for the intraband scattering of the excited electron and hole by optical phonons in the second step involve the deformation potential interaction of the atomic displacements of the LO and TO phonons with the excited electron and

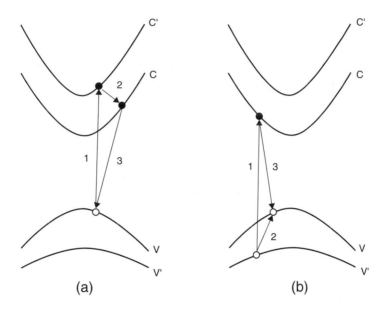

F<small>IG</small>. 6.1. Schematic diagram of the interband transitions which play a role in three-band scattering processes. (a) electron contribution. (b) hole contributions. (The numbers 1, 2 and 3 indicate the time order of the electronic transitions.)

hole and the Fröhlich interaction of the macroscopic Coulomb field of the LO phonon with the excited electron and hole. Loudon (1963) showed that in the long-wave limit (i.e., scattering wavevector $\mathbf{q}_{\mathrm{scat}} = \mathbf{k}_s - \mathbf{k}_i = \mathbf{k}_{\mathrm{LO}} \approx 0$), the relatively large Fröhlich interaction matrix elements for the intraband scattering of the excited electron and hole by the Coulomb field of the LO phonons have the same magnitude but opposite signs and cancel. However, Hamilton (1969) has pointed out that the electron and hole contributions do not cancel when $\mathbf{k}_{\mathrm{LO}}$ is finite and the effective masses of the electron and hole are unequal (i.e., $m_e \neq m_h$). This is a result of the $\mathbf{q}$-dependentfactors in the energy denominators of the Raman scattering tensor and the q-dependence of the Fröhlich interaction matrix element. Under resonance conditions this leads to a sizeable wavevector-dependent contribution to the scattering intensity. Since the intraband scattering of the excited electron and hole by the Coulomb field of the LO phonons does not change the orbital parts of the excited electron and hole wavefunctions, the wavevector-dependent contributions are observed in scattering configurations in which the polarization of the incident light, $\mathbf{e}_i$, is parallel to that of the scattered light, $\mathbf{e}_s$ (i.e., for $\mathbf{e}_i \parallel \mathbf{e}_s$ scattering configurations).

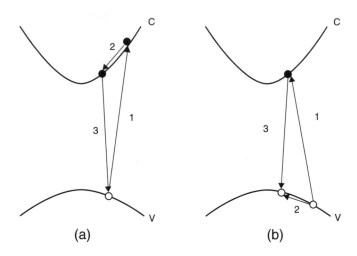

FIG. 6.2. Schematic diagram of the interband and intraband transitions which play a role in two-band scattering processes. (a) electron contribution. (b) hole contribution. (The numbers 1, 2 and 3 indicate the time order of the electronic transitions.)

6.2 Observation of forbidden LO phonon scattering at surface of n-InSb

Pinczuk and Burstein (1968) carried out an investigation of Raman scattering by TO and LO phonons in tellurium doped samples of n-type InSb, using backward-scattering from air cleaved (110) surfaces at $80\,\mathrm{K}$ and $\lambda = 6328\,\mathrm{\AA}$ ($1.96\,\mathrm{eV}$) radiation of a He-Ne laser, which is close to the E_1 energy gap ($1.89\,\mathrm{eV}$) of InSb. The objective was to study Raman scattering by coupled LO phonon–plasmon modes which had been observed by Mooradian and Wright in n-type GaAs samples using the 1.0648 micron radiation of a YAlG laser for which the GaAs is transparent (Mooraddian and Wright, 1966; Burstein *et al.*, 1967). Surprisingly, Pinczuk *et al.* found that the $\mathbf{e}_i \parallel \mathbf{e}_s$ backward-scattering spectra for the n-type InSb samples exhibited a forbidden LO phonon peak at its unscreened frequency, for the scattering configuration, in addition to an allowed TO phonon peak (Fig. 6.3). The intensity of the forbidden LO phonon peak, which was much weaker than the allowed TO peak in the sample with the lowest carrier density, increased with increasing carrier density, becoming much larger than that of the TO peak whose intensity was insensitive to the carrier density, before leveling off and then decreasing. For all carrier densities, the forbidden LO peak appeared

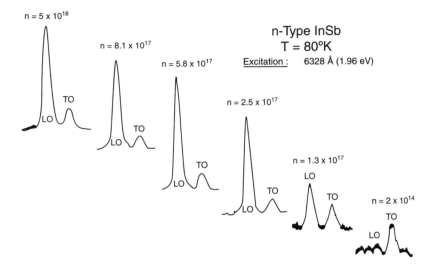

FIG. 6.3. Recorder traces of backward-scattering (6328 Å) Raman spectra of LO and TO phonons at the (110) surface of n-InSb at 80 K. (From Pinczuk and Burstein, 1970, p. 732.) Reprinted figure from [A. Pinczuk and E. Burstein, *Proc. Tenth Int. Conf on Physics of Semiconductors* (ed. by U. S. Atomic Energy Commission). Wash. DC (1970), pp. 727-735], copyright (1970).

at the unscreened frequency, indicating the absence of coupling with plasmons. The unscreened forbidden LO phonon peak and the dependence of its strength on the density of free carriers were attributed to the fact that the Fermi level at the surface was pinned within the energy gap, thereby creating a depletion layer and the absence of free carriers within the skin depth of the incident radiation (Fig. 6.4), and to the effect of the associated surface space-charge electric field E_{sc}.

The observed dependence of the intensity of the forbidden scattering by LO phonons on the density of carriers was attributed to the increase in the electric field in the depletion layer and the associated decrease in width of the depletion layer with increase in the donor density. At low donor density, the skin depth is smaller than the width of the depletion layer and the initial increase of the LO phonon band is due to the increase in the electric field. The leveling off and subsequent decrease of the ratio I(LO)/I(TO) in samples with the highest carrier density is attributed to the decrease in scattering length when the depletion layer width becomes comparable and then smaller than the skin depth of the crystal at the incident and scattered radiation frequencies. In the case of a degenerate sample with $n = 1.4 \times 10^{18}$ cm^{-3}, lowering the temperature to 140 K increased the LO phonon peak by a factor of 6, but did not affect the intensity of the transverse optical (TO) phonon peak. This was attributed to an increase in

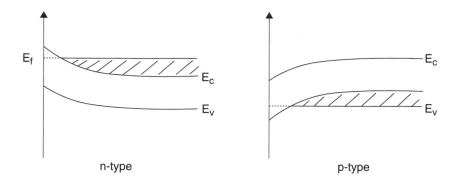

Fig. 6.4. Schematic energy band diagram for the surface space-charge regions of degenerate n- and p-type semiconductor samples in which the Fermi level E_F pinned within the forbidden gap.

resonant enhancement of the LO phonons that originates in the change in the electronic band structure when the temperature of the sample is lowered. The fact that scattering at incident radiation frequencies close to the E_1 and $E_1 + \Delta_1$ gap exhibits intensities of the forbidden LO phonon peak which are much larger than the allowed TO phonons peak was taken as an indication that the major contributions to the forbidden LO phonon scattering originate in two-band processes that were doubly resonant, i.e., $\omega_s = \omega_i + \omega_{LO} \approx \omega_{gap}$, when the energy of the incident photons is close to the E_1 and $E_1 + \Delta_1$ energy gaps (Fig. 6.5). Since similar enhancement are expected for TO and LO phonon bands when the resonance mechanism involves continuum electron–hole pairs in the intermediate state, this was considered to be an indication that excitons (which interact strongly with LO phonons and only weakly with TO phonons) were involved (Burstein *et al.*, 1969).

On lowering the temperature, the E1 energy gap of InSb shifts closer to the energy of the incident radiation photons and the lifetime of excitons at the E_1 gap increases.

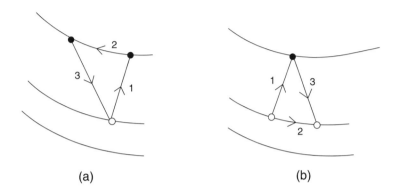

FIG. 6.5. Schematic diagram of transitions that play a role in the two-band
scattering processes at the E_1 and $E_1 + \Delta_1$ gaps. (a) electron contribution;
(b) hole contribution. The numbers 1, 2 and 3 indicate the time order of the
electronic transitions.

6.3   Role of the Franz–Keldysh effect in Raman scattering by LO phonons

Initially, the surface-electric field-induced forbidden Raman scattering by LO
phonons was attributed to the lowering of the crystal symmetry by the electric
field-induced relative displacement of the In and Sb atoms which modifies the
periodic potential and the eigenvectors and energies of the phonons. However, for
the scattering geometry used, i.e., backward scattering from the (110) surface,
the surface space-charge electric field induced $q \approx 0$ LO phonon band is expected
in $\mathbf{e}_i \perp \mathbf{e}_s$, as well as in $\mathbf{e}_i \parallel \mathbf{e}_s$ spectra, whereas the experimental data show that
the forbidden LO phonon band is only observed in the $\mathbf{e}_i \parallel \mathbf{e}_s$ spectra.

Pinczuk and Burstein (1970) suggested, in view of the band-bending associated surface space-charge electric field, that the interband transitions in the
first and third steps of the Raman scattering process correspond to Franz–
Keldysh type (tunneling-assisted) interband transitions that create (or annihi-
late) a spatially-separated excited electron–hole pair, and that the intraband
Fröhlich interaction of the Coulomb field of the LO phonons with the polarized
electron–hole pair is therefore nonzero even in the limit $\mathbf{q}_{scat} \approx 0$. The intraband
Fröhlich interaction matrix element for the interaction of the Coulomb field of

the LO phonons E_{LO} with the electric-dipole of the spatially separated excited electron and hole pair is given by (Burstein and Pinczuk, 1971)

$$(F^{LO}_{\beta\alpha E_{sc}}) = i\left(\frac{eE_{LO}}{q_{LO}\sqrt{V}}\right)\int d^3r\varphi^*_{\beta E_{sc}}(r)\left(e^{i\mathbf{q}\mathbf{r}_e} - e^{i\mathbf{q}\mathbf{r}_h}\right)\varphi_{\alpha E_{sc}}(r)\,. \qquad (6.1)$$

For $\mathbf{q}_{LO} \parallel \mathbf{E}_{sc}$, the leading term of $(F^{LO}_{\beta\alpha E_{sc}})$ is $\mathbf{q}$-independent and is given by

$$(F^{LO}_{\beta\alpha E_{sc}}) \approx i\left(\frac{eE_{LO}}{q_{LO}\sqrt{V}}\right)\int d^3r\varphi^*_{\beta E_{sc}}(r)\mathbf{r}_{E_{sc}}\varphi_{\alpha E_{sc}}(r) \qquad (6.2)$$

$$= i\left(\frac{eE_{LO}}{q_{LO}\sqrt{V}}\right)(\mathbf{r}_{\beta\alpha E_{sc}})\,,$$

where $\mathbf{r}_{E_{sc}}$ is the component of the relative coordinate of the electron and hole along $\mathbf{E}_{sc}$ and $(\mathbf{r}_{\beta\alpha E_{sc}})$ is the $\mathbf{q}$-independent electric-dipole matrix element for $\mathbf{E}_{LO}$-induced intraband transitions of the spatially separated (polarized) excited electron and hole pairs between states α and β whose envelope functions are $\varphi_{\beta E_{sc}}(r)$ and $\varphi_{\alpha E_{sc}}(r)$. The net effect of the surface space-charge field is to open up a strong $\mathbf{q}$-independent channel for Raman scattering by LO phonons. When the electric field is uniform and the Coulomb interaction between the excited electron and excited hole is neglected, the envelope functions $\varphi_{\beta E_{sc}}(r)$ and $\varphi_{\alpha E_{sc}}(r)$ are Airy functions. Since the intraband scattering of the polarized electron–hole pairs by the Coulomb field of the LO phonons does not change the orbital parts of the excited electron and hole wavefunctions, the surface space charge electric field-induced scattering by LO phonons via the Franz–Keldysh mechanism, like the $\mathbf{q}$-dependent forbidden LO phonon scattering, is observed in configurations where the polarization of the scattered radiation is parallel to that of the incident radiation .

6.4   Electric field-induced scattering by odd parity LO phonons

An electric field-induced Raman scattering by odd parity LO phonons in IV-VI compound semiconductors which have NaCl structures was investigated by Brillson *et al.* (1971). Use was made of the fact that although the position of the Fermi level at the free surface of ionic crystals is not strongly pinned, the Fermi level at the surface can be pinned within the bandgap by the presence of a metal film on the crystal surface (Fig. 6.6). In the case of backward Raman scattering from cleaved [100] surfaces of p-type samples of PbTe coated with transparent films of Pb, forbidden LO Raman scattering peaks whose frequencies corresponded to those of unscreened LO phonons were observed. Moreover, the forbidden LO peak exhibited a resonant enhancement at the E_2 gap and was observed only for $\mathbf{e}_i \parallel \mathbf{e}_s$. These peaks did not appear in the absence of the Pb films. As in the case of the III-V semiconductors, the forbidden surface electric field-induced, normally forbidden Raman scattering by the LO phonons was attributed to the polarization of the excited electron–hole pair in the intermediate state by the space-charge field.

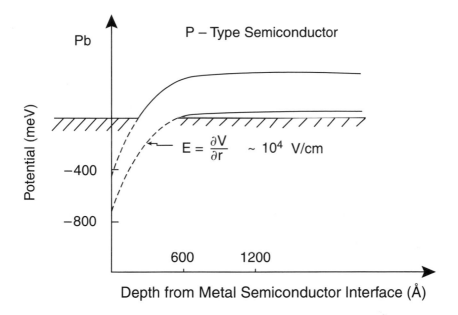

FIG. 6.6. Schematic metal–semiconductor contact diagram for a layer of Pb on a degenerate p-type PbTe sample (from Brillson and Burstein, 1971, p. 321). Reprinted figure from [L. Brillson and E. Burstein, *Proc. Second Int. Conf. on Light Scattering in Solids* (ed by M. Balkanski). Flammarion Science, Paris (1971), pp. 320-325], copyright (1971).

6.5   Microscopic theory of electric field-induced Raman scattering by LO phonons

A microscopic theory of the effect of a uniform electric field on two-band Raman scattering by LO phonons involving polarized coulomb-correlated electron–hole pairs (excitons) in the intermediate state has been formulated by Gay *et al.* (1971) focusing on the electric field induced-polarization of the exciton and on the Fröhlich exciton-LO phonon interaction. The contribution to the Raman scattering efficiency effect due to the polarization of the ground 1s exciton state by a uniform electric field was calculated by perturbation theory. For an exciton of 0.02 eV binding energy and 50 Å radius, the applied electric field f induces a spatial separation of the electron and hole in the 1s ground state of the exciton, creating an electric dipole of magnitude $d(f) = (9/4)ea_0 f$. This corresponds to a charge separation of $\sim 0.25\,a_0$ for a field of 10^4 V/cm. Since a uniform field does not affect the motion of the phonon center of mass, momentum conservation is not destroyed.

The exciton wavefunction is, to first order, independent of f, and the dependence of the Raman scattering tensor $\chi_{11}(q, f)$ on f and q which, to lowest order

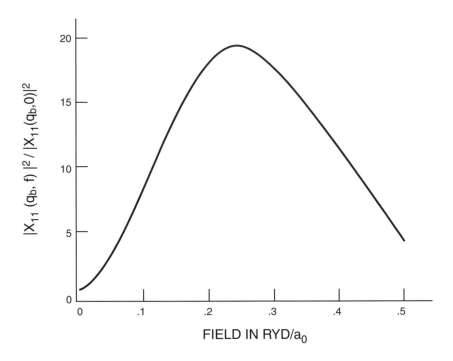

FIG. 6.7. Plot of the ratio $|\chi_{11}(q,f)|^2/|\chi_{11}(q,0)|^2$ as a function of the electric field f. This ratio gives the field-induced enhancement of the contribution to the Raman efficiency from the ground exciton state-LO phonon interaction (from Gay *et al.*, 1971, p. 36). Reprinted figure from [J. G. Gay, J. D. Dow, E. Burstein, and A. Pinczuk, *Proc. Second Int. Conf. on Light Scattering in Solids* (ed by M. Balkanski). Flammarion Science, Paris (1971), 33–38], copyright (1971).

in f and q, comes entirely from the exciton-LO phonon interaction, is given by

$$\chi_{11}(q,f) = \frac{1}{\pi}\left[\frac{m_e^2 - m_h^2}{2M^2}\, q_{\text{LO}} - i\frac{9f}{4}\right]\,, \tag{6.3}$$

where M is the exciton mass, q is in units of the reciprocal of the Bohr radius, and f is in Rydbergs per Bohr radius (Ryd/a_0). The presence of the factor $(m_e^2 - m_h^2)/M^2$ in the q-dependent term is due to the fact that the the macroscopic Coulomb field of the LO phonon interacts only with the charge clouds of the electron and hole of the exciton. In the absence of an electric field, this yields a nonzero contribution when $m_e \neq m_h$ (e.g., when the charge clouds of the electron and hole of the exciton are unequal in magnitude), and a vanishing contribution when $m_e = m_h$. The term linear in q_{LO} is 90° out of phase with the term that is linear in f, a general characteristic of wavevector-dependent scattering.

The wavevector-independent scattering efficiency $|\chi_{11}(q,f)|^2$ is proportional to f^2 at low fields

$$|\chi_{11}(q,f)|^2 = \frac{1}{\pi^2}\left[\frac{(m_e^2 - m_h^2)^2}{4M^4}\,q_{LO}^2 + \frac{81f^2}{16}\right]. \qquad (6.4)$$

$|\chi_{11}(q,f)|$ which has a q-dependent magnitude at zero f, increases linearly with f, reaches a maximum at $f = 0.25$ (Ryd/a_0) $= 10^4$ V/cm, and then decreases (Fig. 6.7), a result that arises from the fact that, with increasing f, the interband coupling of the exciton to the incident and scattered photons decreases more rapidly than the polarized exciton-LO phonon interaction increases.

Gay *et al.* (1971) also carried out a computer perturbation-calculation of the contribution to the electric-field-induced Raman scattering by LO phonons from two band scattering processes involving uncorrelated excited electron and holes in the intermediate state whose envelope functions correspond to Airy functions. As in two-band processes involving excitons, the electric field induces a separation of the electron and hole in the intermediate state and thereby to a wavevector-independent noncancellation of the electron and hole terms in the intraband Fröhlich interaction matrix elements. They note that when the applied field is nonuniform, wavevector is not conserved in the scattering process, since a nonuniform electric field destroys translational invariance symmetry of the crystal. As a consequence, LO phonons with wavevectors that are large compared to the scattering wavevector ($\mathbf{q}_{scat} = \mathbf{k}_i - \mathbf{k}_s$) can participate in the scattering. Non-uniform field effects may be responsible for the fact that the surface electric field induced LO peak in the Raman spectrum of n-type InAs, whose surface is accumulated, shows no screening (Corden *et al.*, 1970). The absence of screening implies that the LO phonons participating in the field induced scattering within the accumulation space-charge region at the surface, have wavevectors that are larger than the Fermi–Thomas wavevector. The fact that the scattering wavevector is smaller than the Fermi–Thomas wavevector, indicates that wavevector is not conserved.

6.6 Raman scattering by LO phonons in crossed electric and magnetic fields

A theoretical formulation of the electric field-induced Raman scattering by LO phonons in crossed electric and magnetic fields for which continuum electrons and holes have well-defined, bounded wavefunctions, was carried out by L. Brillson *et al.* (1972) for a simple two energy band model. The formulation takes into account the effect of $\mathbf{E}_0 \times \mathbf{H}_0$ on the excited electron and hole in the intermediate state, but neglects the tunneling-assisted optical interband transitions in the first and third steps of the two-band Raman scattering process. For sufficiently large applied magnetic fields, relative to the applied electric field, the excited electrons and holes exhibit cyclotron motion (Fig. 6.8) whose centers of orbit depend on the electric field. In the absence of an electric field, the electron and hole created by an interband excitation have the same center of cyclotron orbit. The separation of the average positions of the excited electron and hole, along the direction of the electric field, is given by

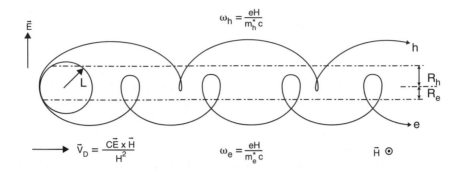

FIG. 6.8. Motion of an electron and hole in crossed electric and magnetic fields for GaAs with $H_0 = 10^5$ G and $E_0 = 10^4$ V/cm. H_0 is out of the paper. V_D is the drift velocity of the electron–hole pair, ω_e and ω_h are the cyclotron frequencies, respectively, of the electron and hole. (From Brillson and Burstein, 1972, p. 2976). Reprinted figure with permission from [L. Brillson and E. Burstein, *Phys. Rev. B* **5**, 2973 (1972)]. Copyright (1972) by the American Physical Society.

$$\mathrm{R}_{eh} = \mathrm{R}_e + \mathrm{R}_h = \frac{m_e^* c^2 E_0}{e H_0^2} + \frac{m_h^* c^2 E_0}{e H_0^2} . \tag{6.5}$$

A relatively simple analytic expression is obtained for the intraband Fröhlich interaction matrix elements of the excited electron and hole in their lowest Landau levels. To first order in the electric field, the wavevector-independent intraband Fröhlich element is diagonal in the orbital electronic states, since only the envelope part is changed. The electric field induced scattering by LO phonons whose wavevector is parallel to the applied electric field is proportional to the electric field-induced separation of the cyclotron-orbit centers of the excited electron and hole in their lowest, respective Landau levels, i.e., proportional to the electric field-induced electric-dipole $|e|R_{eh}$ in the intermediate state. Moreover, the Raman scattering by LO phonons occurs only for parallel polarization of incident and backward scattered radiation. The magnetic field does not affect the polarization selection rules since the orbital part of the excited electron and hole wavefunctions are unaltered. Also, because the Fröhlich interaction does not flip spin, only nonspin flip intraband scattering processes are involved. Moreover, since the allowed interband magneto-optical transitions are between states having the same Landau quantum number l, only $\Delta l = 0$ intraband scattering transitions are involved.

Aronov and Pikus (1967) and Zawadski and Lax (1966) have shown that in the presence of an electric field along y transverse to a relatively strong magnetic field along z, the intermediate continuum states electron and hole have harmonic oscillator wavefunctions in the plane containing the electric field, and Bloch functions along the magnetic field. These functions yield relatively simple analytic expressions for the electron and hole contributions to the Fröhlich intraband scattering matrix element, in the limit $\mathbf{q}_{LO}\mathbf{R}_{eh}$ and $q_{LO}L \to 0$ (where $L = (ch/2\pi eH_0)^{1/2}$ is the radius of the electron and hole in their lowest Landau levels),

$$R^F_{LO} \propto \delta(k_x - k'_x)\delta(k_z - k'_z)|e|\mathbf{R}_{eh}\mathbf{E}_{LO} , \qquad (6.6)$$

where $|e|\mathbf{R}_{eh}$ is the magnitude of the E_0 induced electric-dipole of the excited electron–hole pair in the intermediate state. The wavevector independent Frohlich contribution to R^F_{LO} is proportional to the average separation of the excited electron and hole in the intermediate state and linear in the applied electric field. It is nonzero only when the LO phonon wavevector $\mathbf{q}_{LO}$ has a component along the applied electric field direction and is zero when $\mathbf{q}_{LO}$ is perpendicular to $\mathbf{E}_0$. Thus, as a consequence of the fact that the continuum states in crossed electric and magnetic fields have harmonic-oscillator – type wavefunctions, one obtains a simple analytic expression for the LO phonon Raman scattering tensor, in contrast to the case of electric field-induced scattering in zero applied magnetic field. The Raman tensor is directly proportional to the electric field induced separation of centers of orbit of the electron and hole in the intermediate state. This result, which is similar to electric field-induced Raman scattering by LO phonons via discrete exciton electron–hole pair states is quite general. The Fröhlich interaction contribution to electric field-induced Raman scattering by LO phonons arises from the field-induced separation of the excited electron and hole in the intermediate state.

6.7   Concluding remarks

The surface space-charge electric field induced Raman scattering by LO phonons has proven to be a useful probe of energy band-bending at the surface of degenerate semiconductors. Corden *et al.* (1970) have investigated the Raman scattering by unscreened LO phonons to probe the surface electronic properties of degenerate InAs and GaSb, using the intensity of the LO phonons relative to that of the field-independent TO phonons as a measure of the space-charge electric field. They find a dependence of $I(LO/I(TO)$ on the density of carriers that is in agreement with the Franz–Keldysh-type theory of surface-electric field induced scattering. Buchner *et al.* (1976) have used surface space-charge electric field induced Raman scattering by LO phonons to probe the differences in the character of the A (111) and B (-1-1-1) surfaces of n- and p-InAs. The measurements which were carried out on air-exposed surfaces showed that, in the case of p-InAs, the space-charge electric field of the A surface is considerably smaller than that of the B surface. By contrast, in the case of n-InAs, the space charge electric fields at both surfaces were quite large. The marked difference in the forbidden scattering by LO phonons at the A and B surfaces of n- and p-InAs reflect differences in the pinning of the Fermi levels at the two surfaces. Both the (111) and (-1-1-1) surfaces of n-InAs were shown, by the use of externally applied electric fields, to be depleted of free carriers, in contrast to the (100) surfaces which are accumulated.

The ability to observe electric field-induced Raman scattering by optical phonons has greatly extended the range of semiconductor materials in which normally Raman inactive optical phonons can be investigated by first-order Raman scattering. Moreover, data on the dependence of electric field-induced Raman scattering by optical phonons on the frequency of the incident and scattered radiation has provided a deeper understanding of the elementary processes which play a role in the scattering of light by optical phonons.

The technical help of Celestino Creatore in preparing the figures in this paper is appreciated and gratefully acknowledged.

References

Aronov, A. G. and Pikus, G. E. (1967). *Zh. Eksp. Teor. Phys.* **51**, 505 [(1967). *Sov. Phys. JETP* **24**, 339].

Brillson, L. and Burstein, E. (1971). *Phys. Rev. Lett.* **27**, 808 and *Proc. Second Int. Conf. on Light Scattering in Solids* (ed by M. Balkanski). Flammarion Science, Paris, pp. 320-325.

Brillson, L. J. and Burstein, E. (1972). *Phys. Rev.* B **5**, 2973.

Buchner, S., Ching, L. Y. and Burstein, E. (1976). *Phys. Rev.* B **14**, 4459.

Burstein, E. Pinczuk, A. and Iwasa, S. (1967). *Phys. Rev.* **157**, 611.

Burstein, E., Mills, D. L. Pinczuk, A. and Ushioda, S. (1969). *Phys. Rev. Lett.* **22**, 348.

Burstein, E. and Pinczuk, A. (1971). *The Physics of Optoelectronic Materials* (ed by W. A. Albers). Plenum Publishing Corp. New York, NY, pp. 33-79.

Corden, P., Pinczuk, A. and Burstein, E. (1970). *Proc. Tenth Int. Conf on Physics of Semiconductors* (ed by U.S. Atomic Energy Commission). Wash. DC, pp. 739-745.

Gay, J. G., Dow, J. D., Burstein, E. and Pinczuk, A. (1971). *Proc. Second Int. Conf. on Light Scattering in Solids* (ed by M. Balkanski). Flammarion Science, Paris, pp. 33-38.

Hamilton, D. C. (1969). *Phys. Rev.* **188**, 1221.

Loudon, R. (1963). *Proc. Roy. Soc.* A **275**, 218.

Mooradian, A. and Wright, G. B. (1966). *Phys. Rev. Lett.* **16**, 999.

Pinczuk, A. and Burstein E. (1968). *Phys. Rev. Lett.* **21**, 1073.

Pinczuk, A. and Burstein, E. (1970). *Proc. Tenth Int. Conf on Physics of Semiconductors* (ed by U. S. Atomic Energy Commission). Wash. DC, pp. 727-735.

Zawadzki, W. and Lax, B. (1966). *Phys. Rev. Lett.* **16**, 1001.

7

PHENOMENA IN COLD EXCITON GASES: FROM THEORY TO EXPERIMENTS

L. V. Butov

Department of Physics, University of California at San Diego, USA

Abstract

Leonid Veniaminovich Keldysh created a beautiful world of cold electron–hole systems. The theoretical predictions by Leonid Veniaminovich of the new states – the electron–hole liquid (Keldysh, 1968), exciton Bose–Einstein condensate (Keldysh and Kozlov, 1968), and excitonic insulator (Keldysh and Kopaev, 1965) – generated intensive experimental studies that continued for decades. In this contribution, we overview the phenomena observed in cold exciton gases in quantum well structures.

7.1 Introduction

An exciton is a bound pair of an electron and a hole in a semiconductor. More than three decades ago Keldysh and Kozlov (1968) demonstrated that in the dilute limit ($na_B^D \ll 1$, where a_B is the exciton Bohr radius, n the density, and D the dimensionality) excitons are weakly interacting hydrogen-like Bose particles and are expected to undergo Bose–Einstein condensation (BEC). Since the quantum degeneracy temperature scales inversely with the mass, and the exciton mass, M, is small, even smaller than the free electron mass, exciton BEC should occur at temperatures of about 1 K for experimentally accessible exciton densities, several orders of magnitude higher than for atoms. The discovery of atom BEC is reviewed in (Cornell and Wieman, 2002; Ketterle, 2002). As shown by Keldysh and Kopaev in 1964, in the opposite limit of a dense electron–hole system ($na_B^D \gg 1$) excitons are analogous to Cooper pairs and the exciton condensate, called the excitonic insulator, is analogous to the BCS superconductor state (Keldysh and Kopaev, 1965). Contrary to the BCS superconductor state, the pairing in the excitonic insulator is due to electron–hole interaction, the pairs are neutral and the state is insulating.

Experimental studies of phenomena in cold exciton gases and, in particular, experimental probes of the theoretically predicted exciton condensation rely on implementation of cold exciton gases. Although the semiconductor crystal lattice can be routinely cooled to temperatures well below 1 K in He-refrigerators, after decades of efforts with various materials, it appears to be experimentally challenging to lower the temperature of the exciton gas to even a few kelvin. The problem is the following: The exciton temperature, T_X, determined by the ratio of the energy relaxation and recombination rates, exceeds by far the lattice temperature in most semiconductors. In order to create a cold exciton gas with T_X close to the lattice temperature, the exciton lifetime should considerably exceed the exciton energy relaxation time.

Besides, for experimental implementation of cold and dense exciton gases semiconductors are required in which an excitonic state is the ground state and, in particular, has lower energy than the metallic electron–hole liquid (that is a condensate in real space). The electron–hole liquid is the ground state in Ge and Si and this is the main obstacle for creation of cold and dense exciton gases in these materials (Keldysh and Tikhodeev, 1986).

Over the last two decades the experimental efforts to observe exciton BEC in bulk semiconductors dealt mainly with Cu_2O (Hulin *et al.*, 1980; Snoke *et al.*, 1990; Bulatov and Tikhodeev, 1992; Fortin *et al.*, 1993; Lin and Wolfe, 1993; Mysyrowicz *et al.*, 1996; Goto *et al.*, 1997; O'Hara *et al.*, 1999; Naka and Nagasawa, 2003; Kubouchi *et al.*, 2005), a material whose ground exciton state is optically dipole-inactive and has, therefore, a low radiative recombination rate, as well as with uniaxially strained Ge where stability of the electron–hole liquid is reduced by the strain (Timofeev *et al.*, 1983).

Because of the long lifetime and high cooling rate, the indirect excitons in coupled quantum wells (CQWs) form a system where a cold exciton gas can be created. Examples of semiconductor CQWs are shown in Fig. 7.1(a),(b). The long lifetimes are due to the spatial separation of the electron and hole wells (Lozovik and Yudson, 1976; Fukuzawa *et al.*, 1990), resulting in a radiative lifetime of indirect excitons in CQWs samples that is typically from three to six orders of magnitude longer than that of direct excitons in single QWs. Moreover, the cooling of hot photoexcited excitons down to the temperatures of the cold lattice, which occurs via emission of bulk LA phonons, is about three orders of magnitude faster for excitons in GaAs QWs than that in bulk GaAs. This is due to relaxation of the momentum conservation law in the direction perpendicular to the QW plane. Indeed, for quasi-2D systems the ground-state mode $E = 0$ couples to the continuum of the energy states $E \geq E_0$ rather than to the single energy state $E = E_0 = 2Mv_s^2$ (v_s is the sound velocity) as it occurs in bulk semiconductors (Tikhodeev, 1990; Ivanov *et al.*, 1997; Ivanov *et al.*, 1999) (see Fig. 7.1c).

What temperature is "cold" for the exciton gas? The transition from a classical to quantum gas occurs when bosons are cooled to the point where the thermal de Broglie wavelength $\lambda_{\mathrm{dB}} = \sqrt{2\pi\hbar^2/(Mk_{\mathrm{B}}T)}$ is comparable to the interparticle separation (for instance, BEC takes place when $n\lambda_{\mathrm{dB}}^3 = 2.612$ in 3D systems) and the transition temperature for excitons in GaAs/AlGaAs QWs reaches a value of $T_{\mathrm{dB}} = 2\pi\hbar^2 n_X/(Mgk_{\mathrm{B}}) \approx 3\,\mathrm{K}$ for the exciton density per spin state $n_X/g = 10^{10}\,\mathrm{cm}^{-2}$ (the exciton spin degeneracy $g = 4$ and the exciton mass $M = 0.22\,m_0$ for GaAs/AlGaAs QWs (Butov, 2004), where m_0 is the free electron mass). The estimate is given for a typical exciton density, which is well below the Mott density $n_{\mathrm{Mott}} \sim 1/a_{\mathrm{B}}^2 \sim 2 \times 10^{11}\,\mathrm{cm}^{-2}$ above which the excitons dissociate due to phase-space filling and screening (Schmitt-Rink *et al.*, 1989)

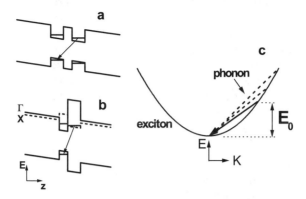

FIG. 7.1. Energy band diagram of GaAs/AlGaAs (a) and AlAs/GaAs (b) CQWs. The indirect transition is indicated by the arrow. (c) Energy diagram for the LA-phonon-assisted relaxation for bulk excitons (solid arrow) and for QW excitons (solid and dashed arrows).

($a_B \sim 20$ nm for the indirect excitons (Dignam and Sipe, 1991)). Peculiarities of condensation in 2D systems were studied in (Popov, 1972; Kosterlitz and Thouless, 1973; Fisher and Hohenberg, 1988; Bagnato and Kleppner, 1991; Ketterle and van Druten, 1996; Prokof'ev *et al.*, 2001).

According to the calculations (Butov *et al.*, 2001), for the bath temperature $T_{bath} = 50$ mK the exciton temperature reaches $T_X = 100$ mK at $t \sim 75$ ns after the excitation pulse while the exciton density at $t = 75$ ns can stay well above 10^{10} cm^{-2} due to the long lifetimes of the indirect excitons (Butov, 2004). Accuracy of the calculations is verified by quantitative agreement between the calculated and measured exciton kinetics. Therefore, the system of indirect excitons in a CQW allows experimental implementation and studies of cold exciton gases with $T_X \sim 100$ mK and $n_X \sim 10^{10}$ cm^{-2}, i.e., with temperatures well below $T_{dB} \sim 3$ K, thus giving an opportunity to experimentally probe what happens in exciton gases when they are cooled below T_{dB} and become quantum. Experimental studies of phenomena in the quantum exciton gases are reviewed in Sections 7.2 and 7.3.

To conclude the introduction we note briefly specific properties of the indirect excitons in a CQW. An indirect exciton is a dipole oriented perpendicular to the QW plane. Interaction between such dipoles is repulsive. The repulsive interaction between the indirect excitons and, more generally, monotonic enhancement of energy with increasing density for the system of spatially separated electron and hole layers, stabilizes the exciton state against formation of metallic electron–hole droplets (Yoshioka and MacDonald, 1990; Zhu *et al.*, 1995; Lozovik and Berman, 1996). Therefore the ground state in the system is excitonic. This is a necessary requirement for the experimental realization of cold exciton gases and ultimately the exciton condensation. Besides, the repulsive interaction results in effective screening of an in-plane disorder potential (Ivanov, 2002). Experimentally, the repulsive interaction is revealed by the enhancement of the exciton energy with increasing density (Butov, 2004). Additionally, the strong density dependence of the indirect exciton energy gives a possibility for experimental measurement of the indirect exciton density: The indirect exciton density can be estimated with a good accuracy from the energy shift δE using the plate capacitor formula $n_X = \epsilon \delta E / (4\pi e^2 d)$, where ϵ is the dielectric constant (de Leon and Laikhtman, 2001).

Note also that an intrinsic property of any QW system is an in-plane disorder potential, which is caused by interface and alloy fluctuations, defects and impurities. High-quality samples are characterized by a small amplitude and large length-scale of the random potential. This is revealed by a small linewidth of exciton photoluminescence (PL). On a large scale, exciton condensate in a random potential is analogous to the "Bose-glass" phase, or to a random Josephson-junction array in superconductors (Ma *et al.*, 1986; Fisher *et al.*, 1989). With the increase of the potential fluctuations the sizes of the condensate lakes (as well as the correlations between the lakes) are reduced, and for a large random potential exciton condensate disappears. Therefore, in order to observe exciton condensate, high-quality samples with small potential fluctuations are required. No condensation effects are observed in CQW samples with a large random potential and the effects disappear with increasing disorder (Butov, 2004). The experiments reviewed below were performed on high-quality CQW samples characterized by the narrowest PL linewidths.

In Sections 7.2 and 7.3 we briefly review phenomena in quantum gases of indirect excitons in CQWs recently observed by our group. We note that a review of our earlier experiments indicating anomalous transport and luminescence of indirect excitons in AlAs/GaAs CQWs (Fig. 7.1b), which are consistent with exciton superfluidity, superradiance of the exciton condensate, and fluctuations near the phase transition (Butov *et al.*, 1994; Butov and Filin, 1998), is not included to this contribution and can be found elsewhere (Butov, 2004). We would also like to refer the reader to a review on studies of indirect excitons by Vladislav Borisovich Timofeev, which is included in the present book.

7.2   Stimulated kinetics of excitons

The scattering rate of bosons to a state **p** is proportional to $(1 + N_p)$, where N_p is the occupation number of the state **p**. At high N_p the scattering process is stimulated by the presence of other identical bosons in the final state. The stimulated scattering is a signature of quantum degeneracy. It is considered to be a spectacular signature for BEC in atomic gases (Ketterle, 2002).

In this and the following section, we review briefly experiments on excitons in high-quality GaAs/Al$_x$Ga$_{1-x}$As CQWs (Fig. 7.1a), where the radiative recombination is dominant. The PL kinetics of indirect excitons are shown in Fig. 7.2(a).

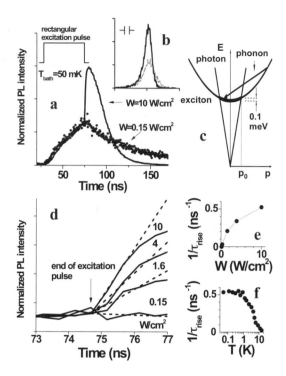

FIG. 7.2. (a) PL kinetics of indirect excitons in GaAs/AlGaAs CQW at high and low excitations W. (b) The PL spectra of indirect excitons at $W = 10\,\mathrm{W/cm^2}$ measured in 5 ns time intervals just before (thin line) and after (thick line) the end of the excitation pulse shown in (a), the scale of energy is 1.545–1.555 eV. (c) Energy diagram for the LA-phonon-assisted relaxation and optical decay of indirect excitons in GaAs/AlGaAs CQWs. The bold sector of the parabolic exciton dispersion indicates the radiative zone for indirect excitons. (d) PL kinetics of indirect excitons near the end of the excitation pulse at different W. The rate $\tau_{\mathrm{rise}}^{-1}$ of the PL rise right after the end of the excitation pulse is presented vs. W at $T_{\mathrm{bath}} = 50\,\mathrm{mK}$ (e), and vs. T_{bath} at $W = 10\,\mathrm{W/cm^2}$ (f). From (Butov *et al.*, 2001). Reprinted figures with permission from [L. V. Butov, A. L. Ivanov, A. Imamoglu, P. B. Littlewood, A. A. Shashkin, V. T. Dolgopolov, K. L. Campman, and A. C. Gossard, *Phys. Rev. Lett.* **86**, 5608 (2001)]. Copyright (2001) by the American Physical Society.

At low excitations, after a rectangular excitation pulse is switched off, the PL intensity of indirect excitons decays nearly monoexponentially. In contrast, at high excitations, right after the excitation pulse is switched off the PL intensity first jumps up and starts to decay only in a few nanoseconds. For delocalized, in-plane free, quasi-2D excitons, only the states with small center-of-mass momenta $|\mathbf{p}| \leq p_0 \approx \left(E_g \epsilon^{1/2}\right)/c$ can decay radiatively by resonant emission of bulk photons (Feldmann *et al.*, 1987; Hanamura, 1988; Andreani *et al.*, 1991; Deveaud *et al.*, 1991; Citrin, 1993; Bjork *et al.*, 1994) (see Fig. 7.2c). Thus the PL dynamics is determined by the occupation kinetics of the optically-active low-energy states $E \leq E_{p_0}$, the radiative zone ($E_{p_0}/k_B = p_0^2/(2Mk_B) \approx 1.2\,\text{K}$ at $B = 0$). The LA-phonon-assisted relaxation of the optically-dark hot photoexcited indirect excitons into the optically-active low-energy states results in a rise of the PL signal, while optical recombination of the low-energy indirect excitons results in a decay of the PL intensity. The end of the excitation pulse is accompanied by a sharp drop in the exciton temperature caused by switching off the generation of hot indirect excitons and, as a result, the PL intensity and the occupation numbers $N_{E \leq E_{p_0}}$ of the optically-active low-energy states abruptly rise within a few nanoseconds right after the trailing edge of the excitation pulse. It is the integrated PL intensity of the indirect excitons that increases after the end of the excitation pulse and the integrated intensity of all PL above the indirect exciton energy is negligibly small (Fig. 7.2b). This indicates that the rise of the indirect exciton PL intensity after the excitation pulse end is due to scattering from optically dark states.

The measured PL kinetics is strongly nonlinear (Fig. 7.2d,e): the observed increase of τ_{rise}^{-1} for delocalized, in-plane free, excitons with increasing exciton concentration (or W) shows that the scattering is stimulated by the final exciton state occupancy which, in turn, is high, $N_{p \leq p_0} > 1$. The observed decrease of τ_{rise}^{-1} with increasing temperature (Fig. 7.2f) results from a thermal reduction of the occupation of the low-energy exciton states $N_{p \leq p_0}$. The numerical simulations of PL dynamics are in quantitative agreement with the experiment (Butov *et al.*, 2001). According to the calculations, the end of the excitation pulse is indeed accompanied by a sharp drop of T_X and the PL jump after the pump pulse is accompanied by $N_{E=0} \gg 1$, i.e., a strongly degenerate Bose gas of indirect excitons builds up. Note that localization effects play an essential role in the kinetics at low exciton densities and should be taken into account.

7.3   Pattern formation in the exciton system

In this section, we review briefly spatially resolved PL experiments revealing a pattern formation in the exciton system.

At the lowest excitation powers, P_{ex}, the spatial profile of the indirect exciton PL intensity practically follows the laser excitation intensity. However, at high P_{ex}, the indirect exciton PL pattern is characterized by a ring structure: the laser excitation spot is surrounded by two concentric bright rings separated by an annular dark interring region (Fig. 7.3a). The rest of the sample outside the external ring is dark. The internal ring appears near the edge of the laser excitation spot and its radius reaches tens of microns, and the external ring can be remote from the excitation spot by more than $100\,\mu\text{m}$. The ring radii increase with P_{ex}. The ring structure follows the laser excitation spot when it is moved over the whole sample area.

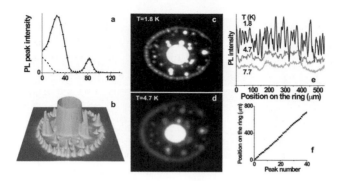

FIG. 7.3. (a) Peak intensity of the indirect exciton PL vs. r, the distance from the excitation spot center, at $T = 1.8$ K, $V_g = 1.22$ V, and the excitation power $P_{ex} = 690\,\mu$W. The excitation spot profile is shown by the dashed line. Spatial pattern of the indirect exciton PL intensity at $T = 1.8$ (c) and 4.7 K (d) for $P_{ex} = 390\,\mu$W. The area of view is $475 \times 414\,\mu$m. (e) The corresponding variation of the indirect exciton PL intensity along the external ring at $T = 1.8$, 4.7, and 7.7 K. The dependence of the position of the indirect exciton PL intensity peaks along the external ring vs. the peak number is nearly linear (f), showing that the fragments form a periodic chain. (b) 3D plot of the PL pattern at $T = 380$ mK, $V_g = 1.24$ V, and $P_{ex} = 930\,\mu$W. From (Butov *et al.*, 2002b; Butov *et al.*, 2004). Adapted figures with permission from Macmillan Publishers Ltd: [L. V. Butov, A. C. Gossard, and D. S. Chemla], *Nature* **418**, 751 (2002)], copyright (2002), and [L. V. Butov, L. S. Levitov, A. V. Mintsev, B. D. Simons, A. C. Gossard, and D. S. Chemla, *Phys. Rev. Lett.* **92**, 117404 (2004)], copyright (2001) by the American Physical Society.

The spatial pattern shows also that the indirect exciton PL intensity is strongly enhanced in certain fixed spots on the sample called localized bright spots (LBS), Fig. 7.3(c). For any excitation spot location and any P_{ex} the LBS are only observed when they are within the area terminated by the external ring, Fig. 7.3(c). In the LBS the indirect exciton PL line is spectrally narrow, full width at half maximum (FWHM) reaches 1.2 meV.

The external ring is fragmented into circular-shape structures that form a periodic array over macroscopic lengths, up to ~ 1 mm (Fig. 7.3c,e). This is demonstrated in Fig. 7.3(f), which shows a nearly linear dependence of the fragment positions along the ring vs. their number. The in-plane potential fluctuations are not strong enough to destroy the ordering (Fig. 7.3f). The fragments follow the external ring either when the excitation spot is moved over the sample area or when the ring radius varies with P_{ex}. Along the whole external ring, both in the peaks and the passes, the indirect exciton PL lines are spectrally narrow with the full width at half maximum ≈ 1.3 meV, considerably smaller than in the center

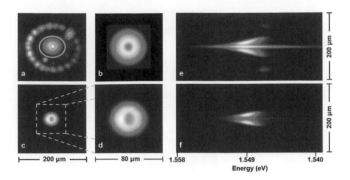

FIG. 7.4. (a), (c)–(f) Experimental and (b) calculated patterns of the PL signal from indirect excitons. In (a), the PL intensity in the area within the green circle is reduced by a constant factor for better visualization. A bright spot in the middle of the inner ring in (a) is due to residual bulk emission. (e) and (f) Image of the PL signal in the E-x coordinates. The external PL ring is also seen for high excitations, both in (a) x-y and (e) E-x coordinates. For (a) and (e) the excitation power is $P_{\text{ex}} = 250\,\mu\text{W}$ and for (b)–(d) and (f) $P_{\text{ex}} = 75\,\mu\text{W}$, respectively, and $T_{\text{b}} = 1.5\,\text{K}$. From (Ivanov, 2005). Adapted figure by permission of EDP Sciences from: [A. L. Ivanov, L. E. Smallwood, A. T. Hammack, Sen Yang, L. V. Butov and A. C. Gossard, *Europhys. Lett.* **73**, 920 (2006)], copyright (2006).

of the excitation spot. The ring fragmentation into the periodic chain appears abruptly at low temperatures below ca. 3 K; this is quantified by the amplitude of the Fourier transform of the PL intensity variation along the ring (Fig. 7.8c). Each fragment contains macroscopic number of excitons which can exceed tens of thousands and the period of the fragmentation varies within macroscopic length-scale, 10-50 microns. We call the exciton state with spatial order on macroscopic length-scale the macroscopically ordered exciton state (MOES).

The features observed in the exciton PL pattern (Butov *et al.*, 2002b) include the inner ring, the external ring, the localized bright spots, and the macroscopically ordered exciton state. One of the features – the external ring – has been observed also in InGaAs/GaAs CQWs (Snoke *et al.*, 2002) and in single GaAs/AlGaAs quantum wells (Rapaport *et al.*, 2004). The inner ring, the external ring, and the localized bright spots are observed up to high temperatures exceeding 10 K. They have been explained within classical framework, attributing the internal rings to nonradiative exciton transport and cooling (Ivanov, 2005) and the outermost rings and bright spots to macroscopic charge separation (Butov *et al.*, 2004; Rapaport *et al.*, 2004). The MOES is observed at low temperatures below ca. 3 K. Understanding the origin of this new state in cold exciton gases is in

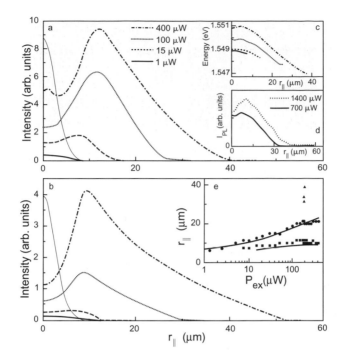

FIG. 7.5. The PL intensity of indirect excitons I_{PL}, (a) measured and (b) calculated, against radius $r_\parallel$ for four optical excitation powers P_{ex}. The Gaussian profile of the optical excitation is shown by the thin solid lines. The cryostat temperature $T_b = 1.5\,\mathrm{K}$. (d) $I_{PL} = I_{PL}(r_\parallel)$ of indirect excitons measured at subbarrier excitation (780 nm) for $P_{ex} = 0.7\,\mathrm{mW}$ and $1.4\,\mathrm{mW}$. Bulk emission is subtracted from the total PL signal in (a) and (d). The energy position of the PL line, $E_{PL} = E_{PL}(r_\parallel)$ for the data shown in (a) is plotted in (c). (e) The measured (square points) and calculated (solid line) inner ring radius $r_\parallel^{rg}$ versus P_{ex} (the triangular points refer to the external PL ring), and the measured (circle points) and calculated (solid line) HWHM spatial extension of the PL signal against $r_\parallel$. From (Ivanov, 2005). Adapted figure by permission of EDP Sciences from: [A. L. Ivanov, L. E. Smallwood, A. T. Hammack, Sen Yang, L. V. Butov and A. C. Gossard, *Europhys. Lett.* **73**, 920 (2006)], copyright (2006).

progress.

7.3.1 *The inner exciton ring*

We now discuss the origin of the observed effects. Under CW photoexcitation, there is a continuous flow of excitons out of the excitation spot mainly due to exciton diffusion and drift. The exciton diffusion originates directly from the exciton density gradient. The exciton drift also originates from the density gradient as the latter results to the gradient of the indirect exciton potential energy because

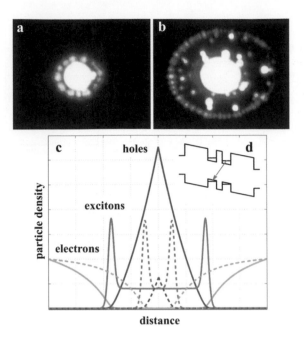

FIG. 7.6. Exciton rings at different photoexcitation intensity: The pattern of indirect exciton PL at $T = 380\,\text{mK}$ and $V_\text{g} = 1.24\,\text{V}$ for (a) $P_\text{ex} = 310\,\mu\text{W}$ and (b) $P_\text{ex} = 930\,\mu\text{W}$. The excitation spot size is ca. $30\,\mu\text{m}$. The area of view is $410 \times 330\,\mu\text{m}$. (c) The carrier distribution as predicted by the transport model at two excitation intensities (the scales are in a.u.). (d) The CQW band diagram. From (Butov *et al.*, 2004). Adapted figure with permission from [L. V. Butov, L. S. Levitov, A. V. Mintsev, B. D. Simons, A. C. Gossard, and D. S. Chemla, *Phys. Rev. Lett.* **92**, 117404 (2004)]. Copyright (2001) by the American Physical Society.

of the repulsive interaction $\delta E \simeq 4\pi e^2 dn_X/\epsilon$ (Fig. 7.5c). The exciton energy decreases with increasing distance from the excitation spot results in an arrow-shaped profile of the exciton PL images in the E-x coordinates (Fig. 7.4e,f). As the excitons travel away from the excitation spot, T_X decreases with increasing radial distance owing to the energy relaxation of the excitons. The reduction of T_X leads to enhancement of the radiative zone occupation (see Section 7.2) and, therefore, increases the PL intensity. This mechanism accounts for the inner ring of radius up to tens of microns observed around the excitation spot. Within this mechanism the excitons travel diffusively in a dark state after having been excited, until slowed down to a velocity below the photon emission threshold, where they can decay radiatively.

A microscopic theory for the long-range transport, thermalization and optical decay of QW excitons is in a quantitative agreement with the experimental data (see Fig. 7.5). The origin of the inner ring has been identified: The inner ring is due to cooling of indirect excitons in their propagation from the excitation spot. Note that the long lifetimes of the indirect excitons allow them to travel to macroscopic distances exceeding tens of microns before recombination (Figs. 7.4 and 7.5). The origin of the inner ring is entirely classical. However, nonclassical statistics occurs at the position of the inner ring, where the exciton gas is already cold but still dense. For instance, in the experiments in (Ivanov, 2005) made at the bath temperature $T_b = 1.5\,\text{K}$, nonclassical occupation numbers of modest values, $N_{E=0}^{\max}(r_\parallel \simeq r_\parallel^{\text{rg}}) \simeq 1$, build up at the position of the inner ring.

7.3.2 The external exciton ring

While the inner exciton ring can be observed both for excitation above the Al-GaAs barrier and for subbarrier excitation, the external rings are observed for the excitation above the barrier only (Ivanov, 2005). The origin of the external ring is the following. Most of the hot electrons and holes, created by the off-resonance excitation, cool down and form excitons giving rise to PL observed within or near the laser spot, in particular in the inner ring. However, charge neutrality of the carriers excited in the CQW by an off-resonance laser excitation is generally violated mainly due to electrons and holes having different collection efficiency to the CQW (Zrenner *et al.*, 1990). Overall charge neutrality in the sample is maintained by opposite charge accumulating in the doped region out-side CQW. We speculate that the current through CQWs from n-doped GaAs layers (Fig. 7.6d) creates an electron gas in the CQW, while excess holes are photogenerated in the laser excitation spot due to higher collection efficiency of photoexcited holes to the CQW which are heavier than electrons.

The holes created at the excitation spot diffuse out and bind with electrons, forming indirect excitons. This process depletes electrons in the vicinity of the laser spot, creating an essentially electron-free and hole-rich region, which al-lows holes to travel a relatively large distance without encountering electrons. At the same time, a spatial nonuniformity in the electron distribution accumu-lates, causing a counterflow of electrons towards the laser spot. A sharp interface between the hole-rich region and the outer electron-rich area forms (Fig. 7.6c), since a carrier crossing into a minority region binds rapidly with an opposite carrier to form an exciton. The origin of the external exciton ring was explained by theories and experiments reported in (Butov *et al.*, 2004; Rapaport *et al.*, 2004). Observation of the external ring up to high temperatures, as high as $\simeq$ 10 K in (Butov *et al.*, 2002b) and $\simeq$ 100 K in (Snoke *et al.*, 2002), and for di-rect excitons which are hot due to their short recombination lifetime (Rapaport *et al.*, 2004), is not surprising since the origin of the ring is classical and does not require degeneracy in the exciton system (Butov *et al.*, 2004; Rapaport *et al.*, 2004).

The mechanism of charge transport can be explored within a simple model in which electrons and holes move in the CQW plane, each species in its own quantum well, governed by the coupled diffusion equations:

$$\frac{\partial n}{\partial t} = D\Delta n - wnp + J(r) \,,$$

$$\frac{\partial p}{\partial t} = D'\Delta p - wnp + J'(r) \,,$$

with $n(r)$ and $p(r)$ the electron and hole concentration, D, D' the diffusion constants, and w the rate at which an electron and hole bind to form an exciton. The source term for holes is localized at the excitation spot, $J'(r) = P_{ex}\delta(r)$, while the electron source is spread out over the entire plane, $J(r) = I(r) - a(r)n(r)$, with $I(r)$ and $a(r)n(r)$ the currents in and out of the CQW. The stationary solution of these equations, with spatially independent parameters I and a, displayed in Fig. 7.6(c), indeed shows a structure of two regions dominated by electrons and holes, and separated by a sharp interface where the exciton density $n_X \propto np$ is peaked.

7.3.3   The localized bright spots

The origin of the localized bright spots was revealed by the defocusing the laser excitation source and illuminating a large area (Fig. 7.7). In this excitation geometry, around the localized spots, small rings appear which mirror the behavior of the outer ring. However, the rings shrink with increasing laser power, indicating that electron–hole contrast here is inverted. This suggests that the spots represent localized sources of electrons (at current filaments crossing CQW) embedded in the hole-rich illuminated area. A model with a bright hole source and a weaker electron source shows the same qualitative behavior (Fig. 7.7c). Increasing hole flux (via P_{ex}), as soon as the electron source is enveloped by the hole rich area, a ring forms around it and then starts to shrink. Therefore the LBS were found to be the collapsed rings which are centered at the localized electron sources. Note that LBS allow accumulating a dense exciton gas in the localized spatial area around the LBS center (Butov *et al.*, 2002a; Lai *et al.*, 2004). However, the core of the LBS is hot as indicated by PL of high-energy direct exciton emission (Fig. 7.7d,e), probably due to the heating by the electron source in the center.

7.3.4   The macroscopically ordered exciton state

The most intriguing feature of the PL pattern is the ring fragmentation into aggregates forming a highly regular, periodic array over macroscopic lengths of up to 1 mm (Figs. 7.3, 7.6 and 7.8). These aggregates are clearly distinct from the localized bright spots originating from the localized sources of electrons: in contrast to the latter, the aggregates move in concert with the ring when the position of the source is adjusted. Moreover, in contrast to the LBS, the aggregates are cold: they do not contain direct excitons (Fig. 7.8a,b).

The external ring is far from the hot excitation spot and, therefore, the conditions in the ring are optimal for the formation of a cold exciton gas: the excitons in the ring are formed from well-thermalized carriers and, more importantly, due to long lifetimes of indirect excitons the binding energy released at exciton formation has little effect on their temperature. The heating source in the ex-

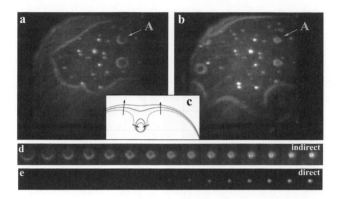

FIG. 7.7. Shrinkage of exciton rings to localized bright spots at nearly homogeneous excitation: Spatial PL pattern for indirect excitons at $T = 1.8\,\text{K}$, $V_g = 1.15\,\text{V}$, and (a) $P_{\text{ex}} = 390\,\mu\text{W}$ and (b) $P_{\text{ex}} = 600\,\mu\text{W}$. With (a,b), the area of view is $690 \times 590\,\mu\text{m}$. (c) Theoretical electron–hole interface for a point source of holes and a weaker point source of electrons. Arrows show interface displacement at increasing hole number. Note the ring shrinking around electron source. The shrinkage of ring A (a,b) of indirect excitons is detailed in (d) for $T = 380\,\text{mK}$, $V_g = 1.155\,\text{V}$, and $P_{\text{ex}} = 77\text{–}160\,\mu\text{W}$ (from left to right) with defocused excitation spot maximum moved to a location directly below ring A. With (d,e), the area of view is $67 \times 67\,\mu\text{m}$ for each image. Ring shrinkage is accompanied by the onset of the direct exciton emission (e) indicating hot cores at the center of the collapsed rings. From (Butov *et al.*, 2004). Adapted figure with permission from [L. V. Butov, L. S. Levitov, A. V. Mintsev, B. D. Simons, A. C. Gossard, and D. S. Chemla, *Phys. Rev. Lett.* **92**, 117404 (2004)]. Copyright (2001) by the American Physical Society.

ternal ring – the binding energy released at exciton formation – is the smallest heating source among those accompanying the exciton generation in CQWs: the binding energy of $\sim 5\,\text{meV}$ (Szymanska and Littlewood, 2003) is considerably smaller than the energy of hot photoexcited carriers in the excitation spot that is hundreds of meV in the experiments on exciton pattern formation. Note that to achieve a high density of indirect excitons, general experiments exploit photoexcitation at or above the direct exciton resonance where the photon absorption coefficient is high; in these experiments, the energies of the hot photoexcited carriers exceed 20 meV typically (Butov, 2004). The external ring is a region where the excess of the temperature of the system – excitons at low T_{lattice} or electron–hole plasma at high T_{lattice} – over the lattice temperature is the smallest. While the external ring is a classical object by itself (Section 7.3.2), it is the region where the coldest exciton gas is created. Indeed, the hot spots on the sample are detected by the positions of the direct exciton emission, and Fig. 7.8(b) shows

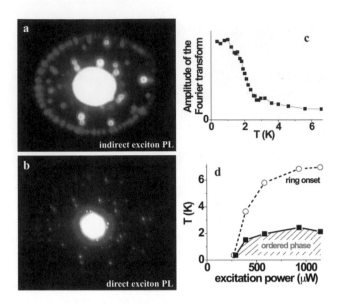

FIG. 7.8. Fragmented ring and localized spots in the (a) indirect and (b) direct
exciton PL image. The area of view is $410 \times 340 \, \mu$m. Note that the hot cores
with high-energy direct excitons, present in the localized spots, are absent
in the ring fragments. (c) Exciton density Fourier transform peak height at
the fragmentation period vanishes continuously at a critical temperature. (d)
The phase diagram of the states of the external ring. The phase boundary
of the modulated state observed at the lowest experimental temperatures
(solid line) along with the ring onset region (dashed line) are marked. From
(Butov *et al.*, 2004). Adapted figure with permission from [L. V. Butov, L. S. Levitov,
A. V. Mintsev, B. D. Simons, A. C. Gossard, and D. S. Chemla, *Phys. Rev. Lett.* **92**,
117404 (2004)]. Copyright (2001) by the American Physical Society.

that the exciton gas is hot in the localized bright spots, LBS, (due to heating by
leakage current through the LBS centers) and there is not observable heating in
MOES.

Cold exciton temperatures in the ring allow observation of the ordering, which
appears abruptly as the temperature is decreased below ∼3 K (Fig. 7.8c). The
phase diagram, obtained from PL distribution along the ring (see Figs. 7.3e and
7.8c), is presented in Fig. 7.8(d).

The exciton ring instability and the formation of the macroscopically ordered
state has not been predicted. The macroscopically ordered exciton state is a low-
temperature phenomenon observed below a few kelvin (Fig. 7.8c,d). Also the
MOES is observed in the external ring, which is far from the hot excitation spot
and where the exciton gas is the coldest.

The exciton density modulation requires a positive feedback. An attractive interaction could be a mechanism for positive feedback and, in particular, pattern formation in electron–hole systems has been long-known in connection with electron–hole droplet formation (Keldysh and Tikhodeev, 1986). However, the CQW geometry is engineered so that droplet formation is energetically unfavorable, the ground state is excitonic, and the interaction between excitons is repulsive: Indirect excitons, formed from electrons and holes confined to different QWs by a potential barrier, behave as dipoles oriented perpendicular to the plane, and an exciton or electron–hole density increase causes an enhancement of energy (Section 7.1). Crucially, the repulsive character of the interaction is observed in the regime of modulational instability. Here the spectral shift observed in PL varies along the circumference of the ring in concert with the intensity, with the largest shift found in the brightest regions. Translating the shift into energy, and noting that the repulsive interaction eliminates the standard mechanism of droplet formation (Keldysh and Tikhodeev, 1986), one is led to conclude that the paradigm of electron–hole droplets is too narrow to account for bead formation and a new mechanism must be sought.

The MOES is a new unpredicted phenomenon and understanding its origin is in progress. The theoretical approach to MOES, available at present, considers it in terms of quantum degeneracy of excitons. Within the model, an instability with a periodic 1D pattern can result from stimulated kinetics of exciton formation that build up near quantum degeneracy below a critical temperature in qualitative agreement with observations (Levitov *et al.*, 2005). Note also that the model predicts both the ring width and modulation wavelength to be of order of the exciton diffusion length. This compares favorably with the measured PL profile at the outer ring (Figs. 7.3, 7.6, 7.8) and is consistent with the exciton diffusion length $l \sim 30\,\mu\mathrm{m}$ estimated in (Ivanov, 2005) from the inner ring radius. The recent experiments on spontaneous coherence (Yang *et al.*, 2006) indicated a strong enhancement of the exciton coherence length at temperatures below a few kelvin. The macroscopic spatial ordering of excitons is correlated with the increase of the coherence length.

Summary The indirect excitons in CQWs form a system where a cold exciton gas with temperatures well below the quantum degeneracy temperature can be created. CQWs provide therefore an opportunity for experimental probe of the theoretical predictions for quantum exciton gases. In this contribution, we reviewed briefly phenomena observed in quantum gases of indirect excitons in CQW semiconductor structures and their understanding at present.

Acknowledgements The experimental studies of cold exciton gases have been inspired by the theoretical works of Leonid Veniaminovich Keldysh. It is a privilege to work in the field created by Leonid Veniaminovich.

The studies of the indirect excitons were performed in collaboration with G. Abstreiter, G. Bohm, K. L. Campman, D. S. Chemla, V. T. Dolgopolov, K. Eberl, A. I. Filin, A. C. Gossard, A. T. Hammack, A. Imamoglu, A. L. Ivanov,

L. S. Levitov, P. B. Littlewood, Yu. E. Lozovik, A. V. Mintsev, A. V. Petinova, A. A. Shashkin, B. D. Simons, L. Smallwood, G. Weimann, Sen Yang, and A. Zrenner and it is a pleasure to thank my collaborators. I am grateful to G. E. W. Bauer, A. B. Dzyubenko, M. Fogler, N. A. Gippius, Yu. M. Kagan, D. H. Lee, L. J. Sham, D. E. Nikonov, S. G. Tikhodeev and, particularly, to Leonid Veniaminovich Keldysh for enlightening discussions.

References

Andreani, L. C., Tassone, F. and Bassani, F. (1991). *Solid State Commun.* **77**, 641.

Bagnato, V. and Kleppner, D. (1991). *Phys. Rev.* A **44**, 7439.

Bjork, G., Pau, S., Jacobson, J. and Yamamoto, Y (1994). *Phys. Rev.* B **50**, 17336.

Bulatov, A. E. and Tikhodeev, S. G. (1992). *Phys. Rev.* B **46**, 15058.

Butov, L. V., Zrenner, A., Abstreiter, G., Bohm, G. and Weimann, G. (1994). *Phys. Rev. Lett.* **73**, 304.

Butov, L. V. and Filin, A. I. (1998). *Phys. Rev.* B **58**, 1980.

Butov, L. V., Ivanov, A. L., Imamoglu, A., Littlewood, P. B., Shashkin, A. A., Dolgopolov, V. T., Campman, K. L. and Gossard, A. C. (2001). *Phys. Rev. Lett.* **86**, 5608.

Butov, L. V., Lai, C. W., Ivanov, A. L., Gossard, A. C. and Chemla, D. S. (2002a). *Nature* **417**, 47.

Butov, L. V., Gossard, A. C. and Chemla, D. S. (2002b). *Nature* **418**, 751.

Butov, L. V., Levitov, L. S., Simons, B. D., Mintsev, A. V., Gossard, A. C. and Chemla, D. S. (2004). *Phys. Rev. Lett.* **92**, 117404.

Butov, L. V. (2004). *J. Phys.: Condens. Matter*, **16**, R1577.

Citrin, D. S. (1993). *Phys. Rev.* B **47**, 3832.

Cornell, E. A. and Wieman, C. E. (2002). *Rev. Mod. Phys.* **74**, 875.

de Leon, S. Ben-Tabou and Laikhtman, B. (2001). *Phys. Rev.* B **63**, 125306.

Deveaud, B., Clerot, F., Roy, N., Satzke, K., Sermage, B. and Katzer, D. S. (1991). *Phys. Rev. Lett.*, **67**, 2355.

Dignam, M. M. and Sipe, J. E. (1991). *Phys. Rev.* B **43**, 4084.

Feldmann, J., Peter, G., Göbel, E. O., Dawson, P., Moore, K., Foxon, C. and Elliott, R. J. (1987). *Phys. Rev. Lett.* **59**, 2337.

Fisher, D. S. and Hohenberg, P. C. (1988). *Phys. Rev.* B **37**, 4936.

Fisher, M. P. A., Weichman, P. B., Grinstein, G. and Fisher, D. S. (1989). *Phys. Rev.* B **40**, 546.

Fortin, E., Fafard, S. and Mysyrowicz, A. (1993). *Phys. Rev. Lett.* **70**, 3951.

Fukuzawa, T., Kano, S. S., Gustafson, T. K. and Ogawa, T. (1990). *Surf. Sci.* **228**, 482.

Goto, T., Shen, M. Y., Koyama, S. and Yokouchi, T. (1997). *Phys. Rev.* B **55**, 7609.

Hanamura, E. (1988). *Phys. Rev.* B **38**, 1228.

Hulin, D., Mysyrowicz, A. and à la Guillaume, C. B. (1980). *Phys. Rev. Lett.* **45**, 1970.

Ivanov, A. L. (2002). *Europhys. Lett.* **59**, 586.

Ivanov, A. L., Ell, C. and Haug, H. (1997). *Phys. Rev.* E **55**, 6363.

Ivanov, A. L., Littlewood, P. B. and Haug, H. (1999). *Phys. Rev.* B **59**, 5032.

Ivanov, A. L., Smallwood, L. E., Hammack, A. T., Yang, Sen, Butov, L. V. and Gossard, A. C. (2006). *Europhys. Lett.* **73**, 920.

Keldysh, L. V. (1968). *Proceedings of the Ninth International Conference on Semiconductors* (ed by S. M. Ryvkin and V. V. Shmaster). Nauka, Moscow, p. 1303.

Keldysh, L. V. (1986). *Contemp. Phys.* **27**, 395.

Keldysh, L. V. and Kopaev, Yu. V. (1964). *Fiz. Tv. Tela* **6**, 2791. [(1965). *Sov. Phys. Solid State* **6**, 2219.]

Keldysh, L. V. and Kozlov, A. N. (1968). *Zh. Eksp. Teor. Fiz.* **54**, 978. [*Sov. Phys. JETP* **27**, 521.]

Ketterle, W. (2002). *Rev. Mod. Phys.* **74**, 1131.

Ketterle, W. and van Druten, N. J. (1996). *Phys. Rev.* B **54**, 656.

Kosterlitz, J. M. and Thouless, D. J. (1973). *J. Phys. C: Solid State Physics* **6**, 1181.

Kubouchi, M., Yoshioka, K., Shimano, R., Mysyrowicz, A. and Kuwata-Gonokami, M. (2005). *Phys. Rev. Lett.* **94**, 016403.

Lai, C. W., Zoch, J., Gossard, A. C. and Chemla, D. S. (2004). *Science* **303**, 503.

Levitov, L. S., Simons, B. D. and Butov, L. V. (2005). *Phys. Rev. Lett.* **94**, 176404.

Lin, J. L. and Wolfe, J. P. (1993). *Phys. Rev. Lett.* **71**, 1222.

Lozovik, Yu. E. and Berman, O. L. (1996). *JETP Letters* **64**, 573.

Lozovik, Yu. E. and Yudson, V. I. (1976). *Zh. Eksp. Teor. Fiz.* **71**, 738. [*Sov. Phys. JETP* **44**, 389]

Ma, M., Halperin, B. I. and Lee, P. A. (1986). *Phys. Rev.* B **34**, 3136.

Mysyrowicz, A., Benson, E. and Fortin, E. (1996). *Phys. Rev. Lett.* **77**, 896.

Naka, N. and Nagasawa, N. (2003). *Phys. Stat. Sol.* (b) **238**, 397.

O'Hara, K. E., Suilleabhain, L. O. and Wolfe, J. P. (1999). *Phys. Rev.* B **60**, 10565.

Popov, V. N. (1972). *Theor. Math. Phys.* **11**, 565. [*Teor. Mat. Fiz.* **11**, 354.]

Prokof'ev, N., Ruebenacker, O. and Svistunov, B. (2001). *Phys. Rev. Lett.* **87**, 270402.

Rapaport, R., Chen, G., Snoke, D., Simon, S. H., Pfeiffer, L., West, K., Liu, Y. and Denev, S. (2004). *Phys. Rev. Lett.* **92**, 117405.

Schmitt-Rink, S., Chemla, D. and Miller, D. A. B. (1989). *Adv. Phys.* **38**, 89.

Snoke, D., Denev, S., Liu, Y., Pfieffer, L. and West, K. (2002). *Nature* **418**, 754.

Snoke, D. W., Wolfe, J. P. and Mysyrowicz, A. (1990). *Phys. Rev. Lett.* **64**, 2543.

Szymanska, M. H. and Littlewood, P. B. (2003). *Phys. Rev.* B **67**, 193305.

Tikhodeev, S. G. (1990). *Zh. Eksp. Teor. Fiz.* **97**, 681. [*Sov. Phys. JETP* **70**, 380.]

Timofeev, V. B., Kulakovskii, V. D. and Kukushkin, I. V. (1983). *Physica B&C* **117-118**, 327.

Yang, Sen, Hammack, A.T., Fogler, M.M., Butov, L.V. and Gossard, A.C. (2006). *Phys. Rev. Lett.* **97**, 187402

Yoshioka, D. and MacDonald, A. H. (1990). *J. Phys. Soc. Jpn.* **59**, 4211.

Zhu, X., Littlewood, P. B., Hybertsen, M. and Rice, T. (1995). *Phys. Rev. Lett.* **74**, 1633.

Zrenner, A., Worlock, J. M., Florex, L. T., Harbison, J. P. and Lyon, S. A. (1990). *Appl. Phys. Lett.* **56**, 1763.

8

COMPOSITE FERMIONS AND THE FRACTIONAL QUANTUM HALL EFFECT IN A TWO-DIMENSIONAL ELECTRON SYSTEM

I. V. Kukushkin[a,b]*, J. H. Smet*[a] *and K. von Klitzing*[a]

[a] Max-Planck-Institut für Festkörperforschung, Heisenbergstraße 1, 70569 Stuttgart, Germany

[b] Institute of Solid State Physics, RAS, Chernogolovka, 142432 Russia

Abstract

Problems involving many strongly interacting bodies are pervasive in physics. In such systems the motion of one particle elicits a response from all others. So it is no wonder that an exact solution of such problems is usually a desperate undertaking. The fractional quantum Hall effect, which occurs in two-dimensional electron systems, is a celebrated example of an interaction phenomenon involving a vast number of electrons. Composite fermions, quasiparticles assembled from one electron and two flux quanta, have emerged as an ingenious dodge to account for this effect in single-particle terms. The remarkable properties of these quasiparticles are investigated in transport experiments as well as with optical techniques.

8.1 Fractional quantum Hall effect

Quantum Hall effects arise when electrons are constrained to move in a plane and are exposed to a perpendicular magnetic field. They are quantum mechanical descendants of a classical effect discovered by Edwin Hall more than a century ago. He observed that a current-carrying conductor in the presence of a magnetic field develops a voltage perpendicular to both the current flow and the field. Ever since, a measurement of the Hall voltage has been a valuable characterization method in solid state physics, because it reveals the number as well as the sign of current-carrying charges. Its application to clean, near-perfect two-dimensional conductors at low temperatures brought a new twist. Here, the Hall voltage does not simply rise linearly with applied field. Instead, it shows plateaux as if the Hall voltage is frozen near specific field values. An example of a Hall measurement is shown in Fig. 8.1. Across the plateaux, the voltage drop in the direction of the current flow vanishes; this is the second hallmark of the quantum Hall effects (von Klitzing *et al.*, 1980; see Das Sarma and Pinczuk (eds.), 1996 for a review).

The magnetic field sends the electrons into circular orbits. Classically, any radius is allowed. Quantum mechanics, however, dictates discrete values of the

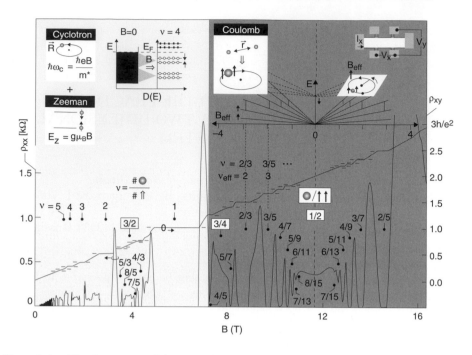

FIG. 8.1. The integer and fractional quantum Hall effects. The longitudinal resistivity measured along the direction of the current flow vanishes whenever
an integer number of Landau levels are completely filled ($\nu = 1, 2, 3, \ldots$). This
integer quantum Hall effect is accompanied by plateaux in the Hall resistivity and can be understood in terms of single-particle physics. The fractional
quantum Hall effect occurs mainly when only the lowest Landau level is occupied at higher magnetic fields for rational values of the filling ν of the form
p/q, where p and q are mutual primes. Despite the phenomenological similarity to the integer quantum effect, the fractional quantum Hall effect has
a very different microscopic origin. It can only be accounted for by considering the Coulomb interaction among the electrons. By introducing suitable
quasiparticles, referred to as composite fermions, it is possible to describe
the fractional quantum Hall effect in single-particle terms. The successive de-
population of composite fermion Landau levels produces an integer quantum
Hall effect of composite fermions and is equivalent to the fractional quantum
Hall effect of the original electrons.

radius, much as it imposes distinct Bohr orbits on an atom. It is an outcome
of the discrete character of magnetic flux: the applied field derives from a large
collection of flux quanta, each contributing the smallest unit of magnetic flux
to the total value. According to the laws of quantum mechanics, only electron
orbits that enclose exactly one such quantum or multiple quanta of magnetic flux
are legitimate. Like the Bohr orbits, each of these orbits has a discrete energy

associated with it – a Landau level. At fixed field, the larger the radius of an orbit, the higher its energy. The electrons are distributed among the orbits or Landau levels so as to minimize the total energy, keeping in mind that each orbit only fits one electron. However, many orbits of equal size (and thus energy) are spread throughout the sample. To be precise, each Landau level can accommodate as many electrons as the total number of flux quanta that thread the sample. As the field is raised and the number of flux quanta increased, Landau levels can take up ever more electrons and levels with higher energy are successively depopulated. The filling factor ν denotes the number of filled Landau levels. When ν takes on an integer value, the system will initially resist the addition of an extra electron, as it has to broach a new Landau level with higher energy. The system is said to be incompressible and this incompressibility is at the heart of the integer quantum Hall effect.

Its cousin, the fractional quantum Hall effect (FQHE), ensues mainly at higher fields when one Landau level is partially occupied and the filling takes on a fraction that can be expressed as a ratio of integers, $\nu = p/q$, with p and q integers (Tsui *et al.*, 1982; Laughlin, 1983). Despite its experimental resemblance, it cannot be accounted for directly in the above picture, considering only the motion of a single electron. After all, if the one occupied level is only partially filled, why the incompressibility? Early on, it was recognized that all electrons must participate to bring about this effect. At many of these fractional fillings, electrons apparently succeed in becoming arranged within the Landau level so as to significantly reduce their mutual repulsion. As we are typically dealing with as many as 10^{10} electrons, all of which interact with each other, developing an intuitive understanding of this effect seems hopeless.

8.2   Composite fermions in a nutshell

Much later than this effect was discovered, it was however suggested that there may be no need to track all electrons to understand this phenomenon (Jain, 1989; Halperin *et al.*, 1993; Heinonen (ed.), 1998). At high fields, compound particles come onto the scene, each assembled from an electron and two flux quanta (or more generally, an even number of them) as depicted in a cartoon-like fashion in Fig. 8.1. This bond between electrons and flux quanta turns out to be a natural way for electrons to avoid each other and the resulting quasiparticles, named composite fermions, may for many purposes be viewed as noninteracting. They too are forced by a field onto circular orbits, which must obey the laws of quantum mechanics. But unlike electrons, they experience only an effective field, greatly reduced from the applied field by an amount equal to the field produced by all the flux quanta of their fellow composite fermions. So, the effective field is zero when two flux quanta thread the sample per electron in the two-dimensional electron system (see the effective field axis in Fig. 8.1). At nonzero effective field, the discrete orbits again have Landau levels associated with them. Filling these gives the integer quantum Hall effect for composite fermions. Sure enough, this integer quantum Hall effect of composite fermions occurs precisely at those external

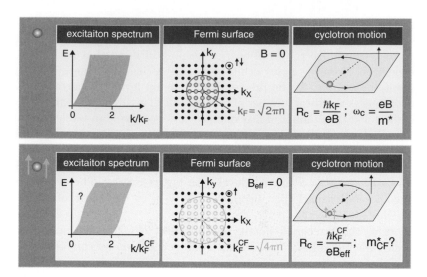

FIG. 8.2. Analogy between electrons and composite fermions (CF). At zero magnetic field, electrons form a metallic state and hence the excitation spectrum is gapless. The Fermi surface is circular and the Fermi wavenumber k_F is determined by the electron density n. Each state in k-space is doubly degenerate due to the spin degree of freedom. In a nonzero field, electrons execute circular orbits with a radius R_c. This radius is determined by k_F, i.e., the density, and shrinks with increasing field. Analogously, a metallic state of composite fermions with a well-defined Fermi surface ensues when the lowest Landau level is half filled (ν equals 1/2). The Fermi wavenumber of this metallic state is identical to that of the electron metallic state at zero external field apart from a factor $2^{1/2}$, since the spin degeneracy at the large external fields where composite fermions are brought alive has been lifted. Composite fermions do not experience the external field, but the drastically reduced effective field. When moving away from half filling, the effective field becomes nonzero and composite fermions are sent into circular orbits much the same way as electrons but the radius of the orbit is controlled by the effective field instead of the external applied field.

applied fields where the fractional quantum Hall effect is routinely observed. This appealing analogy between electrons and composite fermions is summarized in Fig. 8.2.

8.3   Experimental proof for the existence of composite fermions

This analogy has served as a catalyst for experimentalists to dream up experiments seeking to confirm the existence of composite fermions (Goldman *et al.*, 1994; Smet *et al.*, 1996). Examples of such experiments are schematically illustrated in Fig. 8.3. Geometries previously explored for the study of ballistic trans-

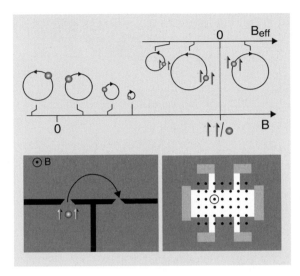

FIG. 8.3. Examples of experimental geometries to prove the existence of composite fermions. Electrons and composite fermions both move on circular orbits. The size of the orbits is however determined by different fields. At high fields, where composite fermions appear, the electron orbits have typically shrunk to a size of less than 10 nm. The bottom left panel illustrates a magnetic focusing experiment. Electrons or composite fermions are injected from a quantum point contact. For the proper value and sign of the external field or the effective magnetic field, electrons or composite fermions should enter the second opening placed at a distance of a micron or less. This current flow can be detected and is turned off when moving to different fields. The composite fermion theory predicts that current flow occurs at the same values of the effective field (apart from a multiplier $2^{1/2}$ due to the lifted spin degeneracy) as those external applied fields where one observes magnetic focusing of electrons. The right panel illustrates a commensurability experiment. It too as well as variations thereof have served as valuable geometries to prove the existence of composite fermions. The basic principle is explained in the text.

port phenomena of electrons at small external magnetic fields were deployed in pursuit of composite fermions. Instead of launching electrons, composite fermions are ejected from an orifice and bent into an adjacent slit arranged along a common boundary in the hope to uncover replica of the transverse magnetic focusing signals for electrons near zero external magnetic field in the vicinity of the field where composite fermions emerge. The discovery of commensurability oscillations of composite fermions in periodic systems would also strengthen the case for composite fermions and would support the claim that the relevant length-scale of motion is set by the composite fermion cyclotron radius and not the

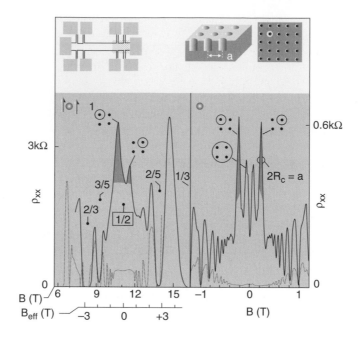

FIG. 8.4. Antidot experiment. Right panel: At certain magnetic fields the electron cyclotron orbit fits exactly around one or a number of antidots. Many carriers are trapped around these antidots and hence no longer contribute to the current flow between the source and the drain contacts. As a result the resistance of the sample develops a maxima at these magnetic fields. The same experiment can be repeated at high magnetic field values, where composite fermions are expected to come onto the scene. Indeed similar features are observed. Composite fermions are more fragile and their mean free path is much smaller than that of electrons and so only the fundamental peak for encircling a single antidot is observed in the experiment.

electron cyclotron radius, which is at least an order of magnitude smaller in this magnetic field regime. Such a commensurability experiment is illustrated in the right panel of Fig. 8.3 and in Fig. 8.4. A periodic lattice of voids (or so-called antidots), where the two-dimensional electron system is entirely depleted, is patterned on top of the sample with electron beam lithography. A constant current is passed through the sample from the source to the drain contact. According to a simple Drude model, the resistivity of the two-dimensional electron system measured along the direction of the current flow (ρ_{xx}) is inversely proportional to the number of charge carriers. When we turn on the magnetic field, the charge carriers start to move on circular orbits. When the orbit fits precisely around one antidot or a set of antidots, a number of charge carriers will be trapped. These carriers no longer contribute to the current flow. The number of current carrying particles is reduced and hence the resistance develops a peak as a larger voltage has to be applied to drive the same current through the sample. The resulting resistance peaks at small fields are illustrated in the right panel of Fig. 8.4. If composite fermions exist, we anticipate replica of such resistance peaks near filling factor $\nu = 1/2$. Such replica are indeed observed in experiment. From their position, one may verify the prediction that the Fermi wavenumber of composite fermions is by $2^{1/2}$ larger than that of the electron Fermi sea. To this date, it is not possible to account for these resistance features at high fields without invoking composite fermions.

8.4   Cyclotron resonance of composite fermions

The previous experiments confirm that composite fermions precess, like electrons, along circular cyclotron orbits, with a diameter determined by the effective field, rather than the external applied field. These experiments owe much of their persuasiveness to their independence on the poorly known composite fermion dispersion. They rely only on the existence of a Fermi surface with the predicted Fermi wavenumber. Unravelling the energy dispersion of composite fermions turned out far more challenging. Indeed, what is the mass of composite fermions? With what frequency do they execute the cyclotron orbit? For electrons this cyclotron frequency is determined by the well-known conduction band mass of the underlying GaAs crystal, in which the two-dimensional electron system forms. It is also relatively straightforward to detect the cyclotron frequency. When microwave radiation with the same frequency is incident on the sample, electrons are accelerated and eventually give up the acquired energy by heating up the sample. This resonant heating can be detected in absorption experiments.

The effective mass of composite fermions is no longer related to the band mass of the original electrons, but is entirely generated from electron–electron interactions and, hence, difficult to predict. The search of the composite fermion cyclotron resonance (Kukushkin *et al.*, 2002a) requires substantial sophistication over conventional methods, used to detect the electron cyclotron resonance, since Kohn's theorem must be outwitted. This theorem states that in a translationally invariant system, homogeneous radiation can only couple to the center-of-mass coordinate and cannot excite other internal degrees of freedom. Phenomena originating from electron–electron interactions will thus not be reflected in the absorption spectrum. An elegant way to bypass this theorem is to impose a periodic density modulation to break translational invariance. The nonzero wavevectors defined by the appropriately chosen modulation may then offer access to the cyclotron transitions of composite fermions, even though they are likely to remain very weak. Therefore, the development of a detection scheme, that boosts the sensitivity to resonant microwave absorption by several orders of magnitude in comparison with traditional techniques, is a prerequisite for these studies.

Luminescence spectra with and without microwave excitation were recorded consecutively by using a CCD-camera, a double-grating spectrometer that provides a spectral resolution of 0.03 meV, and a stabilized semiconductor laser operating at a wavelength of 750 nm and approximately 100 μW of CW-power. The

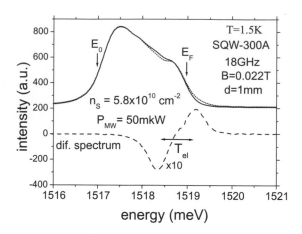

FIG. 8.5. Illustration of the optical scheme to detect resonant microwave absorption for the electron cyclotron–magnetoplasmon hybrid mode at low B-fields. Luminescence spectrum in the presence of (dotted line) and without (solid line) a 50 W microwave excitation of 18 GHz obtained on a disk-shaped two-dimensional electron system (2DES) with a diameter of 1 mm and carrier density $n_s = 5.8 \times 10^{10}$ cm^{-2} at a magnetic field $B = 22$ mT. In the vicinity of the Fermi energy E_F the spectrum is affected significantly under resonant microwave excitation due to heating. The dashed line represents the differential luminescence spectrum. The integration of its absolute value across the entire spectral range yields the microwave absorption amplitude. The width of the differential spectrum reflects the increased electron temperature T_e. From (Kukushkin *et al.*, 2002a). Adapted figure with permission from Macmillan Publishers Ltd: [I. V. Kukushkin, J. H. Smet, K. von Klitzing, and W. Wegscheide, *Nature* **415**, 409 (2002)], copyright (2002).

differential luminescence spectrum is obtained (see Fig. 8.5) when subtracting both luminescence spectra. To improve signal-to-noise ratio, the same procedure was repeated N times ($N = 2$–20). Subsequently, we integrated the absolute value of the averaged differential spectrum over the entire spectral range and hereafter refer to the value of this integral as the microwave absorption amplitude. The same procedure is then repeated for different values of B. This technique offers pre-eminent sensitivity as attested by our ability to observe the electron cyclotron resonance at microwave levels below 10 nW. To establish trustworthiness in this unconventional scheme, we apply it to the well-known case of the electron cyclotron resonance $\omega_{cr} = eB/m^*$, with m^* the effective mass of GaAs ($0.067m_0$). Due to its limited size, the sample also supports a dimensional plasma mode with a frequency ω_p, that depends on both the density n_{2D} and diameter d of the sample, according to $\omega_p^2 = 3\pi^2 n_{2D} e^2 / (2m^* \epsilon_{eff} d)$. The plasma and cyclotron mode hybridize and the resulting resonance frequency of the upper dimensional magnetoplasma–cyclotron mode ω_{DMR} equals $\omega_{cr}/2 + [\omega_p^2 + (\omega_{cr}/2)^2]^{1/2}$. As it is

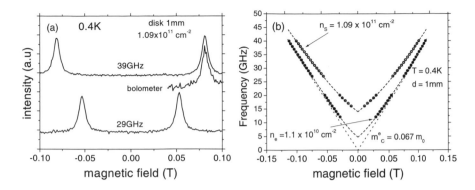

FIG. 8.6. (a) The microwave absorption amplitude at 29 GHz and 39 GHz as a function of B-field by recording differential luminescence spectra for 1 mT field increments at $n_{\rm s} = 1.09 \times 10^{11}$ cm^{-2}. The peaks, symmetrically arranged around zero field, are identified as the dimensional magnetoplasma–cyclotron hybrid mode. Conventional bolometer measurements are also shown. (b) Resonance position for $n_{\rm s} = 1.09 \times 10^{11}$ cm^{-2} (open circles) and 1.1×10^{10} cm^{-2} (solid circles) as a function of incident microwave frequency. The intervals 10–20 GHz and 27–40 GHz were covered. The dashed lines represent the theoretical dependence of the hybrid dimensional magnetoplasma–cyclotron resonance. The dotted line corresponds to the cyclotron mode only. From (Kukushkin *et al.*, 2002a). Adapted figures with permission from Macmillan Publishers Ltd: [I. V. Kukushkin, J. H. Smet, K. von Klitzing, and W. Wegscheide, *Nature* **415**, 409 (2002)], copyright (2002).

clear from Fig. 8.6, the optical method recovers this mode indeed. A comparison with the theoretical expression for $\omega_{\rm DMR}$ yields excellent agreement. No fitting is required, since the density can be independently extracted from the luminescence at higher B-fields, where Landau levels can be resolved. At sufficiently low density, the influence of $\omega_{\rm p}$ on the hybrid mode drops and one recovers at large enough B the anticipated $\omega_{\rm cr} = eB/m^*$-dependence. Additional support for the validity of the detection method comes from a comparison with measurements based on the conventional approach using a bolometer (see Fig. 8.6a). Not only does one find the same resonance position, but also the same line shape. The only difference is the improved signal to noise ratio (30–100 times) for the optical detection scheme.

Disorder and the finite dimensions of the sample in principle suffice to break translational invariance as attested by the interaction of the cyclotron and dimensional plasma mode. However, they provide access to internal degrees of freedom other than the center-of-mass motion of the electrons either at poorly defined wavevectors or too small a wavevector for appropriate sample sizes. Therefore, the imposition of an additional periodic density modulation, that introduces larger and well-defined wavevectors to circumvent Kohn's theorem, is desirable.

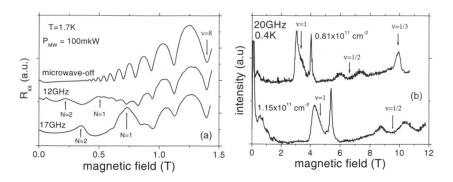

FIG. 8.7. (a) Magnetotransport data without (top curve) and under 100 μW of microwave radiation at 12 (middle curve) and 17 GHz (bottom curve). Curves are offset for clarity. Besides the well-known Shubnikov–de Haas oscillations, additional magnetoresistance oscillations appear under microwave radiation. They are commonly observed in 2DESs on which a static periodic modulation of the density has been imposed. (b) Microwave absorption amplitude at high magnetic fields for $n_s = 0.81 \times 10^{11}$ cm^{-2} and 1.15×10^{11} cm^{-2} and frequency of 20 GHz. The response near $\nu = 1$ and $1/3$ does not shift with frequency. From (Kukushkin *et al.*, 2002a). Adapted figures with permission from Macmillan Publishers Ltd: [I. V. Kukushkin, J. H. Smet, K. von Klitzing, and W. Wegscheide, *Nature* **415**, 409 (2002)], copyright (2002).

Transport experiments in the Hall bar geometry disclosed that additional processing is not required, since the microwaves, already incident on the sample, concomitantly induces a periodic modulation at sufficiently high power. A clear signature is the appearance of commensurability oscillations in the magnetoresistance due to the interplay between the B-dependent cyclotron radius of electrons and the length-scale imposed by the modulation. Examples are displayed in Fig. 8.7(a) and resemble the data measured under modulation produced with the help of surface acoustic wave (SAW)-transducers (Shilton *et al.* 1995). Here, the following scenario is conceivable. Owing to the piezoelectric properties of the Al$_x$Ga$_{1-x}$As-crystal, the radiation is partly transformed into SAW with opposite momentum, so that both energy and momentum are conserved. Reflection from cleaved boundaries of the sample then produces a standing wave with a periodicity determined by the sound wavelength. The involvement of sound waves can be deduced from transport data, since from the minima we expect the modulation period to be approximately 200 and 250 nm for frequencies of 17 and 12 GHz respectively. The ratio of this period to the sound wavelength at these frequencies is 1.12 and 1.15. More detailed investigations of transport properties of 2D-electrons under microwave irradiation indicate transformation of microwaves into surface acoustical waves (Kukushkin *et al.*, 2002b).

Figure 8.7b depicts the microwave-absorption amplitude up to high B-fields.

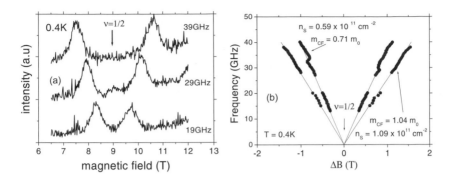

FIG. 8.8. (a) Microwave absorption amplitude in the vicinity of $\nu = 1/2$ (20 mT step size) at three different frequencies and $n_s = 1.09 \times 10^{11}$ cm^{-2}. The peak values are nearly two orders of magnitude weaker and considerably wider (about 30–50 times) than those due to the electron cyclotron resonance. (b) Position of the CF cyclotron mode as a function of the effective magnetic field $B_{\text{eff}} = B - 2n_s \cdot F_0$ (F_0 is the elementary flux quantum) for $n_s = 1.09 \times 10^{11}$ cm^{-2} and $n_s = 0.59 \times 10^{11}$ cm^{-2}. The CF effective mass equals $1.04 \, m_0$ and $0.71 \, m_0$ respectively. The resonances remain visible even if the 2DES condenses in the gapfull fractional quantum Hall states at filling factors 3/5 and 3/7. From (Kukushkin *et al.*, 2002a). Adapted figures with permission from Macmillan Publishers Ltd: [I. V. Kukushkin, J. H. Smet, K. von Klitzing, and W. Wegscheide, *Nature* **415**, 409 (2002)], copyright (2002).

Apart from the strong dimensional magnetoplasma–cyclotron resonance signal at low B-field discussed above, several peaks, that scale with a variation of the density, emerge near filling 1, 1/2 and 1/3. Those peak positions associated with $\nu = 1$ and 1/3 remain fixed when tuning the microwave frequency and are ascribed to heating induced by nonresonant absorption of microwave power. In contrast, the weak maxima surrounding filling 1/2 readily respond to a change in frequency as illustrated in Fig. 8.8(a). They are symmetrically arranged around half filling and their splitting grows with frequency. The B-dependence is summarized in Fig. 8.8(b) for two densities. To underline the symmetry, B_{eff} was chosen as the abscissa. The linear relationship between frequency and field extrapolates to zero at vanishing B_{eff}. We do not expect a deviation at small B_{eff} due to a plasma-like contribution as in Fig. 8.6(b). Excitations for the 1/3, 2/5, 3/7 and other fractional quantum Hall states exhibit in numerical simulations no magnetoplasmon-like linear contribution to the dispersion at small values of k. We conclude that the resonance in Fig. 8.6(a) is the long searched for cyclotron resonance of CFs. Geometric resonances, as they occur in transport at low fields due to the density modulation (Fig. 8.7a), are excluded as an alternative interpretation for the observed features on the following grounds:

I. In the optical data, only the electron cyclotron resonance peak is observed.

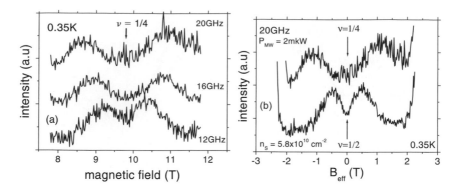

FIG. 8.9. (a) Microwave absorption amplitude in the vicinity of $\nu = 1/4$ (20 mT step size) at three different frequencies and $n_{\mathrm{s}} = 0.59 \times 10^{11}$ cm^{-2}. (b) Comparison of peak positions of cyclotron resonances measured at fixed electron density $n_{\mathrm{s}} = 0.59 \times 10^{11}$ cm^{-2} for composite fermions with two and four flux quanta. From (Kukushkin *et al.*, 2003). Adapted figure with permission from Elsevier Ltd: [I. V. Kukushkin, J. H. Smet, K. von Klitzing, W. Wegscheider, *Physica* E **20**, 96 (2003)], copyright (2003).

Contrary to optical quantities, transport is also sensitive to semiclassical phenomena unrelated to changes in the density of states.

II. Even if the 2DES condenses in a FQHE-state and the chemical potential is located in a gap, the resonance peaks surrounding $\nu = 1/2$ occur (Fig. 8.8b). Commensurability effects are not observable in this regime.

III. The observation of geometric resonances (GRs) requires that the density modulation is temporally static on the time-scale with which CFs execute their cyclotron orbit. For electrons at low fields this condition is met and accordingly transport displays geometric resonances. For the anticipated enhanced mass of CFs, this condition is violated.

IV. Analogous resonance peaks were also detected for the higher order CFs around $\nu = 1/4$ which are shown in Fig. 8.9(a). Since at fixed electron density the CF metallic state is characterized by the same Fermi wavevector GRs would show up at the same distance from $\nu = 1/4$ as they do at $\nu = 1/2$. As it is obvious from Fig. 8.9(b) the observed peaks are located at different positions rendering a commensurability picture untenable.

In contrast to electron cyclotron resonance, the intensity of the CF cyclotron resonance is a strong nonlinear function of microwave-power (Fig. 8.10a). Moreover, its observability only at high power correlates with the first appearance of commensurability oscillations. The drop in intensity at even higher power is most probably due to heating. The intensity diminishes to zero at temperatures

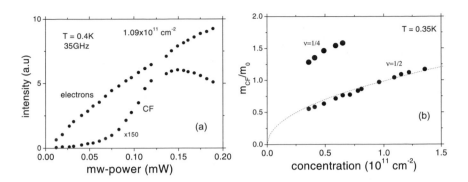

FIG. 8.10. (a) Incident microwave power dependence of the amplitude at the cyclotron resonance measured both for electrons and composite fermions. (b) Density dependence of the CF effective mass measured both for $\nu = 1/2$ and $\nu = 1/4$. The dashed line is a square root fit to the data. From (Kukushkin *et al.*, 2003). Adapted figure with permission from Elsevier Ltd: [I. V. Kukushkin, J. H. Smet, K. von Klitzing, W. Wegscheider, *Physica* E **20**, 96 (2003)], copyright (2003).

above 0.7 K, whereas the electron cyclotron resonance persists up to $T > 2$ K. The slope of the CF cyclotron frequency as a function of $B_{\rm eff}$ in Fig. 8.8(a) defines the cyclotron mass $m_{\rm cr}^{\rm cf}$. This mass is set by the electron–electron interaction scale, so that a square root behavior on density or B-field is forecasted from a straightforward dimensional analysis (Halperin *et al.*, 1993). Numerical calculations predict $m_{\rm cr}^{\rm cf}/m_0 = 0.079 \cdot (B[T])^{1/2}$ for an ideal 2DES, not including Landau level mixing or finite width contributions (Park and Jain, 1998; see Jain and Kamilla, 1998 for a review). The data, shown in Fig. 8.10(b), confirm qualitatively the strong enhancement in comparison with the electron mass (more than 10 times), however a fit to the square root dependence requires a prefactor that is four times larger. It is also clear from Fig. 1.10(b) that attachment of additional flux quanta to electron results to further enhancement of the mass of composite fermion. Note that previously reported mass values based on activation energy gap measurements (Willett, 1998) must be distinguished from the cyclotron mass. The former corresponds to the limit of infinite momentum, whereas here k approaches zero.

8.5 Outlook

Even though we have come a long way in deciphering the properties of composite fermions, some important questions remain. So far, the main thrust has been on ballistic transport experiments exploring the particle-like behavior of composite fermions. What about the particle–wave duality? What is the phase coherence length of composite fermions and is it possible to carry out an Aharonov–Bohm-like experiment for composite fermions? These are some intriguing questions that we wish to address in the future.

References

Das Sarma, S. and Pinczuk, A. eds. (1996). *Perspectives on Quantum Hall Effects*. Wiley, New York.

Goldman, V. J., Su, B. and Jain, J. K. (1994). *Phys. Rev. Lett.* **72**, 2065.

Halperin, B. I., Lee, P.A. and Read, N. (1993). *Phys. Rev.* B **47**, 7312.

Heinonen, O. ed. (1998). *The Composite Fermions: A Unified View of the Quantum Hall Regime*. World Scientific, Singapore.

Jain, J. K. (1989). *Phys. Rev. Lett.* **63**, 199.

Jain, J. K. and Kamilla, R. K. (1998). *The Composite Fermions: A Unified View of the Quantum Hall Regime* (ed by O. Heinonen), p. 1. World Scientific, Singapore.

Kukushkin, I. V., Smet, J. H., von Klitzing, K. and Wegscheider, W. (2002a). *Nature* **415**, 409.

Kukushkin, I. V., Smet, J.H., Falko, V. I., von Klitzing, K. and Eberl, K. (2002b). *Phys. Rev.* B **66**, 121306(R).

Kukushkin, I. V., Smet, J. H., von Klitzing, K. and Wegscheider, W. (2003). *Physica* E **20**, 96.

Laughlin, R. B. (1983). *Phys. Rev. Lett.* **50**, 1395.

Park, K. and Jain, J. K. (1998). *Phys. Rev. Lett.* **80**, 4237.

Smet, J. H., Weiss, D., Blick, R. H., Lutjering, G., von Klitzing, K., Fleischmann, R., Ketzmerick, R., Geisel, T. and Weimann, G. (1996). *Phys. Rev. Lett.* **77**, 2272.

Tsui, D. C., Stormer, H. L. and Gossard, A. C. (1982). *Phys. Rev. Lett.* **48**, 1559.

Shilton, J. M., Mace,D. R., Talyanskii, V. I., Pepper, M., Simmons, M. Y., Churchill, A. C. and Ritchie, D. A. (1995). *Phys. Rev.* B **51**, 14770.

von Klitzing, K., Dorda, G. and Pepper, M. (1980). *Phys. Rev. Lett.* **45**, 494.

Willett, R. L. (1998). *The Composite Fermions: A Unified View of the Quantum Hall Regime* (ed by O. Heinonen), p. 349. World Scientific, Singapore.

9

MICROCAVITIES WITH QUANTUM DOTS: WEAK AND STRONG COUPLING REGIMES

Vladimir D. Kulakovskii[a] and Alfred Forchel[b]
[a] Institute for Solid State Physics, Russian Academy of Sciences,
Chernogolovka, Russia
[b] Technische Physik, Universität Würzburg, Germany

Abstract

The quantum dot/cavity system emission has been investigated in a wide range of exciton-cavity photon mode detunings controlled by temperature variation in the range 10–45 K. A strong Purcell effect has been observed for self-assembled quantum dots (QDs) in small microcavity pillars. The emission rate at fixed cavity finesse increases with reduced pillar diameter and is limited by the transition into the strong coupling regime. The emission lineshape and the relative emission intensities of cavity and QD exciton modes depend sensitively on the primary excitation mechanisms via QD excitons and via the cavity mode. In the case of an excitation via the cavity mode, a well pronounced dip at the QD exciton energy appears in the emission spectrum already in the weak coupling regime, due to an interference effect between the cavity and the QD oscillators. The strong coupling regime has been observed for single enlarged QDs in a 1.5 μm diameter micropillar with finesse $Q = 7500$; the exciton–photon coupling constant $g = 0.055$ meV provides a photoluminescence line splitting of $\Delta = 0.14$ meV at the exciton-cavity resonance.

9.1 Introduction

Since its proposal by Purcell (1946) and work involving atoms in cavities (Weisbuch, 1994), cavity quantum electrodynamics (CQED) has been actively pursued for its potential to investigate fundamental problems in light–matter interaction. Cavity quantum electrodynamics addresses properties of atom-like emitters in cavities. In the cavities, both the energies and modes of photon states, to which the electronic transitions of atoms can couple, are modified by the cavity confinement. The extension to the optical regime requires cavities with sizes of the order of microns; this has become possible due to the epitaxial fabrication of high-quality layered semiconductors (Weisbuch *et al.*, 1992) from which optical cavities have been made providing strong photonic confinement in one direction. Solid state CQED using quantum dots in semiconductor microcavities with three-dimensional optical confinement has recently become a central research field (Yamamoto and Slusher, 1993; Gérard and Gayral, 2001; Vahala, 2003).

Semiconductor quantum dots that exhibit a discrete density of electronic states can be used in the cavities as quasi-atomic light emitters.

A central aspect of cavity research over the years has been the possibility to modify spontaneous emission rates (Purcell, 1946; Hulet *et al.*, 1985; Kleppner, 1981) in the regime of weak coupling between the electronic and photonic excitations, which opens opportunities for basic physics studies as well as in semiconductor optoelectronics. The emission rate can be increased or reduced compared with its vacuum level by tuning discrete cavity modes in and out of resonance with the emitter. These features are of interest both for light-emitting diodes and for quantum computing or quantum cryptography. In the case of quantum cryptography, for example, single-photon devices require a detailed control of spontaneous emission. If the exciton–photon interaction rate is faster than the rate of any dissipative channel in the system, the spontaneous emission of the emitter will create a photon, which has a nonzero probability to re-excite the emitter before leaving the cavity. In this situation the spontaneous emission becomes reversible. This scenario with a qualitatively new emission property is referred to as strong coupling.

In planar semiconductor cavities in the weak coupling regime, only modest changes in emission rates on the order of 10% have been observed (Tanaka *et al.*, 1995; Abram *et al.*, 1998). Thus, the major trend of current semiconductor CQED research is focused towards systems of low dimensionality. Such systems can be realized in three-dimensional (3D) semiconductor microcavities (MCs) containing quantum dots in the active region. In these MCs electrons and holes are localized in the QDs fabricated by a spatial variation of the semiconductor bandgap, whereas photons are confined by a modulation of the semiconductor refractive index, e.g., in planar photonic crystals, micropillars, and microdisks (Burstein and Weisbuch, 1995; Weisbuch and Rarity, 1996; Weisbuch *et al.*, 1992; Zhang *et al.*, 1995; Gérard *et al.*, 1996; Reithmaier *et al.*, 1997; Ohnesorge *et al.*, 1997; Gérard *et al.*, 1998) resulting in the "ultimate" system, where both electrons and photons are confined in all dimensions. Such a system should exhibit peculiar effects characteristic of a two-level atom in interaction with an optical cavity, in particular, a strong enhancement or inhibition of the spontaneous emission rate in the weak-coupling regime (Purcell effect) (Gérard and Gayral, 2001; Kleppner, 1981; Goy *et al.*, 1983; Gabrielse and Dehmelt, 1985; Hulet *et al.*, 1985; Gérard *et al.*, 1998; Bayer *et al.*, 2001; Solomon *et al.*, 2001; Pelton *et al.*, 2002; Santori *et al.*, 2002; Michler *et al.*, 2000).

However, the most striking change of emission properties occurs when the conditions for strong coupling are fulfilled. In this case there is a change from the usual irreversible spontaneous emission to a reversible exchange of energy between the emitter and the cavity mode resulting in vacuum-field Rabi splitting and the formation of dressed states. The strong coupling of individual two-level systems has been observed for atoms in large cavities (Hood *et al.*, 1998; Mabuchi and Doherety, 2002; McKeever, 2003a; McKeever, 2003b). The coherent coupling may provide a basis for future applications in quantum information processing or schemes for coherent control. Due to the rapid progress of the epitaxial growth and nanolithography techniques, the strong coupling has been observed very recently in a number of approaches in semiconductors (Reithmaier *et al.*, 2004; Yoshie *et al.*, 2004; Peter *et al.*, 2005).

In the treatment of coupled modes, the electronic excitations, which are electron–hole pairs or excitons in the case of semiconductors, are usually represented by either two-level systems or by harmonic oscillators (Savona *et al.*, 1994; Andreani *et al.*, 1994). The harmonic oscillator representation does not exhibit driven Rabi oscillations in an external field. However, it provides a vacuum Rabi splitting when coupled to a cavity optical mode and offers greater ease of the mathematical treatment than does the spin representation (Agarwal, 1985). Within these coupled mode treatments in a simple approximation, phenomenological broadening parameters are often used to represent the linewidths of the electronic transition and also the finite quality factor Q of the cavity (Zhu *et al.*, 1990). Although such a representation of the damping is physically attractive for the electronic excitations that couple, for example, to phonons, it is less realistic for the photon. For the latter, the linewidth arises from coupling to the continuum of delocalized photon modes outside the cavity rather than from a true dissipative mechanism. The general problem of coupling of a discrete state to a continuum of states was solved by Fano (1961). Consideration of effects of dissipation and mode broadening in the emission spectra of optical cavities and semiconductor microcavities with the use of the Fano model has shown, however, that the simple approximation describes the main features of the emission spectra (Rudin and Reinecke, 2000).

In a simple approximation, for interacting QD exciton (X) and cavity (C) modes at a resonance $(E_X = E_C \equiv E_0)$, the energy levels of interacting modes are

$$E_{1,2} = E_0 - i(\gamma_C + \gamma_X)/2 \mp \sqrt{g^2 - (\gamma_C - \gamma_X)^2/4}, \tag{9.1}$$

where $\gamma_{C(X)}$ is the half width at half maximum (HWHM) of the cavity (exciton) mode, $2\gamma_C = E_C/Q \gg \gamma_X$, and g is the "vacuum Rabi splitting" parameter, equal to the scalar product of the transition matrix element of the QD dipole moment with the local value of the one photon electric field at the dot position in the cavity (Andreani *et al.*, 1999),

$$g = \sqrt{\frac{\pi e^2 f}{\varepsilon m_0 V_C}}. \tag{9.2}$$

Here ε is the dielectric constant of the cavity material and m_0 is the free electron mass, f/V_C is the exciton oscillator strength per unit volume, and V_C is the effective mode volume. Equation (9.2) shows that g depends on the QD exciton oscillator strength and on the mode volume V_C as $\sqrt{f/V_C}$. In λ-thick GaAs/AlAs micropillar cavities with circular cross-section $V_C \sim 0.5\lambda\pi r_C^2$, where r_C is the radius of the pillar in the cavity plane. The strong coupling regime can be realized only in MCs providing a decay of the cavity mode that is smaller than the exciton–photon coupling that determines the vacuum Rabi splitting. The criterion for the strong coupling regime of the QD exciton and cavity modes can be approximated as $g > (\gamma_C - \gamma_X)/2 \sim \gamma_C/2$ (because $\gamma_C \gg \gamma_X$). Indeed, it follows from equation (9.1) that in the opposite case of $g < \gamma_C/2$, referred to as a weak coupling regime, the interaction results in an enhanced or decreased radiative decay of QD excitons depending on the cavity finesse.

The realization of the strong coupling regime demands a much higher product $Q\sqrt{f/V_C}$ than required for the weak coupling. In this regime the resonant inter-action of cavity electromagnetic modes and electronic excitations should result in a characteristic split-mode ("anticrossing") lineshape of the optical spectra (Zhu *et al.*, 1990; Thomson *et al.*, 1992).

A theoretical consideration of the crossover from weak to strong coupling regimes for 3D confined electronic states in pillar MCs has shown that two classes of QD confined excitons could be promising for getting the strong coupling: exci-tons in self-assembled QDs in highly strained heterostructure layers and excitons bound to monolayer fluctuations in quantum wells, or natural QDs (Andreani *et al.*, 1999). Self-assembled QDs in InAs/GaAs structures are usually characterized with a relatively small size which is comparable to the exciton Bohr radius. As a consequence, they have a rather long radiative recombination time of about 1 ns and small exciton oscillator strength $f \sim 10$. With such single QD excitons the strong coupling regime can be realized only in pillars with very high $Q/\sqrt{V_C}$. The fabrication of MC pillars satisfying this criterion still is quite challenging as the finesse decreases strongly in pillars with radius smaller than 1 micron. In order to reach the strong coupling regime more easily the exciton oscillator strength can be increased by using of QDs of slightly enlarged size. On increasing the QD radius above the exciton Bohr radius, f increases due to an increase of the area of the exciton center of mass wavefunction similar to that of excitons localized on shallow impurities or in microcrystals (Takagahara *et al.*, 1987; Hanamura *et al.*, 1988; Nakamura *et al.*, 1989; Rashba and Gurgenishvili, 1962; Bastard *et al.*, 1984; Citrin, 1993; Bockelmann, 1993; Andreani *et al.*, 1999). The use of enlarged QDs has allowed the first observation of the strong coupling regime in solids by using micropillar MCs with quantum dots (Reithmaier *et al.*, 2004). Strong coupling of QD-cavity modes in solids has been subsequently observed also for other cavity designs. Cavities fabricated from photonic crystals, where the high-value $Q/\sqrt{V_C}$ is achieved due to the small cavity volume (Yoshie *et al.*, 2004) as well as microdisk geometries (Peter *et al.*, 2005) have shown the single dot strong coupling.

The strong coupling regime is usually identified in the emission spectra as an anticrossing of coupled QD exciton and cavity modes. In a simplified analysis a similar filling of the two coupled modes has been assumed. In the present paper we analyze the emission spectra from MCs with QDs both for the weak and strong coupling regimes with taking into account for various excitation condi-tions, namely, under the primary excitation of (i) QD exciton, (ii) cavity mode and (iii) exciton and cavity mode simultaneously. This effect has been suggested

FIG. 9.1. Scanning electron micrograph of a pillar with a diameter of about 0.8 μm. By a combination of electron-beam lithography and reactive dry etching, micropillars with close to vertical and defect-free sidewalls are obtained, as required for high-Q cavities. From (Reithmaier *et al.*, 2004). Adapted figure with permission from Macmillan Publishers Ltd: [J. P. Reithmaier, G. Sęk, A. Löffler, C. Hofmann, S. Kuhn, S. Reitzenstein, L. V. Keldysh, V. D. Kulakovskii, T. L. Reinecke, A. Forchel, *Nature* **432**, 197 (2004)], copyright (2002).

by Keldysh, see in (Keldysh *et al.*, 2006).

9.2 Experiment

The resonators used are based on a one-λ GaAs resonator sandwiched between high-reflectance GaAs/AlAs Bragg mirrors. The Bragg reflectors are composed of 20–26 (23–30) periods of $\lambda/4$ GaAs/AlAs (69 nm/82 nm) layers in the top (bottom) mirrors. The GaAs cavities contain random arrays of In$_x$Ga$_{1-x}$As QDs placed at the antinode of the on-axis resonant fundamental mode of the planar cavity. The quality factor $Q = \lambda_C/\Delta\lambda_C$ of the planar resonators is in the range of several 10 000. All structures are grown by solid source molecular beam epitaxy. High-Q cavities with small mode volumes have been fabricated from these epitaxial samples. Micropillars with circular cross-sections (diameters from 1 to 4 μm) have been processed by electron-beam lithography and reactive ion-etching in inductively coupled or electron cyclotron enhanced Ar/Cl$_2$ plasmas. The strong mode confinement in the micropillars is achieved using a deep etching through almost the total height of the MCs (5 μm), resulting in very smooth pillar side walls as shown in Fig. 9.1. The combination of a high-Q epitaxial cavity with optimized micropillar processing allows us to realize micropillars with maximum Q factors of 25 000, 9000 and 4000 for 2, 1.5 and 1.0 μm diameters, respectively.

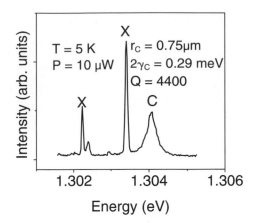

FIG. 9.2. Microphotoluminescence spectrum of a 1.5 μm diameter microcavity ($Q \sim 4400$) with $In_{0.6}Ga_{0.4}As$ QDs recorded at 5 K.

For one type of structure the lateral sidewalls of the resonators were left uncovered after the patterning. For a second type of structure the sidewalls were coated by a gold layer with a thickness of ~150 nm (Bayer *et al.*, 2001). QD emitters form the optically active medium in these cavities. Different types of QDs have been used. First, quasi-zero-dimensional systems have been obtained by placing a strongly inhomogeneously broadened $In_{0.14}Ga_{0.86}As$ quantum well with a nominal width of 7 nm at the center of the cavity. Second, an array of typical self-assembled $In_xGa_{1-x}As$ QDs with a large In content of $x = 0.6$ has been used. These QDs have a relatively small size of 15–20 μm in diameter and a high density of several 10^{10} cm^{-2}. Finally, ensembles of QDs with a smaller In content of $x \sim 0.3$ have been used as active material in the cavities. Before overgrowth, these dots have markedly enlarged dot sizes of 40 to 100 nm. Single dot investigations of *s*- and *p*- shell transitions indicate that the maximum dot dimensions after overgrowth are of the order of 40 to 50 nm.

The emission of the QD layers without cavity effects was studied by removing the upper Bragg reflector. At low temperature (4 K to about 50 K) all QD layers exhibit a broad (about 50 meV) photoluminescence spectrum because of inhomogeneous broadening due to the QD size fluctuations. The spectral position of the cavity mode is detuned to the low-energy side of the QD emission band. The energy of single QDs is tuned on resonance with the cavity mode by using the temperature dependent shift of the bandgap in the dots. With this purpose, the samples have been placed into the optical cryostat with a temperature control. PL of single pillars has been measured using a micro-PL setup based on a spectrometer and a nitrogen cooled CCD-camera with a spectral resolution of 0.05 meV or better, depending on the particular setup. A microscope objective has been used to focus the laser beam into ~ 3 μm diameter spot at the pillar and to collect the pillar emission. The excitation was carried out by second har-

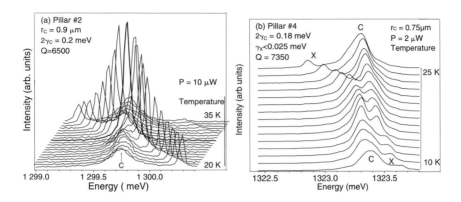

FIG. 9.3. Sets of the PL spectra recorded for pillars with weak (a) and strong
   (b) exciton–photon coupling in the temperature range of 10–35 K. Panel
   (b) from (Reithmaier *et al.*, 2004). Adapted figure with permission from Macmillan
   Publishers Ltd: [J. P. Reithmaier, G. Sęk, A. Löffler, C. Hofmann, S. Kuhn, S. Reitzenstein,
   L. V. Keldysh, V. D. Kulakovskii, T. L. Reinecke, A. Forchel, *Nature* **432**, 197 (2004)],
   copyright (2002).

monics of a CW Nd:YAG laser with a wavelength of 532 nm, or by an Ar-ion
(514 nm).

9.3   MC pillar emission

Figure 9.2 displays PL recorded at 5 K from a single pillar MC #1 with a
diameter of 1.5 μm containing a set of self-assembled $In_{0.6}Ga_{0.4}As$ QDs in the
active layer. A few sharp peaks related to the emission from QDs (X) and a strong
and wider peak connected to the fundamental cavity mode (C) are observed.
The FWHM of the X lines, X, $2\gamma_X$, amounts to about 45 μeV, close to the
spectral resolution of the setup. The strongest line is located near the cavity
mode. We would like to note that even without a dot on resonance, there is
a cavity emission. This emission appears due to excitation of the cavity mode
through, for example, phonon-assisted scattering from radiative recombination
in the dot ensemble with higher energies in the pillar. The cavity mode FWHM
is markedly larger than the dot linewidth. Its FWHM of 0.29 meV corresponds
to a cavity with $Q = 4400$.

The QD lines that are located at low temperature at energies slightly higher
than the energy of the cavity mode can be temperature tuned through the cavity
resonance: the shift of the QD lines due to the temperature induced change in
the QD bandgap is about 1.5 meV in the range of 20–40 K, whereas the shift
of the cavity mode due to the temperature dependence of the refractive index is
much weaker (about 0.25 meV). Thus, a variation of the temperature allows one
to tune the QD emission on and off resonance with the cavity mode.

Figure 9.3 shows two sets of spectra recorded for two pillars in the temperature range of 10–35 K. Pillar #2 (panel a) is a MC with diameter of 1.8 μm,

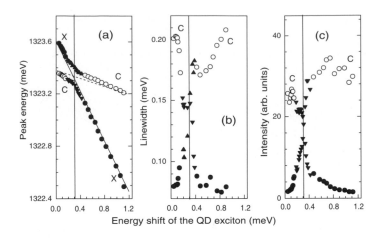

FIG. 9.4. Dependences of the photoluminescence peak energies (a), linewidths (b) and integrated intensities (c) on the temperature-induced energy shift of a reference QD for strong coupling regime. The reference QD-related peak is located about 5 meV above the cavity mode and does not interact with the cavity (bottom scale). Labels C mark the cavity-mode data (open dots); X mark the QD exciton data (filled dots); triangles represent the data on or near the resonance where the strong coupling results in the hybridized modes. From (Reithmaier *et al.*, 2004). Adapted figure with permission from Macmillan Publishers Ltd: [J. P. Reithmaier, G. Sęk, A. Löffler, C. Hofmann, S. Kuhn, S. Reitzenstein, L. V. Keldysh, V. D. Kulakovskii, T. L. Reinecke, A. Forchel, *Nature* **432**, 197 (2004)], copyright (2002).

$Q = 6350$, and self-assembled $In_{0.6}Ga_{0.4}As$ QDs as active material. The spectra consist of two lines, X and C. As shown on the left hand side of the figure, it is seen that the most pronounced effect, observed when tuning the QD through the cavity mode, is a strong increase in the QD emission intensity. No splitting of the cavity and exciton modes is observed at resonance. For an emitter in an optical cavity, the radiation pattern is modified by the cavity and depends on the emission wavelength.

The strong coupling regime was achieved first in pillars with enlarged QDs in the active layer (Reithmaier *et al.*, 2004). A larger QD size was realized by growing QDs with 30% In content instead of 60% In content used previously. Due to the small lattice mismatch between GaAs and $In_{0.3}Ga_{0.7}As$, the QD size has been increased to ≥ 40 nm, which results in a marked increase in the QD exciton oscillator strength. Figure 9.3(b) displays spectra from an MC with a diameter of 1.5 μm, $Q = 7350$, using the enlarged $In_{0.6}Ga_{0.4}As$ QDs as active material (pillar #4). At 25 K the emission of this pillar consists of the cavity mode centered at about 1.32335 eV and a QD exciton emission at slightly higher energy (1.3224 eV). By using lineshape fits of the 25 K spectrum with two Lorentzians, we find that the FWHM of the cavity mode is 0.18 meV for low-excitation conditions. For decreasing temperature down to 10 K, the pillar emission consists of two distinct lines in the whole range of temperatures. However, at 10 K the components of the emission have exchanged their properties: now the low-energy line has an emission intensity and FWHM similar to that of the line located at 25 K at higher energy and therefore has to be assigned to the cavity mode. Simultaneously, the high-energy line displays an emission intensity and FWHM that correspond to the values for the QD exciton mode at 25 K. Over the entire temperature range, the energies of the two contributions to the spectrum are well separated and avoid crossing each other. This is the main indication for an anticrossing of the single QD exciton and cavity-mode dispersions due to strong coupling between these modes. The dependencies of the transition energies, halfwidths and intensities on the exciton-cavity $(X - C)$ detuning are displayed in Fig. 9.4. Approaching the resonance, the PL spectra demonstrate (i) strong broadening of the X line, (ii) strong narrowing of the MC line and, most important, (iii) a well pronounced anticrossing behavior with a splitting $\Delta_0 = 0.14$ meV. During the transformation of the X into the cavity mode, its FWHM increases monotonically from ~ 0.07 meV to 0.19 meV. All these features are characteristics of the strong coupling regime.

Finally, note as well, that the emission from pillars with strong exciton–photon coupling – similar to that from the cavity with weak coupling (Fig. 9.3) – is rather strong both with and without a dot on resonance, which indicates an independent population of the cavity mode. Such an independent excitation of the cavity mode should be taken into account for a correct description of the emission spectra. This problem is considered in Section 9.5.

9.4 Weak coupling regime: Purcell effect

Figure 9.3 shows that the most drastic change in the QD emission line of a 3D QD MC in the weak coupling regime is a strong decrease of the emission intensity with increasing detuning of the cavity and QD modes, $\Delta = |E_X - E_C|$, whereas the measured X linewidth remains nearly constant. The analysis of the linewidth in this regime cannot give any reliable information on the dependence of QD exciton lifetime on the $X - C$ detuning as, in addition to the QD lifetime, the FWHM of the exciton line is influenced by (i) the QD exciton decoherence and (ii) an inhomogeneous line broadening due to charge fluctuations in the vicinity of the QD. It is also very difficult to extract any reliable information on the lifetime from the emission intensity analysis, as the radiation pattern for an emitter in an optical cavity is modified by the cavity and depends on the emission wavelength. The most reliable way to measure the spontaneous emission rate modification is to carry out direct measurements of the QD exciton lifetime.

For this purpose the decay of the QD emission from 3 μm diameter photonic pillars of two types was investigated, with uncoated and metal coated lateral sidewalls (Bayer *et al.*, 2001). Different to other work discussed in this paper, these results have been obtained on cavities in which only the upper Bragg re-

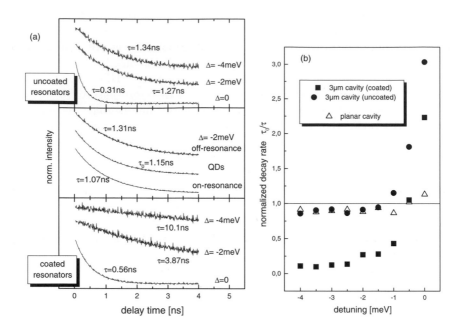

FIG. 9.5. (a) Photoluminescence decay curves of uncoated (top panel) and coated (bottom panel) resonators with a diameter of 3 μm for various detunings between the ground optical mode (indicated at each trace) and single dot transitions. The middle panel shows the decay curves for the QDs with the upper mirror removed (middle trace) and also for the planar cavity on and off resonance. (b) Ratio of the spontaneous emission rates τ_0/τ of QDs in a cavity to that in a homogeneous medium versus detuning from E_C towards lower energies. The data for 3 μm resonators are compared with the data of a planar cavity. From (Bayer *et al.*, 2001). Reprinted figures with permission from [M. Bayer, T. L. Reinecke, F. Weidner, A. Larionov; A. McDonald; and A. Forchel, *Phys. Rev. Lett.* **86**, 3168 (2001)], copyright (2001) by the American Physical Society.

flector was completely etched through. Dots in this case were formed by interface fluctuations in a thin quantum well. The electromagnetic field in the uncoated structures is localized in the pillar resonator due to the large discontinuity of the refractive index between semiconductor and vacuum. This dielectric discontinuity results in a sidewall reflectivity of ~ 0.35 for normal incidence. The coating of the pillar with a gold film increases the reflectivity to a value close to unity. Corresponding calculations for these coated cavities show that the modes are also strongly confined in them and that the energies are essentially the same for the two types of cavities.

For a quantitative comparison, resonators of comparable finesses, $Q \sim 6400$, have been investigated for both structure types. Optical excitation was done by a mode-locked Ti:sapphire laser emitting pulses with a length of 1.5 ps at a repetition rate of 82 MHz. The laser energy was tuned to 1.467 eV to excite the InGaAs only. The radiation was collected, spectrally analyzed by a monochromator and detected by a streak camera with a temporal resolution of the setup of about 10 ps.

Figure 9.5(a) displays the photoluminescence decay curves for uncoated (top panel) and coated (bottom panel) resonators with a diameter of 3 μm for different detunings from the ground photon mode at E_C. The decay curves for the QDs with the top mirror removed from the cavity and for planar resonator are shown in the middle panel. For short delays (< 50 ps) scattered laser light appears in the spectra, so only the data for later times are shown. In all cases the luminescence decays to a good approximation exponentially. For the QDs without cavity the luminescence lifetime is ~ 1 ns. As expected, in the planar resonator only marginal changes of this lifetime are observed, for both on and off resonance, similarly to earlier observations (Tanaka *et al.*, 1995; Abram *et al.*, 1998). Figure 9.5(a) shows that for off resonance in the uncoated resonators, the lifetimes are only slightly increased over the QD value as well. However, for on resonance the lifetime is considerably shortened to ~ 0.3 ns. A similar shortening to ~ 0.5 ns is observed for the coated structures. But, in contrast, for these structures the off-resonance lifetime increases drastically: for $\Delta = 2$ meV, $\tau \sim 3.9$ ns, while for $\Delta = 4$ meV, $\tau > 10$ ns.

Figure 9.5(b) displays the ratio of the spontaneous emission rates of QDs in a cavity to that in a homogeneous medium versus detuning from E_C towards lower energies. The QD emission dynamics for the off-resonance situation in the uncoated patterned resonators is altered by only about 10%, the emission rate still is about 0.9 of that of the QD emission, independent of the detuning. On the other hand, for on resonance it is enhanced by a factor of 3. A somewhat weaker enhancement by a factor of 2 is observed for the coated cavities. For off resonance, however, a considerable reduction of the emission rate is obtained here. For example, for $\Delta < -3$ meV, $\tau_0/\tau \sim 0.1$.

In the dissipative regime the emission kinetics is given by Fermi's Golden Rule: the spontaneous emission rate ratio τ_0/τ for an emitter located at position r in the resonator is given by Gérard *et al.* (1996)

$$\frac{\tau_0}{\tau} = \frac{2}{3} F \left(\frac{E(r)}{E_{\max}} \right)^2 \frac{(\gamma_C/2)^2}{(\Delta^2 + (\gamma_C/2)^2)} + \alpha. \tag{9.3}$$

Equation (9.3) describes two different decay channels: The first term gives the QD emission into the cavity mode that has an electric field distribution E at the QD position r, $E_{\max}$ is a maximum amplitude of the field at the cavity mode. The Purcell factor F is given by

$$F = \frac{3}{4\pi^2} \left(\frac{\lambda}{n} \right)^3 \frac{Q}{V_C}, \tag{9.4}$$

where n is the refractive index of the resonator. The second term in equation (9.3) describes the decay channel due to QD emission into leaky modes.

For a given cavity size, the confinement of the fields of the lowest electromagnetic modes is essentially identical in the coated and uncoated resonators. Thereby, in cavities of comparable finesses, the dependences of the first decay channel should be similar. Then, the difference between the decay rates of the uncoated and coated cavities is given by their different couplings to the leaky modes. The difference for coated and uncoated pillars is equal to 0.75, and this difference does not depend on detuning within the experimental accuracy.

For off resonance, the first decay channel in equation (9.3) can be neglected in comparison to the second one in the uncoated structures, and the emission dynamics is predominantly given by the coupling to the leaky photon modes. The coupling to the leaky modes prevents a strong suppression of the QD spontaneous emission in the resonators. For the coated structures, in contrast, a suppression of this decay channel by an order of magnitude is found for $\Delta < -2$ meV. We note that, based on the detuning dependence of the first term in equation (9.3), an even stronger deenhancement than observed might be expected for larger detunings. In addition, absorption in the gold coatings might play some role. Further technological improvement of the coatings might allow for an even greater suppression.

The experimentally observed ratio $\tau_0/\tau = 3$ in the uncoated cavities is in a good agreement with estimates of the Purcell factor for the investigated 3 μm diameter cavity done by using equation (9.4), $F = 6$, if one takes into account that QDs are located not only in the maximum electric field. Experiments on pillars with various diameters with a similar finesse have shown a two-fold decrease of the emission rates with increasing cavity diameter from 3 to 5 μm. This increase of the emission rate with decreasing cavity diameter results from the Purcell effect, and is given by the reduction of the mode volume. Since the reduction of the cavity finesse is weak and can be neglected in the first approximation, the emission rate is controlled by the confinement of the electrical field, which goes approximately as $1/r_C^2$ and decreases by ~ 2.8 times for a reduction of r_C from 2.5 to 1.5 μm. The decrease in τ_0/τ according to equation (9.3) is ~ 1.8 times, in good agreement with the experimental data.

The strong increase in the emission rate of QD excitons can be used for the fabrication of QD-based single-photon sources with enhanced single-photon emission rates. It follows from equations (9.3) and (9.4) that the QD radiative decay rate can be highly enhanced in high-finesse cavities with a small mode volume. The smaller the cavity volume the higher emission rate can be achieved. Note, however, that these equations are valid for the description of the emission rate from the cavity in the weak coupling regime only, resulting in a limitation of the emission decay rate determined by the cavity finesse, $1/\tau < 2\gamma_C = 2E_C/Q$.

9.5   MC emission: Theory

Before discussion of emission properties of cavities with QDs in the strong coupling regime, let us first consider the theoretical predictions (Keldysh *et al.*, 2006). The spectral density of the electromagnetic field in a cavity for a primary excitation of the cavity mode can be represented as

$$I_C(\omega) = \frac{\omega}{(\omega - \Omega_C - P(\omega))^2 + (\gamma_C + Q(\omega))^2} \quad (9.5)$$

whereas that for a primary excitation through a set of QDs is

$$I_X(\omega) = \sum_j |g_j|^2 \omega \left\{ [(\omega - \Omega_C - P(\omega))(\omega - \Omega_j) - (\gamma_C + Q(\omega))\gamma_j]^2 \right.$$

$$\left. + [\gamma_j(\omega - \Omega_C - P(\omega)) + (\gamma_C + Q(\omega))(\omega - \Omega_j)]^2 \right\}^{-1}. \quad (9.6)$$

Here index j denotes the exciton in the j-th QD; Ω_j and Ω_C are the corresponding resonant frequencies, $2\gamma_j$ and $2\gamma_C$ are the corresponding damping constants; $P(\omega)$ and $-Q(\omega)$ are the real and imaginary parts of the polarizability of the whole ensemble of QDs. In the linear (weak field) regime

$$P(\omega) - iQ(\omega) = \sum_j \frac{|g_j|^2}{\omega - \Omega_j + i\gamma_j}, \quad (9.7)$$

where g_j is the vacuum Rabi splitting parameter for the j-th QD. For the case of a single QD in exact resonance with the cavity mode, the definition of $I_C(\omega)$ and $I_X(\omega)$ practically coincides with formulae (37) and (36) in the paper of Rudin and Reinecke (2000) calculated taking into account the effects of (i) the interaction of the electromagnetic mode with a continuum of electromagnetic modes within the Fano model and (ii) the coupling of the electronic excitations to a dissipative source in the QD in line broadening of the coupled mode spectrum of optical microcavities.

Both functions $I_C(\omega)$ and $I_X(\omega)$ describe the radiation in the cavity mode hybridized with QD resonances. However, they refer to two different mechanisms (channels) of the primary excitation. The distinction considers which of the hybridization partners – cavity mode or quantum dot excitation – is pumped directly by the primary excitation. The function $I_X(\omega)$ corresponds to the process of a primary excitation of QDs (photoexcited exciton is trapped by the QD, then being relaxed to the lowest QD state, which is resonant and therefore hybridized with the cavity mode). The hybridization itself corresponds to the partial transfer of the excitation into the electromagnetic field. In contrast, the function $I_C(\omega)$ refers to a process where the primary excitation occurs in the cavity mode, and then becomes hybridized. As indicated above, this channel may be due to inelastic (e.g., phonon-assisted) recombination of bulk excitons or to a similar process with excitons trapped by nonresonant QDs, i.e., those with

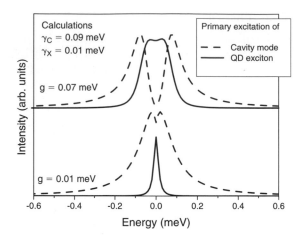

FIG. 9.6. Plots of $I_C(\omega)$ (dashed lines) and $I_X(\omega)$ (solid lines) calculated for the on resonance condition at $g = \gamma_X \ll \gamma_C$ (lower panel) and $\gamma_X \ll g \sim 3/4\gamma_C$ (upper panel).

lowest excited states far from the resonance with the cavity, etc. In a mechanical analogy, one can think about pushing either one or another of two coupled nonidentical oscillators, while always observing the response of the one of them corresponding to the electromagnetic field. Surely, in any case both degrees of freedom become excited, however, with amplitude and phase relations essentially depending on the excitation path. The mathematical manifestation of that is clearly seen in the analytical structure of the spectral functions. Both $I_C(\omega)$ and $I_X(\omega)$ have exactly the same set of poles – hybridized mode frequencies – in the complex plane. However, the shape of the spectra may differ drastically due to the difference in residual values. Even in the simplest case of a single resonant QD, the plots of $I_C(\omega)$ and $I_X(\omega)$ look like mutually complementary. This is shown in Fig. 9.6. At small exciton–photon coupling $g \sim \gamma_X \ll \gamma_C$, the transfer of excitation from the QD to the cavity mode as well as that from the cavity to the QD is relatively weak. As a result, the primary excitation of the QD results in a very narrow line with HWHM $\gamma \sim \gamma_X$. In contrast, the excitation of the cavity mode results in a wide line with $\gamma \sim \gamma_C$ with a relatively weak narrow dip at the QD exciton frequency. The qualitative difference in the spectra continues till the onset of the strong coupling regime, providing the strong exchange between the cavity and QD modes. In particular, Fig. 9.6 shows that even for the case of $g \sim 3/4\gamma_C$, a well pronounced splitting in the emission spectrum is observed only for a primary excitation of the cavity mode.

A more detailed comparison of PL spectra for the two excitation mechanisms and various $X - C$ mode detunings is presented in Fig 9.7. It shows that for the parameter range $\gamma_X \ll |g| \ll \gamma_C$, the dominating feature in the $I_X(\omega)$ plot is a steep increase of a narrow peak of the dot emission in the frequency range a few

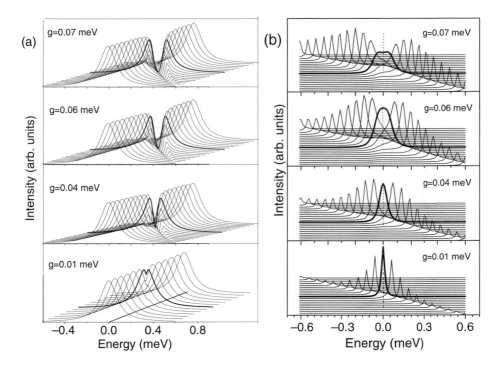

FIG. 9.7. Comparison of the PL spectra for various $X - C$ mode detunings and g/γ_X, g/γ_C ratios for two excitation mechanisms: primary excitation of the cavity mode (a) and QD exciton (b). In all panels $\gamma_X = 0.02$ meV and $\gamma_C = 0.09$ meV.

γ_C around the cavity mode. The cavity mode itself is practically invisible until some dot approaches the resonance. Unlike that, the $I_C(\omega)$ plot is dominated by the strong and relatively broad cavity emission line, cut, however, by a narrow dip – a precursor of the Rabi splitting – at the resonant QD frequency. The intensity of the cavity emission is nearly independent of the $C - X$ detuning. The FWHM of the dip is $\sqrt[4]{|g|^4 + 2\gamma_C\gamma_X|g|^2}$, i.e., exceeds essentially the Rabi splitting in the weak coupling regime. The dip results from the modes interference and exists for $|g|^2 > \gamma_X^3/2\gamma_C$, i.e., long before two coupled modes become resolved. The dip in the $I_X(\omega)$ plot appears only when the usual strong coupling condition $|g|^2 > \frac{1}{2}\left(\gamma_C^2 + \gamma_X^2\right)$ is fulfilled.

Generally, both types of excitation channels may be effective. Then the luminescence spectrum is given by

$$I(\omega) = a\,I_c(\omega) + b\,I_d(\omega), \qquad (9.8)$$

where a and b are the partial excitation rates in corresponding channels. That is the case in experiments, in which both the cavity mode and dot lines are clearly

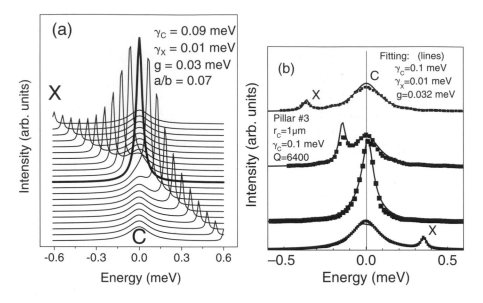

FIG. 9.8. (a) PL spectra calculated for the 1.8 μm diameter pillar (#2) with a weakly coupled QD exciton and cavity modes. (b) Fitting (solid lines) of experimental PL spectra (dots) for a 2 μm diameter pillar with $Q = 6400$ with equations (9.5)–(9.8) for several detunings.

seen at off-resonance conditions. The overlap of both contributions in this case renders a detailed evaluation of the spectra to be rather difficult.

9.6   Fit of the MC emission spectra: Weak and strong coupling regimes

Figures 9.8(a) and (b) illustrate the description of the MC emission spectra taking into account both QD and cavity excitation. The ratio a/b is determined from the fit with equations (9.5)–(9.8) of the set of pillar spectra recorded for various $X - C$ detunings. Figure 9.8(a) displays the spectra calculated for the 1.8 μm pillar (#2) with a weakly coupled QD exciton and cavity modes. The experimental spectra are well reproduced with the use of the ratio $a/b = 0.07$ (compare with Fig. 9.3a). This means that the MC emission is mainly excited via the capture of photoexcited carriers by the QD. A relatively weak direct excitation of the cavity mode does not lead to any marked increase in the FWHM at the resonance. However, the effect of an additional excitation via the cavity mode becomes more pronounced with its increasing. In particular, this leads to an additional broadening of the emission line near the resonance of the cavity and exciton oscillators as shown in Fig. 9.8(b), where the spectra recorded for a 2 μm-diameter pillar with $Q = 6400$ at an elevated excitation are displayed. The QD emission linewidth shows a well pronounced broadening at the resonance. Solid lines in Fig. 9.8(b) show the result of the fit of the emission spectra with

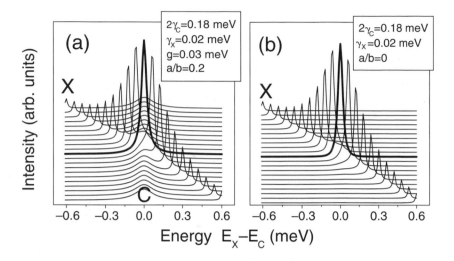

FIG. 9.9. Simulation of the strong coupling spectra with (a) and without (b) accounting for independent cavity excitation in addition to a direct excitation of the QD exciton.

equations (9.5)–(9.8) for several detunings. The spectra are well described with the use of experimental values of $\gamma_C = 0.1$ meV and two adjustable parameters, $g = 0.032$ meV and the ratio of excitations via cavity to QD modes $a/b = 0.3$. Again, the contribution of the primary cavity excitation is rather weak, however, the account for this contribution allows one to describe the emission spectra (line intensities and linewidths) both for zero and nonzero detunings.

The strong coupling regime in MC pillars was achieved first in pillars with enlarged QDs in the active layer. The spectra of such a micropillar (#4) are shown in Fig. 9.3(b). Figure 9.9 shows the simulation of the spectra with the use of equations (9.5)–(9.8). It is seen that, indeed, no marked cavity mode emission is expected out of the resonance if one excites the dot-cavity system through the capture of excitons with QDs. The experimental spectra are qualitatively well reproduced only in the approximation that the primary excitation of cavity mode is about 20% of that of the QDs. The best fit of the set of spectra is obtained using a coupling constant $g = 0.055 \pm 0.005$ meV. This value is markedly smaller than one obtained from a simplified model as one half of line splitting, $\Delta = 0.14$ meV, in the emission spectra recorded at zero detuning, resulting in $g = 0.07$ meV, which indicates the importance of the correct account for both cavity and QD primary excitations.

The value of the exciton–photon coupling in the case of strong coupling, $g = 0.055$ meV, exceeds the value required for weak-to-strong coupling transition, $g = 0.045$ meV, by about 20%. The estimated value of the QD exciton oscillator strength is equal to $f \sim 30$, which is about three times larger than in small self-assembled InGaAs QDs with a large In content.

Vacuum Rabi splitting with a single small QD has been realized in a photonic crystal nanocavity has been recently realized (Yoshie *et al.*, 2004). In these experiments a photonic crystal nanocavity with a markedly smaller cavity volume of $\sim 2(\lambda/n)^3$ with a finesse $f \sim 13\,000$ has been used. The measured zero detuning vacuum Rabi splitting in such a cavity was equal to $2g = 0.17$ meV, which corresponds to a QD exciton radiative lifetime of 1.82 ns assuming that the dot is in the field maximum, whereas ensemble measurements gave $\tau_0 = 1 - 2$ ns.

9.7 Conclusions

In conclusion, micropillars with as high Q-factor as 7500 down to pillar diameters of 1.5 μm with QDs of various compositions and sizes have been fabricated. The QD-cavity system emission has been investigated in a wide range of exciton–photon mode detunings controlled by temperature variation in the range of 10–45 K. A strong Purcell effect has been observed for self-assembled QDs in small microcavity pillars with diameters less than 3 μm. The emission rate at fixed cavity finesse increases with reduced pillar diameter and is limited by the transition into the strong coupling regime. The emission lineshape and relative emission intensities of cavity and QD exciton modes have been shown to depend sensitively on the primary excitation mechanisms via QD excitons and via the cavity mode. In the case of an excitation via the cavity mode a well pronounced dip at the QD exciton energy appears in the emission spectrum already in the weak coupling regime, due to an interference effect between the cavity and the QD oscillators. The strong coupling regime has been observed for single enlarged QDs in a 1.5 μm diameter micropillar with $Q = 7500$ and with an exciton–photon coupling constant $g = 0.055$ meV, providing a PL line splitting at the $X - C$ resonance of $\Delta = 0.14$ meV. The value of Δ exceeds markedly 2g, which indicates the importance of the correct account for both cavity and QD primary excitations.

Acknowledgements

It has always been a pleasure for us and other coworkers in Würzburg, Chernogolovka, Wroclaw and Washington to conduct important parts of this research together with Leonid Keldysh. We hope to continue our collaboration and we wish him all the best in his personal life as well as exciting subjects and results in his great scientist's life.

References

Abram, I., Robert, I. and Kuszelewicz, R. (1998). *IEEE J. Quantum Electron.* **34**, 71.

Agarwal, G. S. (1985). *J. Opt. Soc. Am.* B **2**, 480.

Andreani, L. C., Savona, V., Schwendimann, P. and Quattropani, A. (1994). *Superlattices Microstruct.* **15**, 453.

Andreani, L. C., Panzarini, G. and Gérard, J. M. (1999). *Phys. Rev.* B **60**, 13276.

Bastard, G., Delalande, C., Meynadier, M. H., Frijlink, P. M. and Voos, M. (1984). *Phys. Rev.* B **29**, 7042.

Bayer, M., Reinecke, T. L., Weidner, F., Larionov, A., McDonald, A. and Forchel, A. (2001). *Phys. Rev. Lett.* **86**, 3168.

Bockelmann, U. (1993). *Phys. Rev.* B **48**, 17637.

Burstein, E. and Weisbuch, C. (1995). *Confined Excitons and Photons: New Physics and Devices* (ed by E. Burstein and C. Weisbuch). Plenum, New York.

Citrin, D. S. (1993). *Phys. Rev.* B **47**, 3832.

Fano, U. (1961). *Phys. Rev.* **124**, 1866.

Gabrielse, G. and Dehmelt, H. (1985). *Phys. Rev. Lett.* **55**, 67.

Gérard, J. M., Barrier, D., Marzin, J. Y., Kuszelewicz, R., Manin, L., Costard, E., Thierry-Mieg, Y. and Rivera, T. (1996). *Appl. Phys. Lett.* **69**, 449.

Gérard, J. M., Sermage, B., Gayral, B., Legrand, B., Costard, E. and Thierry-Mieg, V. (1998). *Phys. Rev. Lett.* **81**, 1110.

Gérard, J. M. and Gayral, B. (2001). *Physica* E **9**, 131.

Goy, P., Raimond, J. M., Cross, M. M. and Haroche, S. (1983). *Phys. Rev. Lett.* **50**, 1903.

Hanamura, E. (1988). *Phys. Rev.* B **37**, 1273.

Hood, C. J., Chapman, M. S., Lynn, T. W. and Kimble, H. J. (1998). *Phys. Rev. Lett.* **80**, 4157.

Hulet, R. G., Hilfer, E. S. and Kleppner, D. (1985). *Phys. Rev. Lett.* **55**, 2137.

Keldysh, L. V., Kulakovskii, V. D., Reitzenstein, S., Makhonin, M. N. and Forchel, A. (2006). *Pisma Zh. Eksp. Teor. Fiz.* **84**, 584 [*JETP Lett.* **84**, 494].

Kleppner, D. (1981). *Phys. Rev. Lett.* **47**, 233.

Mabuchi, H. and Doherety, A. C. (2002). *Science* **298**, 1372.

McKeever, J., Boca, A., Boozer, A. D., Buck, J. R and Kimble, H. J. (2003a). *Nature* **425**, 268.

McKeever, J., Buck, J. R., Boozer, A. D., Kuzmich, A., Nagerl, H.-C., Stamper-Kurn, D. M. and Kimble, H. J. (2003b). *Phys. Rev. Lett.* **90**, 133602.

Michler, P., Kiraz, A., Becher, C., Schönfeld, W. V., Petroff, P. M., Zhang, L., Hu, E. and Imamoglu, A. (2000). *Science* **290**, 2282.

Nakamura, A., Yamada, H. and Tokizaki, T. (1989). *Phys. Rev.* B **40**, 8585.

Ohnesorge, B., Bayer, M., Forchel, A., Reithmaier, J. P., Gippius, N. A. and Tikhodeev, S. G. (1997). *Phys. Rev.* B **56**, 4367.

Pelton, M., Santori, C., Vučković, J., Zhang, B., Solomon, G. S., Plant, J. and Yamamoto, Y. (2002). *Phys. Rev. Lett.* **89**, 233602.

Peter, E., Senellart, P., Martrou, D., Lemaître, A., Hours, J., Gérard, J. M. and Bloch, J. (2005). *Phys. Rev. Lett.* **95**, 067401.

Purcell, E. M. (1946). *Phys. Rev.* **69**, 681.

Rashba, E. I. and Gurgenishvili, G. E. (1962). *Fiz. Tverd. Tela* **4**, 1029 [*Sov. Phys. Solid State* **4**, 759.]

Reithmaier, J. P., Röhner, M., Zull, H., Schäfer, F., Forchel, A., Knipp, P. A. and Reinecke, T. L. (1997). *Phys. Rev. Lett.* **78**, 378.

Reithmaier, J. P., Sek, G., Loffler, A., Hoffmann, C., Kuhn, S., Reitzenstein, S.,

Keldysh, L. V., Kulakovskii, V. D., Reinecke, T. L. and Forchel, A. (2004). *Nature* **432**, 197.

Rudin, S. and Reinecke, T. L. (2000). *Phys. Rev.* A **62**, 053806.

Santori, C., Fattal, D., Vuckovic, J., Solomon, G. S. and Yamamoto, Y. (2002). *Nature* **419**, 594.

Savona, V., Hradill, Z., Quattropani, A. and Schwendimann, P. (1994). *Phys. Rev.* B **49**, 8774.

Solomon, G. S., Pelton, M. and Yamamoto, Y. (2001). *Phys. Rev. Lett.* **86**, 3903.

Takagahara, T. (1987). *Phys. Rev.* B **36**, 9293.

Tanaka, K., Nakamura, T., Takamatsu, W., Yamanishi, M., Lee, Y. and Ishihara, T. (1995). *Phys. Rev. Lett.* **74**, 3380.

Thomson, R. J., Rempe, G. and Kimble, H. J. (1992). *Phys. Rev. Lett.* **68**, 1132.

Vahala, K. J. (2003). *Nature* **424**, 839.

Weisbuch, C., Nishioka, M., Ishikawa, A. and Arakawa, Y. (1992). *Phys. Rev. Lett.* **69**, 3314.

Weisbuch, C. (1994). *Cavity Quantum Electrodynamics* (ed by P. R. Berman). Academic Press, New York.

Weisbuch, C. and Rarity, J. (1996). *Microcavities and Photonic Bandgaps: Physics and Applications* (ed by C. Weisbuch and J. Rarity). NATO Advanced Study Institute, Series E: Applied Sciences, Vol. 324, Kluwer, Dordrecht.

Yamamoto, Y. and Slusher, R. E. (1993). *Physics Today* **46**, 66.

Yoshie, T., Scherer, A., Hendrickson, J., Khitrova, G., Gibbs, H. M., Rupper, G., Ell, C., Shchekin, O. B. and Deppe, D. G. (2004). *Nature* **432**, 200.

Zhang, J. P., Chu, D. Y., Wu, S. L., Ho, S. T., Bi, W. G., Tu, C. W. and Tiberio, R. C. (1995). *Phys. Rev. Lett.* **75**, 2678.

Zhu, Y., Ganthier, D. J., Moriu, S. E., Wu, Q., Carmichael, H. J. and Mossber, T. W. (1990). *Phys. Rev. Lett.* **64**, 2499.

10

DYNAMICS OF COLD EXCITONS AND ELECTRON–HOLE ENSEMBLES IN DIRECT-GAP SEMICONDUCTORS STUDIED BY MID-INFRARED PUMP AND PROBE SPECTROSCOPY

Makoto Kuwata-Gonokami
Department of Applied Physics, Graduate School of Engineering, the University of Tokyo, SORST, Japan Science and Technology Agency, 7-3-1 Hongo, Bunkyo-ku, Tokyo 113-8656, Japan

Abstract

We examine the dynamics of high-density, low-temperature electron–hole ensembles and excitons produced by pulsed resonant excitation. Intraband motion, including free carrier motion and exciton internal transition, is probed in the mid-infrared using time-resolved measurements. In CuCl, the creation of a high-density electron–hole plasma via the exciton Mott transition is clearly observed in transient reflection measurements. Analysis of the reflection spectra reveals the formation of voids of exciton gases and colloid-like metallic states. In Cu_2O, excitonic Lyman series transitions up to the $5p$ exciton states are clearly observed from supercooled 1s orthoexcitons generated by two-photon resonant absorption. From the temporal response, a very rapid ortho–para conversion due to the collision of cold orthoexcitons can clearly be seen. Such an efficient generation of long-lived paraexcitons provides new opportunities that may allow us to reach an exciton Bose–Einstein condensation phase.

10.1 Introduction

Collective quantum mechanical phenomena such as superfluidity and superconductivity appear for ensembles of particles at a high density and low temperature. These collective effects can also take place through a precise optical manipulation of the ensemble of particles. The best example is the creation of an ultracold atomic gas prepared by laser cooling. Methods to cool down atoms to sub-microkelvin temperature have been developed and various related quantum mechanical phenomena, including Bose–Einstein condensation (BEC) (Anderson *et al.*, 1995; Davis *et al.*, 1995) and the pairing of fermionic atoms (Regal *et al.*, 2004) have been extensively studied in the past decade.

An ensemble of electrons and holes is also a suitable candidate for the study of collective quantum mechanical effects by optical manipulation through light–matter interaction. Figure 10.1 illustrates the schematic phase diagram of an electron–hole (e-h) ensemble for various e-h pair densities and temperatures. In the dilute region, the electron and hole are combined to form a bound state known

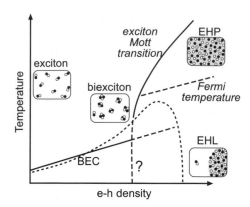

FIG. 10.1. Schematic phase diagram of the photoexcited e-h system in semiconductors. From (Nagai and Kuwata-Gonokami, 2002a). Adapted figure with
   permission from Elsevier Ltd: [M. Nagai and M. Kuwata-Gonokami, *J. Luminescence*
   100, 233 (2002)], copyright (2002).

as an exciton. By increasing the density of excitons, bounded pairs of excitons
appear and form excitonic molecules referred to as biexcitons. Since both excitons
and biexcitons are composed of an even number of fermions, these quasiparticles
obey the Bose quantum statistics. The possibility of BEC of quasiparticles in
semiconductors at low temperature and high density has been studied extensively
(Hanamura and Haug, 1977; Moskalenko and Snoke, 2000).

   At a higher density regime, the ensemble of excitons results in the screening of Coulomb attraction and, finally, in the ionization of excitons. The ionized
e-h ensemble is called e-h plasma (EHP). Such a transformation from an insulator to a metallic phase is referred to as an exciton Mott transition. At a
sufficiently low temperature, the EHP condenses inhomogeneously and forms e-
h droplets (EHD), resembling a spatial phase separation and a first-order phase
transition. In this phase, which is called e-h liquid (EHL), the mean interparticle
distance is less than the Bohr radius of an exciton, and the e-h ensemble has a
high Fermi degeneracy. EHDs have been extensively investigated in indirect gap
semiconductors with a large band degeneracy, and the formation of EHD is well
established both experimentally and theoretically (Jeffries and Keldysh, 1983;
Tikhodeev, 1985).

   Of particular interest is the transformation from excitonic BEC to fermionic
liquid in the quantum degenerate regime at low temperature. With increasing
densities, the ensemble of excitons should evolve continuously from a dilute BEC
into a dielectric superfluid consisting of a BCS-like degenerate two-component
Fermi liquid with Coulomb attraction (Keldysh and Kozlov, 1968; Comte and
Noziéres, 1982). Such a transformation has been extensively discussed theoretically (Noziéres and Schmitt-Rink, 1984; Randeria, 1995).

   Since the initial temperature of an e-h ensemble created by intense optical excitation is very high, it is usually not easy to reach very low temperature within the lifetime of carriers or excitons. However, by carefully selecting the material and the method of optical excitation using stable, short laser pulses to achieve control in the time and frequency domains, a very low temperature can be attained. The recent development of short-pulsed lasers makes it possible to obtain such a precise control of excitation. It might then be possible to observe new phenomena characteristic of optically manipulated quantum systems.

Another important challenge is to find an appropriate way to probe the behavior of high-density carriers. Luminescence spectroscopy has been widely used for such a purpose (Leheny and Shah, 1976; Yoshida *et al.*, 1980; Klingshirn and Haug, 1981). In particular, time-resolved luminescence measurements have been developed to study the dynamics of high-density carriers. However, strong many-body Coulomb correlation and photon-mediated effects modify their spectral line shapes and their interpretation is rather complicated.

The recent development of ultrashort pulsed laser technology has also expanded the frequency range of optical probing, from the far-infrared (THz) to the ultraviolet or soft-X-ray spectral regions. This provides us a new opportunity to probe various aspects of material response under photoexcitation. In particular, pump and probe measurements with low-frequency probe pulses in the mid-infrared (mid-IR) or THz frequency region allow us to probe the intraband motion of carriers (Kira *et al.*, 1998; Nagai *et al.*, 2001; Huber *et al.*, 2001; Kaindle *et al.*, 2003) and also the internal transition of excitons. The nature of e-h ensembles which was hidden in interband luminescence spectra could be revealed with such measurements. In this article, we review our recent time-resolved spectroscopy experiments on the search for high-density, low-temperature states of photogenerated carriers using low-energy pulsed light probing.

10.2 Electron–hole liquid formation via exciton resonant excitation in CuCl

10.2.1 *Search for EHL phase in direct gap semiconductors*

The formation of EHL in direct gap semiconductors with a short carrier lifetime has been studied extensively in the past decades (Shionoya, 1979; Klingshirn and Haug, 1981). In particular, time-resolved emission measurements in GaAs-based III-V compound semiconductors has revealed that EHP created by band-to-band excitation has a high temperature due to excess heating followed by bandgap shrinkage (Tanaka *et al.*, 1980). The lack of an efficient cooling mechanism prevents the e-h ensemble from reaching a low-temperature quantum degenerate state in which the carrier temperature is of the order of electron and hole Fermi energies $E_F^{e,h}$. High-T_c EHD formation has been predicted theoretically in polar II-VI semiconductors, such as CdS and CdSe, in which the phonon-mediated stabilization of the liquid phase takes place (Beni and Rice, 1978). EHD-like behavior in CdS and CdSe has been observed by time-resolved emission measurements (Yoshida *et al.*, 1980). However, despite extensive experimental efforts, no decisive evidence of EHD formation has yet been obtained (Bohnert *et al.*, 1981).

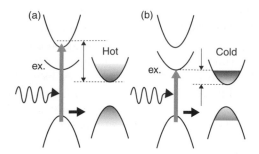

Fɪɢ. 10.2. Photoexcitation scheme of band-to-band excitation (a) and exciton resonant excitation (b). Strong bandgap shrinkage brings excess energy leading the heating of photogenerated carries in (a). For the exciton resonant excitation (b), free electrons and holes are created with a lower effective temperature than in (a) because the exciton binding energy reduces the excess energy of the carriers. From (Nagai and Kuwata-Gonokami, 2002a). Adapted figure with permission from Elsevier Ltd: [M. Nagai and M. Kuwata-Gonokami, *J. Luminescence* **100**, 233 (2002)], copyright (2002).

The resonant excitation of bound e-h pairs (i.e., excitons) could be a method to create a dense ensemble of low-temperature carriers. In this approach, the bandgap energy reduction is nearly compensated by the reduction of exciton binding energy (Bányai and Koch, 1986), reducing the carrier temperature, while the exciton resonance energy is almost unchanged even at high excitation density. When the incident photon flux exceeds the critical density of the exciton Mott transition, a degenerate EHP appears due to screening of the Coulomb attraction between electrons and holes. Schematic diagrams of the band-to-band and resonant excitation schemes are presented in Fig. 10.2.

The resonant excitation scheme was examined in GaAs and self-screening was observed in transmission spectra at a density one order of magnitude above the Mott density (Fehrenbach *et al.*, 1982). However, the obtained carrier temperature was not low enough to reach the quantum degeneracy because the reduction of EHP temperature by the exciton resonant excitation is ineffective for excitons with a small binding energy. CuCl, a direct gap semiconductor with a bandgap energy, $E_G = 3.396$ eV, has a very stable exciton state labeled $1s$-Z_3 exciton that has a binding energy $E_{ex} = 213$ meV and an exciton Bohr radius $a_B = 7$ Å. Because of the small Bohr radius, the Mott transition density is rather high, $n_{Mott} = 5 \times 10^{19}$ cm^{-3}. Thus, CuCl is one of the best candidates among direct gap semiconductors to create a low-temperature EHD using a resonant scheme. In the following sections, we will review our experiments on exciton resonant excitation in CuCl (Nagai *et al.*, 2001; Nagai and Kuwata-Gonokami, 2002a).

10.2.2   *Time-resolved photoluminescence measurements under resonant excitation of excitons above Mott transition density*

First, we briefly review the results on time-resolved emission measurements under intense one-photon resonant excitation of excitons or two-photon excitation of biexcitons in CuCl. We use a high-purity single-crystal CuCl sample (thickness 10 μm), which is cooled to 4 K. The second harmonics of a regenerative amplified mode-locked Ti:sapphire laser (0.3 ps pulse duration and 1 kHz repetition rate) were tuned at 382 nm center wavelength for resonant exciton excitation. Since the absorption coefficient is about 2×10^3 cm^{-1} at a wavelength of 382 nm, which is 200 meV below the bandgap and close to the upper polariton region, the excitation is nearly homogeneous through the crystal. Measurements following band-to-band excitation are performed for comparison by tuning the center wavelength to 360 nm, using the fourth harmonics of signal pulses from an optical parametric generator with β-BaB$_2$O$_4$ crystal. The excitation pulse is focused on the sample with a spot size of 100 μm. The emission is spectrally resolved using a spectrometer equipped with a liquid-nitrogen-cooled CCD camera. For the time-domain emission measurements, the optical Kerr gate method is used with a 0.2 mm-thick high refractive index glass SFL03 as the Kerr medium. The temporal resolution is 0.4 ps, which corresponds to a spectral resolution of 1 nm.

Figure 10.3(a) shows time-integrated emission spectra under band-to-band excitation for different excitation densities. At an excitation density of 0.09 mJ/cm^2, two biexciton emission bands are observed at 391 nm and 392 nm, which are due to transitions from the biexciton to the transverse exciton state (M_T) and to the longitudinal exciton state (M_L), respectively. At a higher excitation of 5 mJ/cm^2, a broad EHP-like emission appears at the low-energy side as reported by Hulin *et al.*, (1985). Figure 10.3(b) shows emission spectra under resonant exciton excitation. At a low excitation density of 0.02 mJ/cm^2, two biexciton emission bands appear, and, at a high excitation density of 5.1 mJ/cm^2, a broad EHP emission band appears. At the intermediate excitation density of 0.9 mJ/cm^2, a new emission band, which does not exist in the case of band-to-band excitation appears at 393 nm.

Figure 10.4 shows results of time-resolved emission measurements for (a) band-to-band excitation at 360 nm with a power density of 5 mJ/cm^2, (b) exciton resonant excitation (382 nm) with a density of 3.5 mJ/cm^2. A broad EHP emission band is observed in all cases.

10.2.3   *Mid-infrared transient reflection spectroscopy*

The collective motion of carriers can be accessed by a measurement of the IR metallic reflection, which is associated with the plasma motion itself. The transient reflection spectra provide direct information of the carrier density and temperature. An outline of the IR reflection methods used to probe carriers is shown in Fig. 10.5. In the presence of free carriers, the dielectric constant ε_b is modified as indicated in the following,

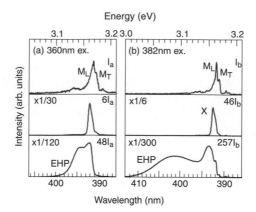

FIG. 10.3. Time-integrated emission spectra in CuCl at different excitation densities under (a) band-to-band excitation (360 nm, $I_a = 0.09$ mJ/cm^2) and (b) exciton resonant excitation (382 nm, $I_b = 0.02$ mJ/cm^2). From (Nagai and Kuwata-Gonokami, 2002a). Adapted figure with permission from Elsevier Ltd: [M. Nagai and M. Kuwata-Gonokami, *J. Luminescence* **100**, 233 (2002)], copyright (2002).

$$\varepsilon_M = \varepsilon_b \left(1 - \frac{\omega_p^2}{\omega^2 + i\gamma\omega} \right), \tag{10.1}$$

where $\omega_p = \sqrt{4\pi n e^2 / \varepsilon_b m}$ is the plasma frequency, n, m, and e are the carriers' density, effective mass (reduced mass of electron–hole pair), and electric charge, respectively, and γ is the plasma damping. Consequently, reflectivity should increase below and decrease above the plasma frequency. In photo-doped semiconductors, the spatial inhomogeneity of the plasma density due to the linear absorption modifies the reflection spectrum in the vicinity of ω_p, so that the spatiotemporal dynamics of the carriers can be probed. The plasma frequency can be scaled with the exciton binding energy, $\omega_p = \sqrt{12/r_s^3} E_{ex}$, where $r_s = \sqrt[3]{3/4\pi n a_B^3}$. In CuCl, ω_p for the degenerate plasma above the Mott transition density $r_s \approx 2.5$ lies in the mid-IR region ($\hbar\omega_p = \sqrt{\pi/4} E_{ex} \approx 0.19$ eV), which is far below the bandgap and well above the LO phonon energy ($\hbar\omega_{LO} = 25.9$ meV).

A platelet shape single-crystal with 24 μm thickness is kept in a cryostat in a strain free state. The probe pulses are obtained by difference frequency generation in a AgGaS$_2$ crystal using the signal and idler of the β-BaB$_2$O$_4$-based optical parametric generator. The probe pulses are focused on the sample with a spot size of 150 μm. The variation in reflectivity is measured using a liquid-nitrogen-cooled HgCdTe mid-IR detector. The spectral resolution is about 30 meV.

Figure 10.6 shows the differential reflectivity spectra $\Delta R/R_0$ for a different

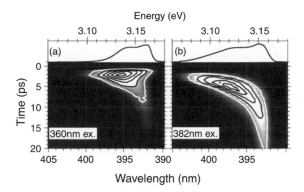

FIG. 10.4. Time-resolved emission spectra in CuCl under (a) band-to-band excitation (360 nm) and (b) exciton resonant excitation (382 nm). The upper curves are the corresponding time-integrated emission spectra. Excitation densities are (a) $5\,\mathrm{mJ/cm^2}$ and (b) $3.5\,\mathrm{mJ/cm^2}$. From (Nagai and Kuwata-Gonokami, 2002a).   Adapted figure with permission from Elsevier Ltd: [M. Nagai and M. Kuwata-Gonokami,   *J. Luminescence* **100**, 233 (2002)], copyright (2002).

pump–probe delay. The excitations correspond to (a) band-to-band transition (360 nm, $4\,\mathrm{mJ/cm^2}$) and (b) resonant exciton transition (383 nm, $4\,\mathrm{mJ/cm^2}$). R_0 is the linear reflectivity, which is 0.14 and does not depend on the probe frequency in the range shown in Fig. 10.6. The right ordinates show the corresponding absolute reflectivity $R_0 + \Delta R$. In Fig. 10.6(a), $\Delta R/R_0$ increases at the low-energy side and decreases at the high-energy side after a 1 ps delay time (reflectivity values at 10 and 1.8 μm are 0.75 and 0.06, respectively), indicating that the strong metallic reflection appears 1 ps after band-to-band excitation. Figure 10.6(b) also shows a similar metallic reflection. The transient reflection spectra are also measured at a lower excitation density of 0.9 $\mathrm{mJ/cm^2}$ under exciton resonant excitation (383 nm) (Nagai *et al.*, 2001). These results clearly indicate that EHP is created via an exciton Mott transition process.

It should be pointed out that there is a difference between the early time delay (1 ps) metallic reflection spectra obtained at band-to-band and exciton resonant excitation (see Fig. 10.6a and b, respectively). In the case of a band-to-band excitation, the short penetration depth of the excitation pulse (about 0.1 μm) results in creation of a very thin plasma layer that makes the reflection spectrum broader and its slope more gentle. Such spatial inhomogeneity effects of the plasma density on the broadening of the plasma resonance have been observed in Si and GaAs (Nagai and Kuwata-Gonokami, 2002b). Assuming the exponential carrier distribution in the sample $n(z) = n_\mathrm{p} \exp(-z/D_\mathrm{p})$ (Vinet *et al.*, 1984), the transient reflection spectrum is calculated in the vicinity of plasma resonance, including the spatial inhomogeneity of the carrier density. Dashed

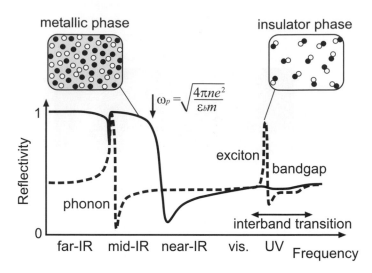

FIG. 10.5. Schematic reflection spectra of a photoexcited semiconductor (solid curve) and of a nonexcited, undoped semiconductor (dotted curve).

curves in Fig. 10.6(a) show calculations with carrier densities $n_p = 2.5$, 1.6, and 1.1×10^{20} cm^{-3}. The dashed curve in Fig. 10.6(b) is calculated taking into account the penetration depth of the exciton resonant excitation condition, and it nearly reproduces the data at a delay time of 1 ps. However, there are two distinct differences between the resonant excitation condition shown in Fig. 10.6(b) and the band-to-band excitation condition of Fig. 10.6(a). The first is the spectral position of $\Delta R/R_0 = 0$, E_0, which decreases from 0.5 to 0.2 eV after a time delay. The second feature is the monotonic decrease of $\Delta R/R_0$ at the low-energy side, typically around 10 μm, with time. A simple plasma model with an exponential profile of carrier density distribution cannot reproduce these features.

It is natural to introduce the effect of phase-separated carrier distribution, in which there are a high-density plasma and a low-density exciton gas region. The free carriers occupy a finite volume fraction f of the crystal. As long as the scale of phase separation is much smaller than the wavelength of the probe light, the effective dielectric constant ε_{eff} can be used in the following Bruggeman's form (Bruggeman, 1934),

$$f \frac{\varepsilon_M - \varepsilon_{\text{eff}}}{\varepsilon_M + 2\varepsilon_{\text{eff}}} + (1 - f) \frac{\varepsilon_b - \varepsilon_{\text{eff}}}{\varepsilon_b + 2\varepsilon_{\text{eff}}} = 0, \qquad (10.2)$$

where ε_M is the dielectric constant of bulk metal. In the case of $f \ll 1$, the effective dielectric constant is simplified as the following equation:

$$\varepsilon_{\text{eff}} = \varepsilon_b \left(1 + 3f \frac{\varepsilon_M - \varepsilon_b}{\varepsilon_M + 2\varepsilon_b} \right). \qquad (10.3)$$

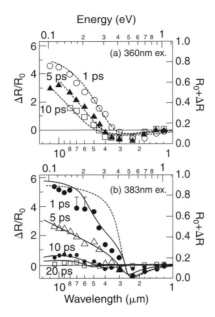

FIG. 10.6. Transient mid-IR reflection spectra in CuCl under (a) band-to-band excitation (360 nm) and (b) exciton resonant excitation (383 nm). Excitation densities are (a) $4\,\mathrm{mJ/cm^2}$ and (b) $4\,\mathrm{mJ/cm^2}$. Dashed curves show the theoretical calculations with the assumption of an exponential profile of the carrier distribution. Solid curves show the Bruggeman's effective-medium theory. From (Nagai and Kuwata-Gonokami, 2002a). Adapted figure with permission from Elsevier Ltd: [M. Nagai and M. Kuwata-Gonokami, *J. Luminescence* **100**, 233 (2002)], copyright (2002).

By substituting equation (10.1) for ε_M, ε_eff exhibits a resonance feature at the frequency $\omega_\mathrm{p}/\sqrt{3}$. Thus, the reflectivity increases at the lower-energy side below $\omega_\mathrm{p}/\sqrt{3}$ and decreases above $\omega_\mathrm{p}/\sqrt{3}$. Because of the small damping $\gamma \ll \hbar\omega_\mathrm{p}$, the reflectivity at the low-energy side is determined by the volume fraction of the plasma phase. By using the material parameters of CuCl, the following equation was obtained: $E_0/\hbar\omega_\mathrm{p} = 1/\sqrt{3}+0.55f$. These procedures, along with the spectra from Fig. 10.6(b), provide the temporal evolution of f and n. Figure 10.7 shows the temporal evolution of the obtained f and n. The solid curves in Fig. 10.6 show calculations simulating the experimental reflectivity spectra using f and n as fitting parameters (Nagai and Kuwata-Gonokami, 2002a).

Since $\Delta R/R_0$ is insensitive to f at $f>0.5$, the obtained volume fraction in Fig. 10.7(b) at a time delay of 1 ps is underestimated, whereas n is overestimated. Figure 10.7(b) shows that, although the value of f always decays monotonically, n rapidly decreases to 2×10^{20} cm^{-3}, 8 ps after excitation and remains almost constant at longer pump–probe delays. This indicates the e-h system reaches a

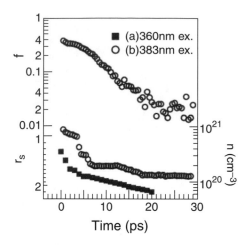

FIG. 10.7. Temporal evolution of volume fractions (upper plots) and plasma density (lower plots) estimated from the transient mid-IR reflection spectra of Fig. 10.6. The center wavelengths of the excitation pulses are 360 nm in (a) (closed squares, band to band excitation) and 383 nm in (b) (open circles, exciton resonant excitation). From (Nagai and Kuwata-Gonokami, 2002a). Adapted figure with permission from Elsevier Ltd: [M. Nagai and M. Kuwata-Gonokami, *J. Luminescence* **100**, 233 (2002)], copyright (2002).

metastable EHD state. The obtained carrier density in the EHD corresponds to $r_s=1.7$, which is in good agreement with theoretical predictions for direct gap semiconductors (Beni and Rice, 1978; Hanamura, 1976).

10.2.4 *Discussion: Electron–hole antidroplet formation*

Our experimental results on time-resolved emission and mid-IR reflectivity clearly indicate the formation of EHD in CuCl within 10 ps, although the numerical simulation shows that EHD cannot grow within the short lifetime of carriers in direct gap semiconductors (Bohnert *et al.*, 1981; Haug and Abraham, 1981). Such a contradiction can be eliminated by taking into account the difference in the initial conditions between our experiment and the theoretical simulation. Specifically, in our experiment, the initial carrier density is set to be much higher than the density of the metastable EHL phase, and a much faster formation time can thus be expected. However, from our numerical simulation following the reported method (Haug and Abraham, 1981), we can observe that EHD does not grow (Lei *et al.*, 2003).

Another aspect of the initial process should be pointed out (Sugakov, 2002). The transient reflectivity spectrum at a 1 ps delay shows a nearly homogeneous high-density carrier distribution. In the next step, it is natural to consider the production of bubbles of excitons in metallic liquid. In such a situation, a new localized plasmon mode at $\sqrt{2/3}\omega_p$ near the void in the metal would appear. This

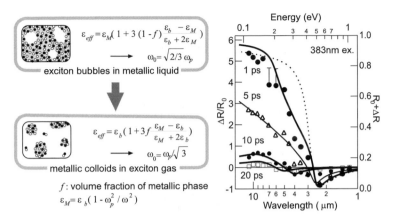

FIG. 10.8. Scenario of the formation of EHDs via exciton Mott transition in CuCl, and corresponding transient mid-IR reflection spectra.

$\sqrt{2/3}\omega_p$ plasmon resonance can be obtained by exchanging ε_b and ε_M in equation (10.3). The solid line at a 1 ps delay in Fig. 10.6(b) shows the calculated curve, with parameters of $n = 3 \times 10^{20}$ cm³ and $f = 0.75$. We clearly see the resonance around 0.3 eV which is associated with this new mode, and a better fit to the experimental curve can be obtained than with the homogeneous plasma model (dashed curve). During the metallic phase of boiling, these bubbles of exciton vapor grow with time. Finally, metallic droplets are formed, and the localized plasmon mode changes into $\omega_p/\sqrt{3}$. This scenario is sketched in Fig. 10.8.

In this section the creation of a cold e-h system by the resonant excitation of excitons and biexcitons was discussed. We perform time-integrated and time-resolved emission measurements along with mid-IR pump–probe measurements, all of which reveal the creation of cold EHP via the exciton and biexciton Mott transition. Both experimental results indicate that EHD with a density of about 10^{20} cm⁻³ forms within a 10 ps delay after excitation in the direct gap semiconductor CuCl.

10.3 Excitonic Lyman spectroscopy in Cu₂O

10.3.1 *Search for exciton BEC*

The observation of the Bose–Einstein condensation (BEC) of neutral atoms or pair of atoms has triggered major advancements in quantum physics of the last decade (Anderson *et al.*, 1995; Davis *et al.*, 1995; Regal *et al.*, 2004). Ensembles of ultracold atoms with high controllability of density, temperature and also interaction strength reveal new aspects of many-body quantum phenomena. Excitons, composite particles in semiconductors made of light-mass fermions, may provide another interesting system offering new insight on quantum many-body physics. Several new experimental results triggered a growing interest to revisit this problem in photoexcited semiconductors (Butov *et al.*, 2002). In usual systems, however, the finite lifetime of excitons, typically of the order of nanoseconds, prevents the observation of BEC at thermal equilibrium condition in contrast to the neutral atoms. According to a recent theoretical investigation, the macroscopic coherence could be built up within a short time (Schmitt *et al.*, 2001). This implies the possibility of detecting exciton BEC in a quasi-equilibrium condition. In such a case, it is crucial to investigate the space-time evolution of excitons by time-resolved measurements.

It has been long recognized that Cu_2O is a material with unique advantages for the observation of BEC of excitons (Mysyrowicz, 1980). Because of the positive parity of the valence and conduction band minima at the center of the Brillouin zone, their radiative recombination is forbidden in the dipole approximation, conferring a long radiative lifetime to the $n = 1$ exciton. The $n = 1$ exciton level is split by exchange interaction in a triply degenerate orthoexciton state and a lower lying singly degenerate paraexciton. Several experiments have shown intriguing results in luminescence and transport suggesting the occurrence of a degenerate quantum fluid at high densities and low temperatures in this material (Fortin *et al.*, 1993; Snoke *et al.*, 1990). However, uncertainties remain on the actual density of particles created by optical pumping. Based on luminescence absolute quantum efficiency measurements, it has been argued that a collision induced annihilation of excitons, which is known as Auger effect, has a giant cross-section. Excitons are destroyed and heated up when the density exceeds 10^{15} cm^{-3} and prevents BEC (O'Hara and Wolfe, 2000). On the other hand, exciton collision could induce ortho- to paraexciton conversion (Kavoulakis and Mysyrowicz, 2000). Since this effect does not destroy excitons but increases the ortho–paraexciton conversion rate to accumulate paraexcitons, the microscopic picture of such collision-induced processes is still under discussion (Jang and Wolfe, 2006). To resolve this controversy, it is important to examine the dynamics of ortho- and paraexcitons as a function of density. More generally, it is important to study the dynamics of paraexcitons at low temperature in order to search for the optimum conditions to achieve BEC. The major difficulty to reach this goal was the lack of a sensitive spectroscopic method to probe optically inactive paraexcitons. We have proposed and demonstrated time-resolved excitonic Lyman spectroscopy, a detection method that probes the internal transitions of excitons, which are in mid-IR spectral region (Kuwata-Gonokami *et al.*, 2004; Kuwata-Gonokami, 2005; Kubouchi *et al.*, 2005; Tayagaki *et al.*, 2005). We clearly detected the temporal evolution of spin forbidden excitons in Cu_2O. In this section, we review our series of experiments on 1s excitons in Cu_2O and discuss prospects for the detection of exciton BEC in this system.

10.3.2   *Excitons in Cu_2O*

The yellow series exciton in Cu_2O is known to be a model system of Wannier excitons with hydrogen-like energy structures (Hayashi and Katsuki, 1952; Gross and Karryev, 1952; Nikitine, 1959). A hydrogenic series of exciton lines with up to 10 terms is clearly observed in linear absorption spectroscopy (Shindo *et al.*, 1974; Fröhlich *et al.*, 1979; Matsumoto *et al.*, 1996). Cu_2O has a particularly simple band structure, with nondegenerate parabolic valence (symmetry Γ_7^+) and conduction bands (symmetry Γ_6^+), with extrema located at the center of the Brillouin zone having the same parity under inversion. This symmetry implies that the yellow series starts with the $n = 2$ term. It corresponds to a so-called second class (weakly allowed) dipole transition, whereby an exciton with an electron and a hole in a p-like internal state of motion are created from the crystal ground state, by absorption of a photon. The $n = 1$ term of the yellow series appears as a weak quadrupole absorption line. The theoretical estimate of the $n = 1$ exciton Bohr radius is small, $a_B = 0.53$ nm (Kavoulakis *et al.*, 1997). The electron–hole exchange interaction further splits the $n = 1$ exciton into a highly forbidden singlet paraexciton of symmetry Γ_2^+ with an internal energy of 2.021 eV at 2 K and a triply degenerate Γ_5^+ orthoexciton (2.033 eV at 2 K). Both the $n = 1$ ortho- and paraexcitons have a long radiative lifetime. The lifetime of the $n = 1$ orthoexciton is limited by ortho- to paraexciton conversion, whereas the optically inactive paraexciton has a lifetime exceeding several tens of microseconds in good quality crystals (Mysyrowicz *et al.*, 1979). Recent studies by high-resolution spectroscopy revealed that the orthoexciton states split into three states due to the long-range k^2 term of electron–hole exchange interaction. Moreover, the anisotropy of the k^2 dependent term in the electron–hole exchange interaction makes the orthoexciton mass anisotropic (Dasbach *et al.*, 2003, 2004).

The $1s$ paraexcitons have been considered to be the most promising candidate to realize exciton BEC because of their extremely long lifetime. Although much work has been devoted to show evidence of BEC and various distinct phenomena have been reported, the existence of exciton BEC is still controversial. Basic properties, such as paraexciton mass and Auger recombination rate, which are crucial for exciton BEC, have not yet been clarified. Photoluminescence measurements are a sensitive method to obtain the steady state information of excitons with a wide dynamic range. However, the measured emission intensity is governed by the relatively long exciton lifetime, which is extremely sensitive to the crystalline quality of the sample. Thus, it is important to develop a method to measure the temporal evolution of exciton population in the picoseconds to microseconds time-scale to solve these basic problems.

10.3.3 *Excitonic Lyman transition*

An alternative method to detect excitons in Cu_2O is the measurement of the transition between the $n = 1$ term and terms of an excitonic series with higher principal quantum numbers (see Fig. 10.9), as suggested theoretically by Haken (1958) and Nikitine (1984). This type of absorption, which changes the angular momentum of the relative electron–hole wave motion from an $l = 0$ (s-like) state to $l = 1$ (p-like) states, is the equivalent of the Lyman-α absorption line in atomic hydrogen. In the case of the yellow series excitons in Cu_2O, these absorption lines appear in the mid-IR region. Because of the large difference in the electron–hole exchange energy between the $1s$ and $2p$ excitons (about 12 meV

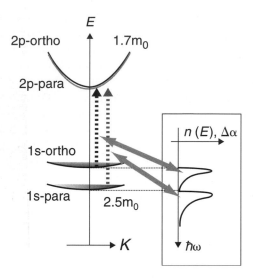

FIG. 10.9. Schematic diagram of excitonic Lyman spectroscopy with energy diagram of the $1s$ and $2p$ exciton states in Cu_2O. Ascending arrows correspond to the measured absorption lines. From (Kuwata-Gonokami, 2005). Adapted figure with permission from Elsevier Ltd: [M. Kuwata-Gonokami, *Solid State Commun.* **134**, 127 (2005)], copyright (2005).

and nearly 0 meV respectively), we can resolve the Lyman-α lines for orthoexcitons and paraexcitons. It should be noted that we can obtain information on the distribution function of excitons from the spectra, since the transition energy depends on the momenta of excitons because of the difference between the $1s$ and $2p$ exciton masses. Recently, Johnsen and Kavoulakis (2001) calculated the line profile for the Lyman-α absorption band and pointed out that it should show a characteristic narrowing when excitons undergo BEC.

An experimental attempt to measure Lyman-α transitions in Cu_2O has been made by Jörger *et al.* (Goppert *et al.*, 2001; Jörger *et al.*, 2005; Karpinska *et al.*, 2005). We have extended this method to time-resolved spectroscopy by using femtosecond mid-IR pump and probe spectroscopy detecting $1s - 2p$ transition to examine the temporal evolution of ortho- and paraexcitons in Cu_2O. We measured time-resolved induced absorption associated with the excitonic Lyman-α transition ($1s$ to $2p$) under pulsed excitation of excitons as a function of the pump–probe time delay. We call this method time-resolved excitonic Lyman spectroscopy. We successfully detected $1s$ paraexcitons as well as $1s$ orthoexcitons and found that a significant amount of $1s$ paraexcitons were generated (Kuwata-Gonokami, 2005; Kubouchi *et al.*, 2005; Tayagaki *et al.*, 2005).

10.3.4 *Time-resolved excitonic Lyman spectroscopy*

The experimental setup is shown in Fig. 10.10. We used a Ti:sapphire based regenerative amplifier (150 fs pulses at a central wavelength of 775 nm or 80 fs, 800 nm) with an energy per pulse of 1 mJ and a repetition rate of 1 kHz. Part of the pulse is sent to an optical parametric amplifier (OPA), where a tunable probe pulse with a wavelength around 10 μm is produced by difference frequency generation (DFG). The rest of the primary pulse is converted to a tunable pump source in the visible part of the spectrum (energy per pulse ~ 2 μJ), focused onto a sample area of 1.3×10^5 μm^2. High-purity, naturally grown single crystals cut along the (100) or (110) planes with thicknesses ranging between 220 and 400 μm are used. Samples are mounted on the sample holder and cooled by contact with a copper block maintained at liquid helium temperature. The delayed probe pulse traverses the pumped region of the sample under a small incidence angle with respect to the pump beam. It is then analyzed by a monochromator (resolution ~ 0.1 meV) followed by a HgCdTe detector. The inset of Fig. 10.10 shows the absorption spectrum of a Cu$_2$O crystal in mid-IR region. We observe an absorption band associated with the multiphonon excitation. Fortunately, windows transparent between 110 and 135 meV are available, allowing the observation of the transitions from the $n = 1$ ortho- and paraexcitons to their $2p$ states.

Figure 10.11 shows a sequence of time-resolved absorption spectra for the ortho- and paraexciton Lyman-α transitions. The pump pulse was tuned to 600 nm, inside the orthoexciton $1s$ phonon-assisted absorption sideband. The mid-IR absorption reveals the presence of two lines peaked at 116 meV and 129 meV respectively.

Referring to Fig. 10.11, it can be recognized that the low-energy edge of each line is given by E_{1s-2p}, the separation between the respective $1s$ and $2p$ excitons at rest. There is no ambiguity for the position of the low-energy edge E_{1s-2p} of the $1s - 2p$ orthoexciton Lyman-α line, because the internal energy of the $1s$ and $2p$ orthoexcitons is well known from the literature (Shindo *et al.*, 1974; Fröhlich *et al.*, 1979; Matsumoto *et al.*, 1996). One finds $E_{1s-2p} = 116$ meV, in good agreement with the present results. The internal energy of the 1s paraexciton is also well known, but there is no available data on the position of the $2p$ paraexciton. However, it is expected to lie very close to the $2p$ orthoexciton resonance since the electron–hole exchange energy (responsible for the ortho–para splitting) should be negligibly small for $2p$ states. The electron–hole exchange interaction is proportional to the spatial overlap of electrons and holes in the exciton wavefunction. This overlap is minimized by the larger Bohr radius and the p-like relative motion. One then expects the paraexciton Lyman line $E_{1s-2p} = E_{2p;\text{para}} - E_{1s;\text{para}}$ to be 128.5 meV. As can be seen, the agreement with the experiment is excellent and confirms that the electron–hole splitting is small for the $2p$ states. Therefore the two lines can be readily identified. The line at 116 meV is due to the Lyman $1s - 2p$ transition of the orthoexciton, while the line at 129 meV is due to the $1s - 2p$ paraexciton absorption. For small time

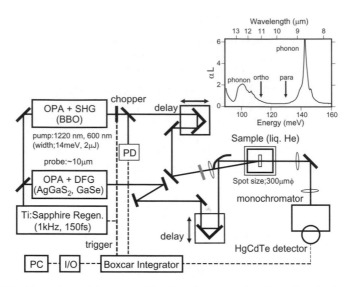

FIG. 10.10. Experimental setup for the time-resolved excitonic Lyman spectroscopy of Cu_2O. Inset shows a linear mid-IR absorption spectrum of Cu_2O. The observed lines around 100 and 145 meV correspond to phonon overtones of the crystal. From (Kuwata-Gonokami, 2005). Reprinted figure with permission from Elsevier Ltd: [M. Kuwata-Gonokami, *Solid State Commun.* **134**, 127 (2005)], copyright (2005).

delays, a blue-shift as well as a broadening are observed as shown in Fig. 10.11.

The broadening and high energy shift of the lines at an early time apparent in Fig. 10.11 can be explained by the cooling of an initial hot exciton cloud towards the lattice temperature. Upon pumping at a 600 nm wavelength, an excess energy of about 20 meV is transferred to the created orthoexciton gas. The majority of the excess energy is quickly dissipated by emission of optical phonons. The remaining excess energy is dissipated at a slower rate by acoustic phonons. From the data, one concludes that the initial ortho- and paraexciton temperature is of the order of 80 K, and decreases to a few kelvin within 200 ps when the pump is tuned to the orthoexciton phonon side band absorption.

Figure 10.12 shows the paraexciton $1s - 2p$ line measured as a function of crystal temperature at a probe delay time of 4 ns. At higher temperatures, the induced absorption band becomes broad in accordance with the thermal distribution of excitons. Solid lines show the fitted results with Maxwell–Boltzmann distribution of excitons. To perform a simulation of the data, one needs to know the translational mass of the $1s$ and $2p$ excitons, their distribution in momentum space as well the homogeneous linewidth of the initial and final states. In this figure, we used the exciton mass parameter of the $n = 1$ orthoexciton, $m = 2.7m_e$, as estimated by resonant Raman scattering experiments where m_e is the free electron mass (Caswell *et al.*, 1981). Its value differs from the sum of the indi-

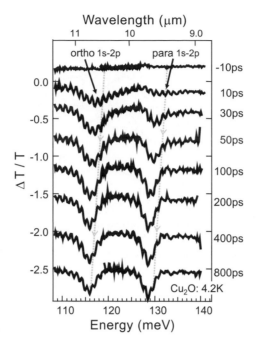

FIG. 10.11. Time-resolved differential absorption spectra in Cu₂O held at 4.2 K
and recorded at pump-pulse delays of −10, 10, 30, 50, 100, 200, 400 and
800 ps. The two absorption bands of each spectrum correspond to $1s - 2p$
transitions for the ortho- (116 meV) and paraexciton (129 meV). The pump
pulse wavelength is tuned at 600 nm, corresponding to the phonon-assisted
absorption band of the orthoexcitons. It is incident perpendicular to the (100)
crystal plane and has an energy of 2 μJ for a pulse duration of 150 fs. The
observed fringe pattern is due to multiple reflections of the probe pulse in the
sample. The sample thickness is 170 μm. From (Kuwata-Gonokami, 2005).
Adapted figure with permission from Elsevier Ltd: [M. Kuwata-Gonokami, *Solid State
Commun.* **134**, 127 (2005)], copyright (2005).

vidual electron and hole masses, $m = 1.68m_e$ measured by cyclotron resonance
(Hodby *et al.*, 1976). This difference is due to the nonparabolicity of electrons
and holes forming 1s excitons with a small Bohr radius. The $n = 1$ exciton Bohr
radius is comparable to the lattice constant $a_l = 0.426$ nm. On the other hand,
the Bohr radius of the $2p$ state is 4.4 nm, much larger than a_l. The wavefunctions of the electron and hole are confined around the center of the Brillouin
zone. The $n = 2$ exciton mass is then simply the sum of the individual electron
and hole masses. From a line shape analysis of the paraexciton $1s - 2p$ band in
an equilibrium condition, we obtain a value around $(2.5 - 2.7)\ m_e$.

It is interesting to prepare the initial state of orthoexcitons directly at low
temperature by resonant optical excitation. Drawing an analogy with the two-

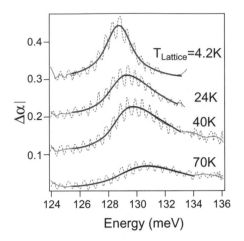

FIG. 10.12. Lineshape of the $1s - 2p$ paraexciton absorption measured at different sample temperatures, as indicated. The solid lines are calculated lineshapes using translational exciton masses of $2.7m_e$ for the $1s$ state, $1.68m_e$ for the $2p$ state and the sample temperature (see text). The fringe pattern due to multiple reflection of the probe light by the surfaces has been subtracted by Fourier filtering, as shown by the grey lines. From (Kuwata-Gonokami *et al.*, 2004). Reprinted figure with permission from [M. Kuwata-Gonokami, M. Kubouchi, R. Shimano, and A. Mysyrowicz, *J. Phys. Soc. Jpn.* **73**, 1065 (2004)], copyright (2004) by the Physical Society of Japan.

photon excitation of biexcitons in CuCl with ultrashort laser pulses (Kuwata-Gonokami *et al.*, 2002), we can generate orthoexcitons at a very low initial temperature by direct two-photon absorption (TPA) with short-pulse lasers. Despite the broad laser bandwidth, the small group velocity dispersion at the TPA laser frequency ($\hbar\omega = 1.0164$ eV) makes it possible to generate orthoexcitons at a state with an initial momentum spread even smaller than in the case of biexcitons in CuCl. A conservative estimate yields an initial orthoexciton temperature much less than 10^{-3} K. Figure 10.13 shows a schematic diagram of the Lyman spectroscopy with resonant two-photon excitation scheme (Tayagaki *et al.*, 2005).

Figure 10.14(b) shows an induced absorption spectrum under the resonant two-photon excitation of $1s$-orthoexcitons at low-density excitation condition (pump pulse energy 3 mJ/cm^2) measured after a pump–probe delay of 5 ps. Four lines appear around 116, 129, 133 and 136 meV, respectively. A linear absorption spectrum in the region of np orthoexciton states is shown in Fig. 10.14(a). The curve is shifted by the energy of the 1s orthoexciton $E_{1s;\,ortho} = 2033$ meV (Uihlein *et al.*, 1981) to match the spectral position of the induced absorption signal of Fig. 10.14(b). There is a good agreement between both sets of resonances (Uihlein *et al.*, 1981; Gross, 1956). Thus we can assign the series of Fig. 10.14(b)

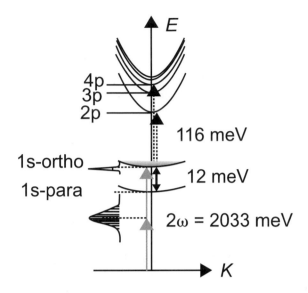

FIG. 10.13. Schematic energy diagram for the yellow series 1s and np exciton states in Cu₂O and their Lyman series transitions from the super cooled orthoexcitons generated by pulsed resonant two-photon absorption. From (Tayagaki *et al.*, 2005). Reprinted figure with permission from [T. Tayagaki, A. Mysyrowicz, and M. Kuwata-Gonokami, *J. Phys. Soc. Jpn.* **74**, 1423 (2005)], copyright (2005) by the Physical Society of Japan.

to the transition from the 1s orthoexciton to the np orthoexciton ($n = 2, 3, 4, 5$) of the Lyman series.

Figure 10.14(c) shows the transient absorption spectrum measured at a probe delay time of 4 ns. The orthoexciton $1s - 2p$ signal has diminished and the higher energy narrow lines have disappeared, because of the decay of orthoexcitons into paraexcitons, while a broader signal now appears around 128 meV. This new line is assigned to the $1s - 2p$ transition of the paraexcitons (Kubouchi *et al.*, 2005).

The excitonic Lyman series shows a systematic n-dependence of the linewidths and absorption strengths. A similar behavior was observed in the np absorption spectra and has been discussed in terms of n-dependent LO-phonon scattering rate from np to 1s state. Figure 10.15 shows the linewidths of the $1s - np$ transitions (circles). The total width of the lines has several contributions: the energy distribution of 1s excitons f_{1s}, the intrinsic linewidths of the 1s state and the width of the final np states. The finite energy distribution of 1s excitons, f_{1s}, contributes to the broadening because of the difference in the mass parameters of 1s and np excitons, as schematically shown in Fig. 10.9. The widths of the Lyman lines can be expressed as $(\Delta E_{1s-np})^2 \sim (\Delta E_{np})^2 + [(m_{1s}/m_{2p} - 1) f_{1s}]^2$. The recent high-resolution spectroscopy revealed that the intrinsic linewidth of the 1s exciton state is much narrower ($\gamma_{1s} \ll 1\,\mu$eV) (Dasbach *et al.*, 2003, 2004) than

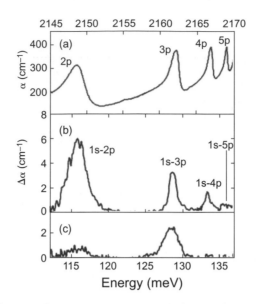

FIG. 10.14. (a) Linear absorption spectrum. Induced absorption spectra of (b) the $1s$ orthoexciton generated instantaneously by the two-photon excitation and (c) the $1s$ paraexciton measured 4 ns after the excitation. From (Tayagaki *et al.*, 2005). Reprinted figure with permission from [T. Tayagaki, A. Mysyrowicz, and M. Kuwata-Gonokami, *J. Phys. Soc. Jpn.* **74**, 1423 (2005)], copyright (2005) by the Physical Society of Japan.

the spectral resolution of the present experiment and is negligible in our analysis. The linewidths of the np states are known to show a n-dependence which can be deduced from an analysis of the usual linear absorption spectrum. We estimate the linewidths of the np states by fitting the linear absorption data lines shown in Fig. 10.9(a) with asymmetric Lorentzian curves following the model of Toyozawa (Toyozawa, 1964). The obtained results are also plotted in Fig. 10.15 (triangles).

We can estimate the upper limit of the energy spread of initial $1s$ orthoexcitons, from the difference in the width between the $1s-4p$ and $4p$ lines < 0.1 meV. The obtained upper limit is 0.36 meV$/[(3.0m_e/1.7m_e) - 1] \sim 0.5$ meV. It should be pointed that this is narrower than the lattice thermal energy at 4.2 K, $k_B T = 0.7$ meV, indicating that the initial supercooled $1s$-orthoexcitons are still at a temperature below that of the lattice even after a delay of 4 ps.

We now discuss the $1s$ state. It is known that the spectral positions of np exciton states ($n \geq 2$) are well described with the Rydberg series $E_n = 17525 - 786/n^2$ cm^{-1} (Nikitine, 1969). This implies that the wavefunctions of np states are well described by the hydrogen-like model. However, the $1s$ exciton state shows a significant deviation from this Rydberg series which has been discussed by several authors. More recently, Artoni *et al.* (2002) analyzed the ac Stark

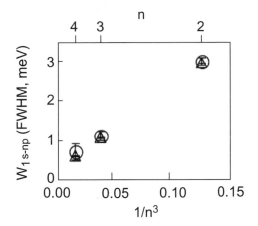

FIG. 10.15. Linewidth of $1s - np$ line (circles) and np absorption (triangles).
From (Tayagaki *et al.*, 2005). Reprinted figure with permission from [T. Tayagaki,
A. Mysyrowicz, and M. Kuwata-Gonokami, *J. Phys. Soc. Jpn.* **74**, 1423 (2005)], copyright
(2005) by the Physical Society of Japan.

effect under resonant pumping of $1s - 2p$ transition of orthoexciton and adopted
a value of 6.3 eÅ (30 D) for μ_{1s-2p} by a direct analogy with hydrogen atoms.
Kavoulakis *et al.* (1997) examined the shrinkage of the Bohr radius of the $1s$
exciton state caused by the central cell correction due to the nonparabolicity
of the valence and conduction bands. They took into account the quadratic
terms of the wavevector dependence of the bands under the constraint that
the known value of the translational mass of $1s$ orthoexciton, $3m_e$, must be
reproduced. This analysis gave a Bohr radius of $a_{1s} = 0.53$ nm and a binding
energy $E = 164.9$ meV. They found that the wavefunction of the $1s$ exciton is
well approximated with such a modified $1s$ function (Uihlein *et al.*, 1981). It was
pointed out that such a small effective Bohr radius of $1s$ exciton reduces the
overlap between the $1s$ and $2p$ exciton wavefunctions, yielding a revised dipole
moment $1s - 2p$ of 1.64 eÅ (Jörger *et al.*, 2005).

The systematic analysis of the excitonic Lyman absorption provides us a
unique opportunity to determine the "unknown" $1s$ exciton wavefunction from
a set of projection to the known np states through the dipole matrix element
$\mu_{1s-np} = e \langle u_{1s} |z| u_{np} \rangle$.

The integrated area of the excitonic Lyman absorption line $1s - np$ can be
expressed as follows,

$$S_{1s-np} = \int_{1s-np} \Delta\alpha(E)dE = \frac{n_{1s}4\pi^2}{\hbar c\sqrt{\varepsilon_b}} E_{1s-np}|\mu_{1s-np}|^2, \qquad (10.4)$$

where E_{1s-np}, μ_{1s-np}, n_{1s} and ε_b are the $1s - np$ transition energy, dipole moment, $1s$ exciton density and background dielectric constant at probe frequency,

respectively. We determine the background dielectric constant by measuring the interference fringe pattern of the absorption spectrum in the mid-IR region. We obtain $\varepsilon_b \sim 6$ in the probe frequency region of the excitonic Lyman spectroscopy. Knowing the density of the $1s$ exciton and from the measured integrated absorption $\int_{1s-np} \Delta\alpha(E)dE$, we can determine the dipole moment μ_{1s-np}. In practice, however, it is difficult to estimate the absolute density of the $1s$ exciton from the absorbed photon number, which strictly depends on the detailed pumping condition especially in the case of a two-photon excitation with short pulses. Therefore in order to evaluate the $1s$ exciton state, we rely on the relative intensity of the integrated absorption of $1s - np$ lines S_{1s-np}/S_{1s-2p}, which are density independent and insensitive to the pump condition. Indeed, in the low-density excitation limit, equation (10.4) yields the ratio $|\mu_{1s-np}/\mu_{1s-2p}|^2 \times (E_{1s-np}/E_{1s-2p})$.

We find that the integrated areas of the $1s - 3p$ and $1s - 4p$ transitions are proportional to the integrated area of the $1s - 2p$ line, i.e., the $1s$ exciton density. From the fitting of the result, we obtain the ratio $S_{1s-3p}/S_{1s-2p} = 0.26 \pm 0.01$ and $S_{1s-4p}/S_{1s-2p} = 0.11 \pm 0.01$. With our analysis, we obtain $S_{1s-np}/S_{1s-2p} = |\mu_{1s-np}/\mu_{1s-2p}|^2 \times (E_{1s-np}/E_{1s-2p})$ where $\mu_{1s-np} = e\langle u_{1s}|z|u_{np}\rangle$ and E_{1s-np} are the transition dipole moment and resonance energy of the $1s - np$ transition. From the observed value of S_{1s-np} / S_{1s-2p}, we obtain the best fit value of $1s$ Bohr radius of $a_B = 7.9$ Å, which gives a dipole moment of $\mu_{1s-2p} = 4.2$ eÅ. For comparison, the $1s$ Bohr radius extracted from the np ($n \geq 2$) exciton series $E_n = 17525 - 786/n^2$ cm^{-1}, $n = 2, 3, \ldots$ (Nikitine, 1969) yields a Bohr radius of $a_{np} = 11.1$ Å n^2.

Equation (10.4) implies that we can determine the value of the transition dipole moment from the absorption strength if we know the density of excitons. To cross check the validity of our procedure to determine the dipole moment and Bohr radius based on the ratio S_{1s-np}/S_{1s-2p}, we also measured the induced Lyman absorption under one-photon excitation at 600 nm wavelength corresponding to the phonon side band of the $1s$ orthoexciton, where we can expect almost 100% quantum efficiency for the photon-exciton conversion. With an absorbed photon density of 2×10^{15} photons/cm^3, we measured an induced absorption signal at 50 ps delay of S ~ 53 cm^{-1}meV in the spectral region from 112 meV to 137 meV, which includes the contributions from the $1s - np$ orthoexciton and the $1s - 2p$ paraexcitons. At such short delay time, we can neglect the decay of the orthoexcitons. We obtain $\mu_{1s-2p} = 4.8$ eÅ. The value is slightly overestimated since the contribution of the $1s - 3p$ transition is neglected. Thus the obtained value is consistent with the estimation from the ratio S_{1s-np}/S_{1s-2p}.

Within the hydrogenic model, the main origin for the deviation of the $1s$ exciton level from the Rydberg series is the dynamical effect on the electron–hole screening. Here, we introduce an ac dielectric constant at the frequency close to the exciton binding energy. The measured dielectric constant, obtained from the interference fringe pattern in the absorption spectrum in the range from 100 to 150 meV, is $\varepsilon_b \sim 6$. The effective Bohr radius is derived from the binding energy

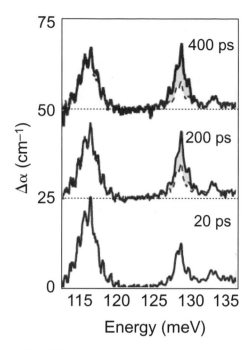

FIG. 10.16. Time-resolved Lyman absorption spectra recorded at different delays from the pump pulse, which is tuned at 1220 nm. The sample, held
at 1.8 K, is a single Cu$_2$O crystal with a platelet shape and a thickness of
220 μm along the (100) plane. Hatched areas show the paraexciton components (Tayagaki *et al.*, 2006). Adapted figure with permission from [T. Tayagaki,
A. Mysyrowicz, and M. Kuwata-Gonokami, *Phys. Rev.* B **74**, 245127 (2006)], copyright
(2006) by the American Physical Society.

E_{bin} and dielectric constant ε with $a_{\mathrm{B}} = e^2/(2E_{\mathrm{bin}}\varepsilon)$. $E_{\mathrm{bin}} \sim 150.8$ meV and
$\varepsilon \sim 5.9$ in turn gives a value $a_{\mathrm{B}} \sim 7.9$ Å, which agrees with our estimate. This
indicates that the hydrogen-like model with only the correction due to screening
by the ac dielectric constant works well for the $1s$ excitons in Cu$_2$O.

10.3.5   *Exciton cold collision and ortho- to paraexciton conversion*

We now discuss the temporal evolution of induced absorption spectra under resonant two-photon excitation of orthoexcitons as shown in Fig. 10.16 (Tayagaki *et
al.*, 2006). From the Lyman series spectra we can extract the density of paraexcitons and orthoexcitons. Dashed areas show the contribution from paraexcitons.
In our previous paper, we did not take into account the overlap between the
$1s - 3p$ orthoexciton and $1s - 2p$ paraexciton signals (Kubouchi *et al.*, 2005)
so that we overestimated the paraexciton density. It is clear, however, that the
paraexciton signal grows within 400 ps which is much faster than the ortho–
para conversion via the participation of a transverse acoustic phonon (Jang *et*

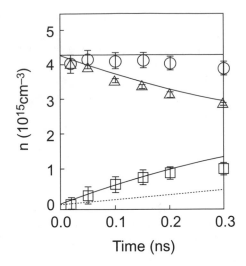

FIG. 10.17. Temporal evolution of ortho- (triangles), paraexcitons (squares) and total density of excitons (circles) extracted from the time-resolved Lyman spectra shown in Fig. 10.16.

al., 2004). This process takes about 3 ns to be completed. Figure 10.17 shows the temporal evolution of ortho- and paraexcitons and their sum. The dashed curve shows that the paraexciton grows as expected from the phonon mechanism. The total number of excitons is almost conserved. The density-dependent fast ortho–para conversion was clearly observed. Kavoulakis and Mysyrowicz have proposed a fast collision-induced conversion effective at high densities (Kavoulakis and Mysyrowicz, 2000). It corresponds to an electron spin exchange between two orthoexcitons in a relative singlet spin configuration, resulting in their conversion into two paraexcitons. This mechanism scales as,

$$\frac{dn_o}{dt} = -Cn_o^2, \tag{10.5}$$

where n_o is the orthoexciton density and C is the constant associated with the collision cross-section between two orthoexcitons. The solid curves in Fig. 10.17 shows the calculation with $C = 2.6 \times 10^{-16}$ cm^3/ns. This clearly indicates that a collision-induced ortho–para conversion process dominates for the cold excitons at a density above 10^{15} cm^{-3}. Metastable biexcitons could resonantly enhance the scattering similar to the Feshbach resonance of cold atoms. For a detailed comparison of the collision cross-section between theory and experiment, further experiments and an improved theoretical treatment which takes into account the low-temperature state of the orthoexcitons are necessary.

In this section, we reviewed recent developments on a method developed by our group to detect excitons in Cu$_2$O with mid-IR time-resolved pump–probe measurement of $1s - np$ internal transitions. Results clearly indicate that this method is a powerful tool to investigate the space-time evolution and energy distribution of ortho- and paraexcitons. We observe the Lyman series up to the 5p term of the yellow series exciton in Cu_2O under resonant two-photon excitation of orthoexcitons with short pulses in a very narrow spread in momentum space. From a systematic analysis of the Lyman series absorption strength, we obtain a Bohr radius of the 1s exciton of $a_B = 7.9$ Å and a $1s - 2p$ transition dipole moment of $\mu_{1s-2p} = 4.2$ eÅ.

We observed that $1s$ paraexcitons are efficiently created within 200 ps via the cold collision of orthoexcitons which overcomes the Auger decay process. The observed enhanced production of spin forbidden paraexcitons from cold orthoexcitons provides a unique opportunity to reach the BEC state with excitons.

With this method, we could clarify some of the basic issues crucial for the eventual observation of exciton BEC, such as quantitative estimations of the Auger coefficient and of the ortho- to para- transformation rate, and spatial diffusion effects. Further refinements of the method are required for the observation of exciton BEC, namely a quantitative estimation of the $1s - 2p$ dipole moment.

10.4 Conclusion

In this article, we summarize our recent experimental results on the study of high-density excitation in semiconductors by time-resolved probing of the intraband motion of electron–hole ensembles, including the free carrier metallic optical response and internal transitions of excitons. We first discussed experiments on the generation of high-density and low-temperature states of electron–hole ensembles by strong pulsed resonant excitation of excitons in CuCl. A rapid build-up of strong metallic reflection in the mid-infrared region was observed. Transient reflection spectra clearly reveal the creation of high-density electron–hole liquid via the exciton Mott transition. Analysis of the reflection spectra reveals that the cold liquid phase has voids of exciton gases and is transformed into a metallic colloid-like state within 10 picoseconds.

Second, we showed results on the mid-infrared spectroscopy of internal exciton transitions in Cu_2O. Supercooled orthoexcitons created via resonant two-photon excitation with femtosecond pulses, reveal the fine structure of the excitonic Lyman series transitions of orthoexitons. At a high-density regime, we observe a very rapid ortho–para conversion due to the collision of cold orthoexcitons. These experiments reveal that with a well-designed optical excitation and detection scheme, opportunities exist to explore new aspects of high-density and low-temperature states of electron–hole ensembles in a quantum degenerate regime.

Acknowledgements

The results reviewed in this article have been achieved by a fruitful collaboration with students and colleagues at the University of Tokyo. The author would like to acknowledge valuable contributions by R. Shimano, M. Nagai, T. Tayagaki, M. Kubouchi and K. Yoshioka. This work has benefited crucially from long-term collaborations with A. Mysyrowicz and Y. P. Svirko. Fruitful discussions with J. B. Héroux and N. Naka are appreciated.

We acknowledge important support through the ERATO cooperative excitation project (1997–2002) and SORST program from the Japan Science and Technology organization and partial support from KAKENHI (S) and JSPS.

References

Anderson, M. H., Ensher, J. R., Mattews, M. R., Wieman, C. E. and Cornell, E. A. (1995). *Science* **269**, 198.

Artoni, M., La Rocca, G. C., Carusotto, I. and Bassani, F. (2002). *Phys. Rev.* B **65**, 235422.

Bányai, L. and Koch, S. W. (1986). *Z. Physik*. B **63**, 283.

Beni, G. and Rice, T. M. (1978). *Phys. Rev.* B **18**, 768.

Bohnert, K., Anselment, M., Kobbe, G., Klingshirn, C., Haug, H., Koch, S. W., Schmitt-Rink, S. and Abraham, F. F. (1981). *Z. Physik* B **42**, 1.

Bruggeman, D. A. G. (1934). *Ann. Phys. (Leipzig)* **24**, 636.

Butov, L. V., Gossard, A. C. and Chemla, D. S. (2002). *Nature* **418**, 751.

Caswell, N., Weiner, J. S. and Yu, P. Y. (1981). *Solid State Commun.* **40**, 843.

Comte, C. and Noziéres, P. (1982). *J. de Physique* **43**, 1069.

Dasbach, G., Fröhlich, D., Stolz, H., Klieber, R., Suter, D. and Bayer, M. (2003). *Phys. Rev. Lett.* **91**, 107401.

Dasbach, G., Fröhlich, D., Klieber, R., Suter, D., Bayer, M., and Stolz, H. (2004). *Phys. Rev.* B **70**, 045206.

Davis, K. B., Mewes, M.-O., Andrews, M. R., van Druten, N. J., Durfee, D. S., Kurn, D. M. and Ketterle, W. (1995). *Phys. Rev. Lett.* **75**, 3969.

Fehrenbach, G. W., Schiffer, W., Treusch, J. and Ulbrich, R. G. (1982). *Phys. Rev. Lett.* **49**, 1281.

Fortin, E., Fafard, S. and Mysyrowicz, A. (1993). *Phys. Rev. Lett.* **70**, 3951.

Fröhlich, D., Kenklies, R. and Uihlein, Ch. (1979). *Phys. Rev. Lett.* **43**, 1260.

Goppert, M., Becker, R., Maier, C., Jörger, M., Jolk, A. and Klingshirn (2001). *Int. J. Modern Physics* B **15**, 3615.

Gross, E. F. and Karryev, N. A. (1952). *Dokl. Akad. Nauk SSSR* **84**, 261 [Sov. Phys. Dokl. **84**, 471].

Gross, E. F. (1956). *Nuovo Cimento Suppl.* **3**, 672.

Haken, H. (1958). *Förtschr. Phys.* **38**, 271.

Hanamura, E. (1976). *J. Luminescence* **12-13**, 119.

Hanamura, E. and Haug, H. (1977). *Phys. Rep.* **33**, 209.

Haug, H. and Abraham, F. F. (1981). *Phys. Rev.* B **23**, 2960.

Hayashi, M. and Katsuki, K. (1952). *J. Phys. Soc. Jpn.* **7**, 599.

Huber, R., Trauser, F., Brodschelm, A., Bichler, M., Abstreiter, G. and Leitenstorfer, A. (2001). *Nature* **414**, 286.

Hulin, D., Mysyrowicz, A. Migus, A. and Antonetti, A. (1985). *J. Luminescence* **30**, 290.

Hodby, J. W., Jenkins, T. E., Schwab, C., Tamura, V. and Trivich, D. (1976). *J. Phys.* C **9**, 1429.

Jang, J. I., O'Hara, K. E. and Wolfe, J. P. (2004). *Phys. Rev.* B **70**, 195205.

Jang, J. J. and Wolfe, J. P. (2006). *Solid State Commun.* **137**, 91.

Jeffries, C. D. and Keldysh, L. V. (Eds.) (1983). *Electron–Hole Droplets in Semiconductors*, North Holland, Amsterdam.

Johnsen, K. and Kavoulakis, G. M. (2001). *Phys. Rev. Lett.* **86**, 858.

Jörger, M., Fleck, T. and Klingshirn, C. (2005). *Phys. Rev.* B **65**, 245209.

Kaindle, R. A., Carnahan, M. A., Hägele, D., Lövenich, R. and Chemla, D. S. (2003). *Nature* **423**, 734.

Karpinska, K., Van Loosdrecht, P. H. M., Handayani, I. P. and Revcolevschi, A. (2005). *J. Luminescence* **112**, 17.

Kavoulakis, G. M., Chang, Y-D. and Baym, G. (1997). *Phys. Rev.* B **55**, 7593.

Kavoulakis, G. M. and Mysyrowicz, A. (2000). *Phys. Rev.* B **61**, 16619.

Keldysh, L. V. and Kozlov, A. N. (1968). *Zh. Eksp. Teor. Fiz.* **54**, 978. [*Sov. Phys. JETP* **27**, 521].

Kira, M., Jahnke, F. and Koch, S. W. (1998). *Phys. Rev. Lett.* **81**, 3263.

Klingshirn, C. and Haug, H. (1981). *Phys. Rep.* **70**, 315.

Kubouchi, M., Yoshioka, K., Shimano, R., Mysyrowicz, A. and Kuwata-Gonokami, M. (2005). *Phys. Rev. Lett.* **94**, 016403.

Kuwata-Gonokami, M., Shimano, R. and Mysyrowicz, A. (2002). *J. Phys. Soc. Jpn.* **71**, 1257.

Kuwata-Gonokami, M., Kubouchi, M., Shimano, R. and Mysyrowicz, A. (2004). *J. Phys. Soc. Jpn.* **73**, 1065.

Kuwata-Gonokami, M. (2005). *Solid State Commun.* **134**, 127.

Leheny, R. F. and Shah, J. (1976). *Phys. Rev. Lett.* **37**, 871.

Lei, J., Ming-Wei, W., Nagai, M. and Kuwata-Gonokami, M. (2003). *Chin. Phys. Lett.* **20**, 1833.

Matsumoto, H., Saito, K., Hasuo, M., Kono, S. and Nagasawa, N. (1996). *Solid State Commun.* **97**, 125.

Moskalenko, S. A. and Snoke, D. W. (2000). *Bose–Einstein Condensation of Excitons and Biexcitons and Coherent Nonlinear Optics with Excitons*, Cambridge University Press, Cambridge.

Mysyrowicz, A., Hulin, D. and Benoit a la Guillaume, C. (1979). *Phys. Rev. Lett.* **43**, 1123.

Mysyrowicz, A. (1980). *J. de Physique* **41**, 281.

Nagai, M., Shimano, R. and Kuwata-Gonokami, M. (2001). *Phys. Rev. Lett.* **86**, 5795.

Nagai, M. and Kuwata-Gonokami, M. (2002a). *J. Luminescence* **100**, 233.

Nagai, M. and Kuwata-Gonokami, M. (2002b). *J. Phys. Soc. Jpn.* **71**, 2276.

Nikitine, S. (1959). *Phil. Mag.* **4**, 1.

Nikitine, S. (1969). *Optical Properties of Solids* (ed by Nudelman, S. and Mitra, S. S.), Plenum, New York.

Nikitine, S. (1984). *J. Phys. Chem. Solids.* **45**, 955.

Noziéres, P. and Schmitt-Rink, S. J. (1984). *Low Temp. Phys.* **59**, 195.

O'Hara, K. E. and Wolfe, J. P. (2000). *Phys. Rev.* B **92**, 12909.

Randeria, M. (1995). *Bose–Einstein Condensation* (ed by Griffin, A., Snoke, D. W. and Stringari, S.), Cambridge University Press, Cambridge, p. 355.

Regal, C. A., Greiner, M. and Jin, D. S. (2004). *Phys. Rev. Lett.* **92**, 040403.

Schmitt, O., Tran Thoai, D. B., Bànyai, L., Gartner, P. and Haug, H. (2001). *Phys. Rev. Lett.* **68**, 2839.

Shindo, K., Goto, T. and Anzai, T. (1974). *J. Phys. Soc. Jpn.* **36**, 753.

Shionoya, S. (1979). *J. Luminescence* **18-19**, 917.

Snoke, D. W., Wolfe, J. P. and Mysyrowicz, A. (1990). *Phys. Rev.* B **41**, 11171.

Sugakov, V. I. (2002), *private communication.*

Tanaka, S., Kobayashi, H., Saito, H., and Shionoya, S. (1980). *J. Phys. Soc. Jpn.* **49**, 1051.

Tayagaki, T., Mysyrowicz, A. and Kuwata-Gonokami, M. (2005). *J. Phys. Soc. Jpn.* **74**, 1423.

Tayagaki, T., Mysyrowicz, A. and Kuwata-Gonokami, M. (2006). *Phys. Rev.* B **74**, 245127.

Tikhodeev, S. G. (1985). *Usp. Fiz. Nauk*, **145**, 3 [*Sov. Phys. Usp.* **28**, 1].

Toyozawa, Y. (1964). *J. Phys. Chem. Solids* **25**, 59.

Uihlein, Ch., Fröhlich, D. and Kenklies, R. (1981). *Phys. Rev.* B **23**, 2731.

Vinet, J. Y., Combescot, M. and Tanguy, C. (1984). *Solid State Commun.* **51**, 171.

Yoshida, H., Saito, H., Sionoya, S. and Timofeev, V. B. (1980). *Solid State Commun.* **33**, 161.

11

EXCITON COHERENCE

P. B. Littlewood
Cavendish Laboratory, University of Cambridge, JJ Thomson Ave., Cambridge
CB3 OHE, UK

Abstract
In a multiband metal, the Coulomb interaction between carriers in different bands can under some circumstances be strong enough to bind electrons and holes together and destroy the metallic state. Exchange interactions between the excitons will favor coherence between the excitons, just as in Bose–Einstein condensation (BEC) of atomic quantum liquids. Excitonic BEC provides a parallel description to the excitonic instability of the metal in the same way that BEC of bound fermion pairs is the strong coupling limit of the Fermi-surface BCS instability of superconductivity. It turns out that many different physical systems – excitons, polaritons, quantum Hall bilayers, dimer spin systems, ultracold atomic Fermi gases – share the same underlying physics, and this article provides a brief review of the underlying ideas, the physical phenomena, and some recent results.

11.1 Introduction

In a celebrated paper Keldysh and Kopaev (1965) explored the conditions for spontaneous instability of a semimetal to binding of the electron–hole pairs by the Coulomb interaction. A little earlier, Moskalenko (1962) and Blatt *et al.* (1962) had raised the idea of the Bose condensation of excitons, but the paper of Keldysh and Kopaev (soon to be followed by the paper of Keldysh and Koslov (1968) containing a study of the collective excitations) established the full mean field theory of the phenomenon, over the complete range of interesting density. The state so formed at this instability has been variously called an *excitonic Bose–Einstein condensate*, an *excitonic insulator*, an *excitonic dielectric* and a *phase-coherent condensate*. One of the purposes of this short review will be to explain why such a state might have so many apparently incompatible labels, hinting at a character of an insulator that is nearly a superfluid. In this article, I will use the shorthand of excitonic insulator (EI), as being the most accurate and most comprehensive.

A brief cartoon of the instability is shown in Fig. 11.1. Consider a semimetal with a pocket of holes and a pocket of electrons, and the ability to tune the bandgap so that the pockets overlap in energy. For simplicity, we will assume that the two pockets are spherically symmetric, and set the chemical potential to have

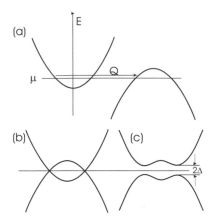

FIG. 11.1. Cartoon of the excitonic insulator instability. (a) shows a semimetal-
    lic state, with two pockets of electrons and holes. In (b) these bands have been
    shifted by the nesting wavevector Q, and in (c) the collective binding of pairs
    produces a self-consistent gap.

equal numbers of electrons and holes. Starting with an (indirect) gap, we tune
the bandgap to zero, and then overlapping. Suppose that the overlap in energy
is very small and the density of electrons and holes is tiny. We know then that
electrons and holes will then bind together in pairs with the attractive Coulomb
interaction to form a new particle that we call an exciton. In the simplest case to
analyze, the band masses are light and the wavefunction is essentially hydrogenic,
so the binding energy is the exciton Rydberg (because of a light mass and large
dielectric constant in many semiconductors, the Rydberg is often measured a few,
or few tens, of meV). Clearly, if the Rydberg is larger than the kinetic energy in
the Fermi pockets, binding is to be favored over free electron and hole gases. The
pair binding will produce a gap in the quasiparticle spectrum for electron–hole
pair excitations which is why we call this state an excitonic insulator. But there
is not necessarily an overall gap in the spectrum – if we can ignore the interaction
of the excitons with the underlying lattice the center of mass is free to move.
And further if the excitons are put into a collective coherent state, then there is
a long-range phase rigidity of the wavefunction that leads us to enquire into its
superfluid character.

    In the forty years since, there has been continuing experimental activity to
search for this state of matter, in various systems and in various guises. Depending on one's perspective about what the EI state is, it may be said to have been
discovered many times, or not at all. As I will explain below, the charge-density
wave state (CDW), first identified in transition metal dichalcogenides (Wilson *et
al.*, 1974), is a particular example of a generalized EI that is accompanied by a
density modulation. Recently two new systems have emerged as likely candidates
for EI: quantum Hall bilayers near Landau-level filling ν (Fertig, 1989; Eisen-

stein *et al.*, 1992; Spielman *et al.*, 2000; Kellogg *et al.*, 2004; Tutuc *et al.*, 2004) and "triplon BEC" in some insulating spin systems (Nikuni *et al.*, 2000; Ruegg *et al.*, 2003; Jaime *et al.*, 2004). The relation between these systems and the EI concept will be discussed later. Some early work focussed on semimetals such as Bi (where the instability does not in fact occur) and this phase of work is reviewed in Halperin and Rice (1968a, 1968b).

There has been much effort on nonequilibrium excitonic systems: optically excited excitations in semiconductors, with work particularly on Ge, CuCl, Cu_2O, and more lately III-V and II-VI semiconductor systems, particularly in reduced dimensionality $Ga_{1-x}Al_xAs$. The work on Ge is notable for having led to an unexpected discovery of the beautiful phenomenon of electron–hole droplets (Hensel *et al.*, 1977; Rice, 1977) (but no EI); the work on CuCl (Kuwata-Gonokami *et al.*, 2002) demonstrated a coherent EI induced by pumping (though not spontaneous instability); cuprous oxide has long been of interest (Wolfe *et al.*, 1994) because of the long-lived tightly bound excitons and high-quality crystals, and despite some recent difficulties associated with strong Auger processes (O'Hara *et al.*, 1999; Snoke and Negoita, 2000), it remains attractive to study; there is progress in GaAs quantum wells, particularly coupled quantum wells where the electrons and holes are physically separated, a system first proposed and theoretically studied by Lozovik and Yudson (1975), Shevchenko (1976), and Zhu *et al.* (1995). Coupled quantum wells have some particular attractions because the excitons can be made to have very long lifetimes, the interaction between the excitons is repulsive, and the coupling of a two-dimensional exciton system to a three-dimensional phonon bath allows more efficient cooling. A number of recent experiments have given evidence of degenerate statistics (Butov *et al.*, 2001), the preparation of cold excitons by electron–hole recombination far from the source (Butov *et al.*, 2003; Snoke *et al.*, 2003), and trapping by stress (Snoke *et al.*, 2004) or by hot-exciton pressure from optical pumping (Hammack *et al.*, 2006).

With the development of high-quality semiconductor microcavities, a recent advance is the study of exciton-polariton systems where the elementary excitations are a coherent mixture of exciton and photon (Dang *et al.*, 1998; Baumberg *et al.*, 2000; Savvidis *et al.*, 2000; Deng *et al.*, 2002, 2003, 2006; Weihs *et al.*, 2003; Kasprzak *et al.*, 2006). What the optical work brings is the nonequilibrium aspect. The difficulty of reaching thermal equilibrium on the time-scales available (determined by the exciton lifetime) provides the principal challenge to experiment. While now a challenge, in the longer term, the ability to have nonequilibrium control will surely produce a level of richness and new phenomena that are not yet appreciated. As a general matter, the history of work on excitonic systems is that new and interesting phenomena continue to be revealed as part of the search for the EI state.

Another important impact of the work has been on what is called the BCS to BEC crossover. The Keldysh–BCS wavefunction for fermion pairing is a variational one. In the high-density limit, it describes a weak Fermi-surface instability akin to a weak-coupling superconductor. In the dilute limit, a wavefunction of the same form describes instead a phase-coherent superposition of tightly bound fermion pairs – excitons. Following the investigation of this crossover in the exciton problem by Comte and Nozières (1982) and Nozières and Comte (1982), the understanding was adapted to the superconducting problem (Leggett, 1980; Nozières and Schmitt-Rink, 1985; Randeria, 1994) to describe the crossover from weak to strong coupling. Tuning through the crossover has recently become possible in ultracold Fermi gas systems, with the tuning parameter being provided by the proximity to a Feshbach resonance (Regal *et al.*, 2004; Zweirlein *et al.*, 2004). We will not comment extensively on the ultracold gas phenomena here – except to mention that the use of multicomponent systems and optical lattices allows a realization of many models of interest in the condensed matter arena. A specific proposal for an excitonic insulator state in an optical lattice has recently been made (Ho, 2006) and we may expect much more news from this arena.

It is not possible in this short review to give anything approaching a complete account of the experimental systems and the experimental work, but a few topics will be discussed at greater length below. There are a number of books (Griffin *et al.*, 1994; Moskalenko and Snoke, 2000; Kavokin and Malpuech, 2003) and reviews (Halperin and Rice, 1968a; Halperin and Rice, 1968b; Hanamura and Haug, 1977; Littlewood and Zhu, 1996; Littlewood *et al.*, 2004) of the fundamental material to which the interested reader is referred for details. The aim of this short article is to give an overview of the different physical systems that are linked together by the basic idea of exciton condensation, and in particular to explicate the mathematical connection between the various topics that is embodied in the mean-field wavefunction written down by Keldysh and Kopaev. Within these topics, the aim is then to provide a link to that literature, including the many extensions to the basic mean field ideas, and also the extra physics that comes with the different physical systems.

11.2 Excitonic instability of a semiconductor: Elementary concepts

Now we return to put some flesh on the simple model alluded to in Fig. 11.1. The Hamiltonian consists of the single-particle band structure $\epsilon_{ck} = E_{\text{gap}} + \epsilon_k$; $\epsilon_{vk} = -\epsilon_k$, and the Coulomb interactions between excited particles (electrons and holes). Written in second-quantized notation, and shifting the valence band momenta by Q for simplicity

$$H = H_\text{o} + H_{\text{Coul}}, \tag{11.1}$$

where

$$H_\text{o} = \sum_k \left[\epsilon_{ck} a_{c,k}^\dagger a_{c,k} + \epsilon_{vk} a_{v,k}^\dagger a_{v,k} \right], \tag{11.2}$$

and

$$H_{\text{Coul}} = \frac{1}{2} \sum_q \left[V_q^{ee} \rho_q^e \rho_{-q}^e + V_q^{hh} \rho_q^h \rho_{-q}^h - 2 V_q^{eh} \rho_q^e \rho_{-q}^h \right]. \tag{11.3}$$

A complete analysis of the mean field theory can be found in many places, e. g., (Comte and Nozières, 1982; Littlewood and Zhu, 1996), and here we stress the major results.

The single-exciton wavefunction can be written in a form

$$\Phi^\dagger(q) = \sum_k \phi_q(k) a^\dagger_{c,k+q/2} a_{v,k-q/2} |0\rangle . \tag{11.4}$$

Here $a^\dagger_{c,k}$ and $a^\dagger_{v,k}$ are creation operators for electrons in the conduction and valence bands; q is the center of mass wavevector; $|0\rangle$ the state with a filled valence band and empty conduction band; and $\phi(k)$ the relative wavefunction of a pair. For a hydrogenic Wannier exciton, $\phi(k)$ is just the Fourier transform of the hydrogenic wavefunction that falls off with the exciton Bohr radius, and it also contains the Bloch parts of the wavefunction in a periodic solid. In the general case where the conduction and valence band minima are offset (say by wavevector Q), the lowest energy pair will be near $q \approx Q$ (or more generally, a linear combination of all equivalent pockets). The natural units are the exciton Rydberg, $\mathrm{Ry}^* = (\mu e^4)/(2\epsilon^2 \hbar^2) = [\mu/(m\epsilon^2)]\mathrm{Ry}$, and the exciton Bohr radius, $a^* = (\epsilon \hbar^2)/(\mu e^2) = \epsilon(m/\mu)a_o$. It is conventional to measure the exciton density n in terms of the average spacing r_s between excitons, measured in units of the Bohr radius: $(4\pi/3)(r_s a^*)^3 = 1/n$, with obvious generalization to physical dimensions other than 3.

11.2.1 *Keldysh mean field theory*

It is intuitively clear that if the binding energy is larger than the single-particle gap, the binding of an exciton is more favorable than the empty vacuum. Although the ground state now must be thought of as containing many exciton pairs (in a way to be specified later), the state is still an insulator for quasiparticle excitations. (The lowest excitation energy is changed from E_{gap} to $E_{\mathrm{gap}} + \mathrm{Ry}^*$.) Suppose however, that the gap is negative (so that in the absence of Coulomb, we have two carrier pockets). Now, the excitonic bound state will turn out to produce a gap at the Fermi level, so the excitation spectrum is changed from that of a metal to an insulator. Of course, in order to see that, one has to write down a many-exciton wavefunction, and here a natural candidate is the *coherent state* (Keldysh and Kopaev, 1965), inspired by the BCS wavefunction for superconductivity

$$\Psi = A e^{\Phi^\dagger(Q)} |0\rangle \tag{11.5}$$

$$= A \exp\left(\sum_k \phi_{Q(k)} a^\dagger_{c,k+Q/2} a_{v,k-Q/2}\right) |0\rangle$$

$$= A \prod_k \exp\left(\phi_{Q(k)} a^\dagger_{c,k+Q/2} a_{v,k-Q/2}\right) |0\rangle$$

$$= \prod_k \left[u(k) + v(k) a^\dagger_{c,k+Q/2} a_{v,k-Q/2} \right] |0\rangle .$$

Here A is a normalization, satisfied in the last equation by the requirement $|u|^2 + |v|^2 = 1$, and $v(k)/u(k) = \phi_Q(k)$. The transformation from the exponentiated form of a coherent state for bosons to the product form for fermions is produced by a Taylor series expansion of the exponential, and the realization that all terms higher than quadratic disappear due to Pauli exclusion.

Notice that while equation (11.4) is a formally exact solution of a two-body problem, the coherent state of equation (11.5) is approximate, and certainly *not* a good approximation (with the hydrogenic $\phi(k)$) to the dense system. If we generalize $\phi(k)$ to be a *variational* function, then we expect to do better. The bosonic form will remain at low densities, (so in the fermionic representation $u \sim 1$, and $v \propto \phi(k)$) and at high densities $r_s \ll 1$, the form of the wavefunction will evolve to the Fermi function: $v(k) = f_k = 1/[e^{\beta(\epsilon_k - \mu)} + 1]$; $u(k) = 1 - f_k$ (see Fig. 11.2). It turns out that the coherent state wavefunction is in fact a solution of a BCS-like mean field theory, expected to be a good approximation in weak coupling, or when the range of the interaction is much longer than the interparticle spacing.

The Keldysh wavefunction is in fact a very good approximation to the ground state over the whole range of density, describing a weak "Fermi surface" instability in the high-density limit (where there are no real bound pairs) and condensation in a dilute Bose gas in the low-density limit, where the excitons retain their integrity. Of course in the dilute limit, the mean field theory is not a good theory of the transition *at finite temperature* because it does not include the fluctuations of the order parameter phase. It is straightforward to incorporate such fluctuations to derive the conventional theory of BEC in the ultradilute limit, and we will not go into it here because it has been much discussed elsewhere (Keldysh and Koslov, 1968; Nozières and Schmitt-Rink, 1985; Côté and Griffin, 1988; Bauer, 1990; Littlewood and Zhu, 1996; Littlewood *et al.*, 2004).

The wavefunction has a broken symmetry, with an order parameter

$$\rho_{cv}(Q) = \sum_k \left\langle a^\dagger_{c,k+Q/2} a_{v,k-Q/2} \right\rangle \quad . \tag{11.6}$$

$u(k)$ and $v(k)$ are in general complex, but when they all have the *same* phase, this order parameter is nonzero. However, it is easy to see that the *overall* phase of the order parameter equation (11.6) is undetermined, a fact that can be traced to a conservation law in equation (11.1) – the Hamiltonian separately conserves the numbers of electrons and holes. In that case, the densities

$$\rho_{cc}(Q) = \sum_k \left\langle a^\dagger_{c,k+Q/2} a_{c,k-Q/2} \right\rangle = n \quad , \tag{11.7}$$

$$\rho_{vv}(Q) = \sum_k \left\langle a^\dagger_{v,k+Q/2} a_{v,k-Q/2} \right\rangle = n \quad , \tag{11.8}$$

are unchanged.

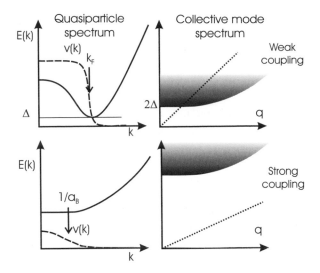

FIG. 11.2. Form of the wavefunction at low density (lower panels) and high density (upper panel), and the corresponding excitation spectrum. The left panels show the occupance $v(k)$ and the single-particle excitation spectrum $E(k)$; the right panels give the collective excitations, which consists of the Bogoliubov mode (dashed line) and the particle–hole pair excitations (gapped continuum). At high density ($r_s \ll 1$), the occupancy $v(k)$ approaches a Fermi function and the quasiparticle spectrum has a small gap confined to momenta very near the Fermi surface: this is the "BCS" limit. At lower density, the wavefunction is better described as a dilute gas of bosons (excitons) in a phase coherent state. Then $v(k)$ approximates the hydrogenic wavefunction, with a characteristic momentum scale of the inverse Bohr radius a_B^{-1}. Except for the Bogoliubov mode, the excitation spectrum is then just that of ionization of the particle–hole pair. (Note that in this approximation the excited bound states are neglected.)

11.2.2 *Pseudospins and two-level system analogies*

It is often useful to make a pseudospin analogy to the fermion problem, one that is also quite familiar from the semiconductor Bloch equations. For specificity, consider an exciton in a small quantum dot, so that the relevant states are just no excitation (full valence band and empty conduction band) and a single electron hole pair on any dot. If we assume strong repulsion of excitons that strongly overlap (mostly from Pauli exclusion) and thereby in a lattice model we disallow higher occupation, then we can represent the spectrum as a two level system, where the fermion operators $a_j^\dagger$, $(b_j^\dagger)$ create electrons in the lower (upper) levels on the jth dot. Conservation requires $b_j^\dagger b_j + a_j^\dagger a_j = 1$, and because of this constraint, we can represent the two level system by an SU(2) pseudospin S. The average occupancy is

$$n = 1/N \sum_{j=1}^{N} \left[b_j^\dagger b_j - a_j^\dagger a_j \right] = 1/N \sum_j \left[S_j^z + 1/2 \right] \qquad (11.9)$$

and the operators to create (destroy) excitons are the raising (lowering) operators

$$S_j^+ = b_j^\dagger a_j \ , \qquad (11.10)$$

$$S_j^- = a_j^\dagger b_j \ . \qquad (11.11)$$

The coherent state wavefunction of equation (11.5) is then written as

$$\Psi = \exp\left(\sum_j w_j S_j^+ \right) |0\rangle \ . \qquad (11.12)$$

When the w_j have all the same phase, this describes a quantum ferromagnet in the pseudospin variables. Here the $|0\rangle$ state is of course a set of completely filled levels of a and empty of b, i.e.,

$$|0\rangle = \prod_i a_i^\dagger |\text{vac}\rangle \ , \qquad (11.13)$$

so the coherent magnetic state is also

$$\Psi = \prod_j \left[a_j^\dagger + w_j b_j^\dagger \right] |\text{vac}\rangle \ . \qquad (11.14)$$

11.3   Some other physical systems

Not surprisingly, with this generality there are a number of other systems which fit the spin analogy, and especially here we will refer to some quantum spin systems, and the quantum Hall effects.

11.3.1   *BEC of quantum magnets*

The compound $BaCuSi_2O_6$ is an insulating blue pigment with a singlet ground state. The singlet arises because the crystal structure contains dimers of Cu^{2+} ions (that are $3d^9$ configuration and therefore spin-1/2) which are (antiferromagnetically) coupled. The lowest-lying triplet state is separated by a gap from the ground state, and the magnetic susceptibility is exponentially small at low temperatures because of the gap. In a magnetic field, the triplet splits, and the lowest triplet state shifts down in energy approximately linearly with magnetic field. If the dimers are very weakly coupled (quite a good approximation), at a field H_c the energy level will cross below the singlet, so that (at $T = 0$) the ground state will be the lowest spin-polarized triplet state ($m^z = 1$, using the notation m for spin to avoid confusion with the pseudospin S above). At this point the magnetization will jump.

Of course in the presence of interactions, the magnetization will not jump discontinuously to the full saturated moment. Instead, one will find a state at intermediate fields that is a linear combination of the two levels: singlet and triplet ($m^z=1$). Because of intersite interactions the triplet excitations ("tripletons") will disperse, and typically will have a minimum in the excitation spectrum at $q = 0$, dispersing up quadratically. Once dispersion and interactions are present, this mixed combination will acquire some broken symmetry as a result of coupling between sites, perhaps either a (spatially) periodic structure where the mixing varies from site to site, or possibly a spatially homogeneous but phase coherent mixture of these two. There are now several systems where the proponderance of the evidence is for phase coherence (Nikuni *et al.*, 2000; Ruegg *et al.*, 2003; Jaime *et al.*, 2004). Interestingly, there are some closely related compounds that apparently have a ground state with periodic order (Kageyama *et al.*, 1999), which makes the important point that all the phase coherent condensed states are actually in competition with solid phases.

The phase-coherent ground state can be written in the notation of equation (11.10) where the physical singlet state ($m_j = 0$) is equivalent to $S_j = |\downarrow\rangle$, and the lowest of the triplets $m_j^z = 1$ is $S_j = |\uparrow\rangle$. The solution of a realistic model for this system yields a coherent ground state of the form of equation (11.12), and the calculated physical properties are in very good agreement with the experimental data (Jaime *et al.*, 2004). In a different language, the phase-coherent ground state can instead be regarded as the BEC of "tripletons", with the transition occurring when the bottom of the tripleton band reaches zero energy, and acquires a macroscopic population.

11.3.2 *Quantum Hall bilayers*

The quantum Hall effect arises for two-dimensional electronic systems because of the quantizing effects of a large magnetic field. In a large field, the density of states is split into Landau levels, each of which contains a density of states $n_o = 1/2\pi l^2$ per unit area, where $l = (\hbar c/eB)^{1/2}$ is the magnetic length. Thus with an areal density of electrons n, the system is said to have a Landau level filling of $\nu = n/n_o$. The quantum Hall effect is associated with integer filling, when the Hall conductance is quantized in units of e^2/h and the diagonal conductance vanishes.

Now consider a system of parallel bilayers, where the electrons have the possibility to tunnel between the layers. The effect of tunneling is to establish two mixed states of the "up" $|\uparrow\rangle$ and "down" $|\downarrow\rangle$ layers, $\psi_\uparrow \pm \psi_\downarrow$. If the tunneling were strong, then the splitting is large, and the lowest Landau level is then the composite layer: in particular at a filling of $\nu = 1/2$ *per layer* the lowest Landau level will be completed, and perhaps unsuprisingly the quantum Hall effect at integer filling (i.e., $1/2 + 1/2$) is observed (Eisenstein *et al.*, 1992). However, it is possible to manipulate the tunneling between the layers by the intervening barrier, and it was seen that *even if the tunneling was made very small, the $1/2 + 1/2$ quantum Hall effect could persist*, provided that the layers where still physically close enough that the Coulomb interaction between the layers remain large.

Some recent experiments have demonstrated the underlying origin of the phenomenon, by making separate electrical contact to the individual layers and applying electrical currents in the two layers in opposite directions. If the layers were uncoupled, one would then expect Hall voltages of the opposite sign, but in the regime where the composite $1/2 + 1/2$ Hall effect is seen these counterflow Hall experiments (Kellogg *et al.*, 2004; Tutuc *et al.*, 2004) yield *zero* Hall effect in each layer – suggesting that the particles carrying the current are neutral, so that the Lorentz force vanishes. The transport is in fact by interlayer excitons, and the zero Hall affect can be seen to be a consequence of their condensation into a phase coherent state (Fertig, 1989; Wen and Zee, 1992).

Excitons can be formed in this system despite that the carriers in both layers are electrons and all the interactions are *repulsive*. The important effect of the field is that within a single Landau level the kinetic energy of an electron is quenched. Thus an equally valid picture of a half-filled Landau level of electrons is to regard it as a filled Landau level populated by positively charged holes with a hole filling also of $1/2$. This particle–hole transformation costs no kinetic energy. The Coulomb binding of a hole in one layer, and an electron in the other makes an exciton, and the cooperative transport owes to the condensation of excitons into a collective state.

This sleight of hand is rather short on insight as to the physics behind the formation of an exciton condensate, and the parallel is clearest if we look at variational wavefunctions. As above, we will define a pseudopspin variable that is in this case a layer index: thus $c_{k\uparrow}^{\dagger}$ creates an electron in the state k of the upper top well. The Coulomb energy is minimized by equal occupations of the layers, and a coherent state that will do this is (Fertig, 1989)

$$|\Psi\rangle = \prod_{k}\left[c_{k\uparrow}^{\dagger} + e^{i\phi}c_{k\downarrow}^{\dagger}\right]|0\rangle . \tag{11.15}$$

Notice that because the filling is $\nu = 1$, the sum is over all states k in the Landau level. The phase coherence of the paired state is promoted by the exchange energy, just as in a conventional ferromagnet. In the absence of tunneling between the layers, the energy is independent of the phase ϕ. The parallel to equation (11.12) is immediate.

11.3.3 *Polaritons*

Returning to more conventional exciton systems, we now consider the effect of strong coupling to light, as obtained in an optical microcavity with a photon mode made nearly resonant with the exciton. Since the exciton has a much heavier mass than the photon, it is a good approximation to neglect the exciton kinetic energy and to replace it by a set of localized levels. But in order to keep the effects of a strong local repulsion between excitons (largely from Pauli exclusion) we then exclude double occupancy of the sites, which is formally accomplished by treating the excitons as two-level systems.

The model that results is the Dicke model of atomic physics (Dicke, 1954):

$$H_{2\text{level}} = \sum_q \omega(q)\psi_q^\dagger \psi_q + \sum_{j=1}^N \frac{\epsilon_j}{2}(b_j^\dagger b_j - a_j^\dagger a_j) + \frac{g}{\sqrt{N}}\sum_{jq}(b_j^\dagger a_j \psi_q + \psi_q^\dagger a_j^\dagger b_j). \quad (11.16)$$

$H_{2\text{level}}$ describes an ensemble of N two-level oscillators with an energy ϵ_j dipole coupled to one cavity mode. b and a are fermionic annihilation operators for an electron in an upper and lower states respectively (with a local constraint $b_j^\dagger b_j + a_j^\dagger a_j = 1$ so that there is an electron either in the lower level or in the upper level) and ψ is a photon bosonic annihilation operator. The operator that counts the number of excitations in the system, $N_{\text{ex}} = \sum_q \psi_q^\dagger \psi_q + \frac{1}{2}\sum_j(b_j^\dagger b_j - a_j^\dagger a_j)$, commutes with $H_{2\text{level}}$ so is conserved.

The mean field coherent state wavefunction is now

$$|\lambda, u, v\rangle = e^{\lambda \psi_0^\dagger} \prod_j (v_j b_j^\dagger + u_j a_j^\dagger)|\text{vac}\rangle, \quad (11.17)$$

with the (real) variational parameter λ and variational functions $v_j = v(\epsilon_j)$. (The vacuum state is here defined to be empty of both levels.) The constraint is satified by setting $u_j^2 + v_j^2 = 1$, and the variational functions are obtained by minimizing $H_{2\text{level}} - \mu N_{\text{ex}}$. For detailed results see (Eastham and Littlewood, 2000, 2001). This coherent state is now a superposition of excitons and photons, a generalization of the previous wavefunctions while retaining exactly the same form. As usual in the limit of low density the wavefunction is just a coherent superposition of weakly interaction bosons (in this case polaritons). But the wavefunction is very good at higher density because the wavelength of light is much larger than the typical separation between exciton states – mean field theory is of course promoted by long-range interaction.

The BCS-BEC crossover in this system is a little different to the exciton system because of the very light mass of the cavity photon (Keeling *et al.*, 2004, 2005). From the dispersion of the polaritons, effectively only modes of very long wavelength are coupled and the mass of the polariton (near $q = 0$) is very light – typically about $m^* \simeq 10^{-4}$–10^{-5} of the bare exciton mass. The light mass of course means that a dilute gas of polaritons will begin to establish quantum degeneracy at a temperature $T_0 \propto n/m^*$ (in two dimensions) that is substantial even at low densities, and polariton condensation has been studied starting from a dilute Bose gas picture (Kavokin and Malpuech, 2003; Sarchi and Savona, 2004). The dilute Bose gas approach will be less accurate once the polaritons have substantial overlap, or equivalently when $k_B T_0/g$ becomes not small. Beyond this density, the long-range interaction mediated by photons makes the BCS-like description a very accurate one, and equation (11.17) is rather good, whether or not one starts from a picture of localized excitons. But whereas in exciton systems the crossover occurs when excitons overlap (namely around $r_s \approx 1$), in polaritons the density of the crossover will be shifted down by many orders of magnitude because of the light polariton mass. In practice, we have argued elsewhere that it will be difficult to obtain polariton BEC in the density regime appropriate for a dilute gas treatment (Keeling *et al.*, 2004).

Experiments have shown continuous progress over several years. The first experimental indication that spontaneous quantum degeneracy could occur in the strong coupling regime in microcavities were obtained nearly a decade ago (Dang *et al.*, 1998; Savvidis *et al.*, 2000) and there have been a number of experiments demonstrating nonlinear gain (Deng *et al.*, 2002), and also second-order coherence (Deng *et al.*, 2002). However, in many of these experiments the distribution of polaritons is quite clearly nonequilibrium, sometimes deliberately so in the case of a driven cavity (Baumberg *et al.*, 2000). Only very recently has there been evidence for quasithermal equilibrium population of excitons in microcavities (Deng *et al.*, 2006; Kasprzak *et al.*, 2006), and only the latter experiment has also combined the results on nonlinear emission with a BEC-like population growth at low energies, first-order phase coherence, and also broken symmetry of the polarization.

11.4 Further symmetry breaking, superfluidity, and decoherence

In the last section of this article we will step back again to ask a few questions about the condensed state in principle. For the models we have used, our states are superfluid, because of the invariance of the energy with respect to the phase of the wavefunction. This of course implies that a small gradient of the phase can be applied at little cost in energy; in the usual way, a gradient in the macroscopic phase will give rise to a (super)current, and we have the recipe for a true superfluid.

Unfortunately, it turns out that in all the cases we have discussed above there are (possibly extremely small) terms in the Hamiltonian that have been neglected that will destroy this neat picture. The basic point was laid out long ago (Guseinov and Keldysh, 1972) and is simply that excitons are not conserved particles: there will always be processes that allow the electron and hole to recombine, and such a process will *pin* the phase of the order parameter to a special value.

11.4.1 *Lattice effects and relation to charge density waves*

The pinning of the phase is easiest to understand in the simple case of coupled quantum wells, and to separate out the effect of dynamical photon generation, imagine that we have electrons and holes either side of the barrier maintained at the same chemical potential. This in fact occurs at an InAs/GaSb interface (Lakrimi *et al.*, 1997) where the valence band on one side of the interface lies above the energy of the conduction band on the other, so there is a natural charge transfer to form a dipole layer. Tunneling through the barrier is surely allowed, and therefore the conduction and valence band states will be mixed by a term

$$H_{\text{tunnel}} = \sum_k t(k) a^\dagger_{c,k} a_{v,k} + h.c. \; . \tag{11.18}$$

Suppose this is small, so that it can be treated in first-order perturbation theory. Once there is an order parameter established, $\rho_{cv} = |\rho_{cv}| e^{i\phi}$ (equation 11.6), this term can be evaluated

$$\langle H_{\text{tunnel}} \rangle = t|\rho_{cv}| \cos \phi \; , \tag{11.19}$$

where the origin for the phase ϕ has been fixed by reference to the phase of t, by choosing the latter parameter to be real.

Clearly, this term breaks the rotational XY symmetry and will introduce a gap in the Bogoliubov mode, thereby destroying the superfluid response and turning the material instead into an insulator.

The existence of terms of this form is generic for excitons, and marks a distinction between an excitonic insulator and a superconductor. The analogous term in a superconductor would be

$$H_{\text{tunnel}} = \sum_k t(k) a^\dagger_{\uparrow k} a^\dagger_{\downarrow,k} + h.c. \; , \tag{11.20}$$

namely a term that destroys or creates *pairs* of electrons. Obviously this is disallowed inside a solid, but the analogous term does exist as a Josephson coupling between two bulk superconductors that locks the relative phases of superconductors *to each other*, viz $H_J = \Delta^\dagger_1 \Delta_2$. The overall phase of the two superconductors remains free.

Now let us go back and consider excitonic condensation between pockets shifted in momentum space by some wavevector Q, which is not a reciprocal lattice vector (including not zero). Then there are no single-particle terms like equation (11.18). The most general two-body term coupling the two bands is of the form

$$H_2 = \sum_{mnm'n'} \sum_{kk'q} V(nk; mk + q; n'k'; m'k' - q) a^\dagger_{nk} a_{mk+q} a^\dagger_{n'k'} a_{m'k'-q} \; , \tag{11.21}$$

where m, n, m', n' are band indices. The excitonic insulator pairing that we have treated before arises from the terms in equation (11.21) – direct Coulomb interaction between charge densities in different bands (say n, n'), namely

$$\sum_{nn'} \sum_{kk'q} V_{\text{Coul}}(q) a^\dagger_{nk} a_{nk+q} a^\dagger_{n'k'} a_{n'k'-q} \; , \tag{11.22}$$

which then factorizes in mean field to give an expectation value in the EI state proportional to $\rho^\dagger_{cv}(Q)\rho_{cv}(Q)$ which is manifestly invariant under change of the phase of the order parameter. However, in a general band-structure there will be nothing to prevent terms that mix the band indices in a less symmetric way, yielding eventually couplings between the diagonal and off-diagonal components of the density matrix, for example $\rho^\dagger_{cv}(Q)\rho_{cc}(Q)$. This means, not surprisingly, that an EI instability will drive also a modulation of the overall charge density at the same wavevector Q, namely it corresponds to a periodic *charge density wave*. Furthermore the phases of the CDW and EI are locked, and since the CDW phase is nothing more than the position of the density wave relative to the underlying lattice, we will eventually find Umklapp terms that lead to pinning of the CDW (if Q is a rational fraction of a reciprocal lattice vector), or if not, weak disorder will do the job.

This is a complicated way to produce a gap in the spectrum – although the physical origin is simple – and indeed the gaps may be small because the processes are high-order. The phenomena produced are just that of a pinned CDW, where indeed if the pinning is weak CDW will slide in an electric field large enough to overcome the disorder (Ong and Monceau, 1977), and the gaps are in the microwave region of the spectrum. But nevertheless, CDW's are insulators and not superconductors. It is not the purpose of this article to expatiate on the fascinating phenomena of CDW dynamics, so interested readers are referred to the book of Grüner (1994).

11.4.2 *Dynamics and decoherence*

The last section dealt with pinning effects where the electron and hole subsystems are in equilibrium at the same chemical potential. But in many of the systems we have been discussing, the electrons and holes are out of equilibrium and maintained in separate populations. To the extent that strict separation can be preserved, the individual conservation laws apply, but the moment we allow recombination, and escape of particles from the system, we now have an open system that will show decay – and therefore need to be pumped from the outside. This is quite obviously of great relevance to the problem of polariton condensation, because the lifetime of cavity polaritons is quite short, even in very high-quality microcavities.

There are many sources of decoherence: Because the mirrors are not perfect, light will leak out of the lasing mode and the excitation (in steady state) must be replaced by incoherent pumping of excitons; Excitons themselves may decay spontaneously into photon modes other than the cavity mode; Phonons and disorder inside the material can scatter the excitons, and produce pairbreaking and dephasing. All of these may be modeled by coupling of the internal degrees of freedom to (bosonic) baths of dynamic fluctuations $B_\gamma(r,t)$, which in the Dicke model will be of three generic types:

$$H_{\mathrm{SB}} = \sum_{j=1}^{N}(b_j^\dagger b_j - a_j^\dagger a_j)(B_{1j}^\dagger + B_{1j})$$

$$+ \sum_{j=1}^{N}(b_j^\dagger b_j + a_j^\dagger a_j)(B_{2j}^\dagger + B_{2j})$$

$$+ \sum_{jq}(b_j^\dagger a_j B_{3j}^\dagger + a_j^\dagger b_j B_{3j})\,. \qquad (11.23)$$

This is already a simplification in that we have kept just diagonal (on-site) terms. The three terms in equation (11.23) correspond to neutral, pairbreaking, and phase-breaking scattering respectively. Their treatment in a quasi-equilibrium situation is discussed in (Szymanska and Littlewood, 2002; Szymanska et al., 2003, 2006).

The first term in equation (11.23) represents dynamic or static fluctuations of the excitation energy ϵ_j. Provided these fluctuations are slow and weak enough, they are unimportant: the wavefunction is robust against static disorder in the energy levels, in the same way that a singlet superconductor is insensitive to weak charge disorder.

The second term is more dangerous, and if a static potential plays just the same role as magnetic impurities in a superconductor (Zittartz, 1967; Szymanska and Littlewood, 2002; Szymanska et al., 2003; Marchetti et al., 2004). This corresponds to scattering that breaks up the electron–hole pair (in order for it to be relevant, one must relax the two-level constraint). If the pairbreaking is weak, it is relatively harmless, but strong pairbreaking drives a crossover into the conventional regime of an electron–hole plasma laser. Here the electron–hole system becomes gapless, and the coherence is entirely supported by the photon field.

The last term in equation (11.23) does not exist in a superconductor where it would be forbidden by symmetry. This is an XY-like random-field term (coupling to S_x, S_y if we represent two-level systems as a spin model). It is sensitive to the phase of the local order parameter, and the difference from the tunnel interlayer coupling above is that this term is dynamical. We already know that at a classical level, a static random field term of this kind is relevant in dimensions less than 4 and will destroy long-range order of the phase (Imry and Ma, 1975). As a dynamical field, the effect of this term is that at the very longest length-scales the Bogoliubov mode becomes diffusive (Szymanska et al., 2006). In recent experiments that show the best evidence of polariton condensation (Kasprzak et al., 2006), narrowing of the polariton emission was seen at the onset of nonlinearity as predicted by the theoretical models, and it seems plausible that the remanent homogeneous linewidth represents exactly the effects this kind of decoherence.

11.5 Conclusion

This concludes our brief review of the excitonic insulator story. An idea, introduced some 40 years ago, has had enormous and unexpected applications across a wide swath of condensed matter physics – in charge-density waves, quantum Hall physics, and quantum magnets – and we now seem very close to the elusive identification of the original motivator of excitonic Bose–Einstein condensation.

Acknowledgements

All of the author's research in this area was performed in collaboration with others, especially Paul Eastham, Alex Ivanov, Jonathan Keeling, Francesca Marchetti,

Ben Simons, and Marzena Szymanska. Many illuminating discussions with Leonid Butov, Le Si Dang, David Snoke, and Roland Zimmermann are gratefully acknowledged. This work was supported under the EU programme "Photon-Mediated Phenomena in Semiconductor Nanostructures", and by the Engineering and Physical Sciences Research Council.

References

Bauer, G. E. W. (1990). *Phys. Rev. Lett.* **64**, 60.

Baumberg, J. J., Savvidis, P. G., Stevenson, R. M., Tartakovskii, A. I., Skolnick, M. S., Whittaker, D. M. and Roberts, J. S. (2000). *Phys. Rev.* B **62**, R16247.

Blatt, J., Brandt, W. and Boer, K. (1962). *Phys. Rev.* **126**, 1691.

Butov, L. V., Ivanov, A. L., Imamoglu, A., Littlewood, P. B., Shashkin, A. A., Dolgopolov, V. T., Campman, K. L. and Gossard, A. C. (2001). *Phys. Rev. Lett.* **86**, 5608.

Butov, L. V., Gossard, A. C. and Chemla, D. S. (2003). *Nature* **418**, 751.

Comte, C. and Nozières, P. (1982). *J. Phys. (Paris)* **43**, 1069.

Côté, R. and Griffin, A. (1988). *Phys. Rev.* B **37**, 4539.

Dang, L. S., Heger, D., André, R., Bœuf, F. and Romestain, R. (1998). *Phys. Rev. Lett.* **81**, 3920.

Deng, H., Wiehs, G., Santori, C., Bloch, J. and Yamamoto, Y. (2002). *Science* **298**, 199.

Deng, H., Weihs, G., Snoke, D., Bloch, J. and Yamamoto, Y. (2003). *PNAS* **100**, 15318.

Deng, H., Press, D., Götzinger, S., Solomon, G. S., Hey, R., Ploog, K. H. and Yamamoto, Y. (2006). *Cond-mat//0604394*.

Dicke, R. H. (1954). *Phys. Rev.* **93**, 99.

Eastham, P. R. and Littlewood, P. B. (2000). *Solid State Commun.* **116**, 357.

Eastham, P. R. and Littlewood, P. B. (2001). *Phys. Rev.* B **64**, 235101.

Eastham, P. R., Szymanska, M. H. and Littlewood, P. B. (2003). *Solid State Commun.* **127**, 117.

Eisenstein, J. P., Boebinger, G. S., Pfeiffer, L. N., West, K. W. and Song, H. (1992). *Phys. Rev. Lett.* **68**, 1383.

Fertig, H. (1989). *Phys. Rev.* B **40**, 1087.

Griffin, A., Snoke, D. W. and Stringari, S. (1994). *Bose–Einstein Condensation.* Cambridge University Press, Cambridge.

Grüner, G. (1994). *Density Waves in Solids.* Addison-Wesley Publishing Company.

Guseinov, R. R. and Keldysh, L. V. (1972). *Zh. Eksp. Teor. Fiz.* **63**, 2255 [*Sov. Phys. JETP* **36**, 1193].

Halperin, B. I. and Rice, T. M. (1968a). In *Solid State Physics.* Academic Press Inc., New York.

Halperin, B. I. and Rice, T. M. (1968b). *Rev. Mod. Phys.* **40**, 755.

Hammack, A. T., Griswold, M., Butov, L. V., Smallwood, L. E., Ivanov, A. L. and Gossard, A. C. (2006). *Phys. Rev. Lett.* **96**, 227402.

Hanamura, E. and Haug, H. (1977). *Physics Reports* **33**, 209.

Hensel, J. C., Phillips, T. G. and Thomas, G. A. (1977). *Solid State Physics* **32**, 87.

Ho, A. F. (2006). *Cond-mat//0603299.*

Imry, Y. and Ma, S. (1975). *Phys. Rev. Lett.* **35**, 1399.

Jaime, M., Correa V. F., Harrison, N., Batista, C. D., Kawashima, N., Kazuma, Y., Jorge, G. A., Stern, R., Heinmaa, I., Zvyagin, S. A., Sasago, Y. and Uchinokura, K. (2004). *Phys. Rev. Lett.* **93**, 087203.

Kageyama, H., Yoshimura, K., Stern, R., Mushnikov, N. V., Onizuda, K., Kato, M., Kosuge, K., Slichter, C. P., Goto, T. and Ueda, Y. (1999). *Phys. Rev. Lett.* **82**, 3168.

Kasprzak, J., Richard, M., Kundermann, S., Baas, A., Jeambrun, P., Keeling, J. M.J., Marchetti, F. M., Szymanska, M. H., Andre, R., Staehli, J. L., Savona, V., Littlewood, P. B., Deveaud, D. and Dang, L. S. (2006). *Nature* **443**, 409.

Kavokin, A. and Malpuech, G. (2003). *Cavity Polaritons*. Elsevier North Holland, Elsevier, Amsterdam.

Keeling, J., Eastham, P. R., Szymanska, M. H. and Littlewood, P. B. (2004). *Phys. Rev. Lett.* **93**, 226403.

Keeling, J., Eastham, P. R., Szymanska, M. H. and Littlewood, P. B. (2005). *Phys. Rev. B* **72**, 115320.

Keldysh, L. V. (1964). *Zh. Eksp. Teor. Fiz.* **47**, 1515 [(1965). *Sov. Phys. JETP* **20**, 1018].

Keldysh, L. V. and Kopaev, Yu. V. (1964). *Fiz. Tv.Tela* **6**, 279 [(1965). *Sov. Phys. Solid State* **6**, 2219].

Keldysh, L. V. and Kozlov, A. N. (1968). *Zh. Eksp. Teor. Fiz.* **54**, 978 [*Sov. Phys. JETP* **27**, 521].

Kellogg, M., Eisenstein, J. P., Pfeiffer, L. N. and West, K. W. (2004). *Phys. Rev. Lett.* **93**, 036801.

Kuwata-Gonokami, M., Shimano, R. and Mysyrowicz, A. (2002). *J. Phys. Soc. Japan* **71**, 1257.

Lakrimi, M., Khym, S., Nicholas, R. J., Symons, D. M., Peeters, F. M., Mason, N. J. and Walker, P. J. (1997). *Phys. Rev. Lett.* **79**, 3034.

Leggett, A. J. (1980). In *Modern Trends in the Theory of Condensed Matter*. Springer-Verlag, Berlin.

Littlewood, P. B. and Zhu, X. J. (1996). *Physica Scripta* **T68**, 56.

Littlewood, P. B., Eastham, P. R., Keeling, J., Marchetti, F. M., Simons, B. D. and Szymanska, M. H. (2004). *J. Phys.: Condens. Matter* **16**, S3597.

Lozovik, Yu. E. and Yudson, V. I. (1975). *Pisma Zh. Eksp. Teor. Fiz.* **22**, 556 [*JETP Lett.* **22**, 274].

Marchetti, F. M., Simons, B. D. and Littlewood, P. B. (2004). *Phys. Rev. B* **70**, 155327.

Marchetti, F. M., Keeling, J., Szymanska, M. H. and Littlewood, P. B. (2006). *Phys. Rev. Lett.* **96**, 066405.

Moskalenko, S. A. (1962). *Fiz. Tv. Tela* **4**, 276 [*Sov. Phys. Solid State* **4**, 199].

Moskalenko, S. A. and Snoke, D. W. (2000). *Bose–Einstein Condensation of Excitons and Biexcitons and Coherent Nonlinear Optics with Excitons.* Cambridge University Press, Cambridge.

Nikuni, T., Oshikawa, M., Oosawa, A. and Tanaka, H. (2000). *Phys. Rev. Lett.* **84**, 5868.

Nozières, P. and Comte, C. (1982). *J. Phys. (Paris).* **43**, 1083.

Nozières, P. and Schmitt-Rink, S. (1985). *J. Low Temp. Phys.* **59**, 195.

O'Hara, K. E., Suilleabhain, L. O. and Wolfe, J. P. (1999). *Phys. Rev. B* **60**, 10565.

Ong, N. P. and Monceau, P. (1977). *Phys. Rev. B* **16**, 3443.

Randeria, M. (1994). In *Bose–Einstein Condensation.* Cambridge University Press, Cambridge.

Regal, C. A., Greiner, M. and Jin, D. S. (2004). *Phys. Rev. Lett.* **92**, 040403.

Rice, T. M. (1977). *Solid State Physics* **32**, 1.

Ruegg, Ch., Cavadini, N., Furrer, A., Gudel, H. U., Kramer, K., Mutka, H., Wildes, A., Habicht, K. and Vorderwisch, P. (2003). *Nature* **423**, 62.

Sarchi, D. and Savona, V. (2004). *Cond-mat*/0411084.

Savvidis, P. G., Baumberg, J. J., Stevenson, R. M., Skolnick, M. S., Whittaker, D. M. and Roberts, J. S. (2000). *Phys. Rev. Lett.* **84**, 1547.

Shevchenko, S. I. (1976). *Fiz. Nizk. Temp.* **2**, 505 [*Sov. J. Low Temp. Phys.* **2**, 251].

Snoke, D. W. and Negoita, V. (2000). *Phys. Rev. B* **61**, 002904.

Snoke, D. W., Denev, S., Liu, Y., Pfeiffer, L. and West, K. (2003). *Nature* **418**, 754.

Snoke, D. W., Voros, Z., Liu, Y., Pfeiffer, L. and West, K. (2004). *Cond-mat*/0410298.

Spielman, I. B., Eisenstein, J. P., Pfeiffer, L. N. and West, K. W. (2000). *Phys. Rev. Lett.* **84**, 5808.

Szymanska, M. H. and Littlewood, P. B. (2002). *Solid State Commun.* **124**, 103.

Szymanska, M. H., Littlewood, P. B. and Simons, B. D. (2003). *Phys. Rev. A* **68**, 13818.

Szymanska, M. H., Keeling, J. and Littlewood, P. B. (2006). *Cond-mat*/0603447.

Tutuc, E., Shayegan, M. and Huse, D. (2004). *Phys. Rev. Lett.* **93**, 036802.

Weihs, G., Deng, H., Huang, R., Sugita, M., Tassone, F., and Yamamoto, Y. (2003). *Semicond. Sci. Technol.* **18**, S386.

Wen, X. G. and Zee, A. (1992). *Phys. Rev. Lett.* **69**, 1811.

Wilson, J. A., Di Salvo, F. J. and Mahajan, S. (1974). *Phys. Rev. Lett.* **32**, 882.

Wolfe, J. P., Lin, J. L. and Snoke, D. W. (1994). In *Bose–Einstein Condensation* (ed by A. Griffin, D. W. Snoke and S. Strinfari). Cambridge University Press, Cambridge.

Zhu, X. J., Hybertsen, M. S., Littlewood, P. B., and Rice, T. M. (1995). *Phys. Rev. Lett.* **74**, 1633.

Zittartz, J. (1967). *Phys. Rev.* **164**, 575.

Zwierlein, M. W., Stan, C. A., Schunck, C. H., Raupach, S. M. F., Kerman, A. J. and Ketterle, W. (2004). *Phys. Rev. Lett.* **92**, 120403.

12

INELASTIC LIGHT SCATTERING BY LOW-LYING EXCITATIONS OF QUANTUM HALL FLUIDS

Vittorio Pellegrini[a] and Aron Pinczuk[b]
[a] NEST CNR-INFM and Scuola Normale Superiore, Piazza dei Cavalieri 7, I-56126 Pisa, Italy
[b] Department of Physics, Department of Applied Physics and Applied Mathematics, Columbia University, New York, New York 10027, USA
Bell Labs, Lucent Technologies, Murray Hill, New Jersey 07974, USA

Abstract

Strongly-correlated low-dimensional electron fluids occur in artificial semiconductor heterostructures of high crystal perfection. When subjected to quantizing magnetic fields at low temperatures in the millikelvin range these structures are hosts to two-dimensional (2D) quantum fluids. The states of the quantum Hall effects are archetypes of the novel behaviors. The quantum Hall fluids support dispersive low-energy collective excitations that represent time- and space-dependent oscillations in the charge and spin degrees of freedom. These collective modes manifest fundamental interactions that are responsible for electron correlation. Studies of low-lying collective excitation modes play pivotal roles in the low-temperature phases of the electron liquids and offer insights on energetics, coherence, magnetization, instabilities and quantum phase transitions of the 2D system. We review here recent light-scattering results obtained from 2D quantum Hall fluids in semiconductor quantum structures under extreme conditions of low temperature and large magnetic field. Inelastic light-scattering methods currently offer unique access to elementary excitations of 2D fluids and of systems with lower dimensions (quantum dots with few electrons). The studies reviewed here suggest that light-scattering methods will continue to provide access to the physics of quantum fluids and of states in novel low-dimensional electron systems.

12.1 Introduction

When 2D systems in artificial semiconductor heterostructures of high perfection are subjected to quantizing perpendicular magnetic fields and at low temperatures in the mK range they give rise to quantum Hall (QH) states at integer and some peculiar fractional values of the Landau level filling factor $\nu = nhc/eB$ where h is the Planck constant, e is the electric charge and n is the carrier density (von Klitzing *et al.*, 1979; Tsui *et al.*, 1982; Laughlin, 1983). The remarkable, sometimes bizarre, behaviors of electrons in the quantum Hall regimes are among the most striking consequences of fundamental interactions that occur in contemporary physics (Sarma and Pinczuk, 1997).

Quantum Hall fluids display characteristic collective excitations linked to changes in degrees of freedom such as confined-motion perpendicular to the layer, in-plane motion (free-electron-like), and spin. The collective modes represent time- and space-dependent oscillations of the charge or/and the orientations of spin of the many-electron system. Despite the presence of a quantizing magnetic field, the modes have characteristic energy versus in-plane wavevector dispersions (Lerner and Lozovik, 1980; Kallin and Halperin, 1984).

These collective modes manifest fundamental interactions that are responsible for electron correlation. For this reason studies of low-lying collective excitation modes play pivotal roles in the investigation of low-temperature phases of the electron liquids. The impact of these studies is significant when they offer insights on physics that currently is not accessible by the more conventional low-temperature magnetotransport experiments. Experimental explorations of low-lying collective excitation modes offer venues to study energetics, coherence, states of spin polarization, instabilities and quantum phase transitions of quantum Hall fluids.

We review here recent light-scattering results obtained from quantum Hall fluids in semiconductor quantum structures under extreme conditions of low temperature, large magnetic field and low electron density. We also review recent light-scattering results in nanofabricated semiconductor quantum dots. These experiments exploit the resonant enhancement of the inelastic light-scattering cross-sections. This method allows probing the fundamental behavior due to interactions, with a sensitivity that will keep light-scattering studies at the frontiers of research of quantum fluids in low-dimensional electron systems.

12.2   Inelastic light scattering by low-dimensional electron systems

It was pointed out in 1978 that the resonant enhancements of the inelastic light-scattering cross-sections may provide the high sensitivity required to measure spin and charge excitations of 2D electron systems (2DES) confined in semiconductor heterostructures (Burstein *et al.*, 1979). Contemporary research of fundamental behavior in collective quantum phases of electrons in 2D indeed relies on the remarkably large and sharp resonant enhancement of inelastic light-scattering intensities that occur in semiconductor heterostructures of high crystal perfection.

It is useful to recall the kinematics of light-scattering processes that is based on conservation laws. For the energy we have $\omega_L - \omega_S = \omega(\mathbf{q})$, where L, S refer to the incident and scattered photons and $\omega(\mathbf{q})$ is the collective mode energy for the in-plane wavevector $\mathbf{q}$. Conservation of in-plane wavevectors, $\mathbf{k}_L - \mathbf{k}_S = \mathbf{k} = \mathbf{q}$, where $\mathbf{k}_L$ and $\mathbf{k}_S$ are the in-plane wavevectors of the laser and scattered light respectively, occurs when the 2D system has full translational invariance. The conservation laws are displayed in Fig. 12.1(a), while Fig. 12.1(b) shows a typical set-up for light-scattering measurements at low-temperatures and high magnetic fields.

Numerous experiments, some of them highlighted in this chapter, have revealed that breakdown of the in-plane wavevector is pervasive and occurs even in the highest quality samples. With breakdown of wavevector conservation, light-scattering experiments probe excitations corresponding to finite values of $\mathbf{q} \neq \mathbf{k}$ at which there are maxima in the density of states. This is a major characteristic of resonant inelastic light-scattering experiments in which photon energies are in resonance with optical transitions (or excitons) of the semiconductor quantum structures.

A mechanism for wavevector conserving light-scattering processes is represented in time-dependent perturbation theory by a third-order process in which the incoming photon at ω_L creates an intermediate valence-to-conduction band (or interband) electron–hole state. This state then relaxes to a lower-energy interband state by creation or annihilation of the collective excitation of the 2DES. In the third step there is creation of the scattered photon at ω_S. Descriptions of resonant inelastic light-scattering processes that break wavevector conservation could be represented by higher-order time-dependent perturbation processes that explicitly incorporate carrier relaxation due to weak residual disorder.

In resonant light-scattering experiments, a finely tunable laser is used to generate the incident polarized radiation while the scattered light is spectroscopically resolved by double or triple grating spectrometers that are often equipped with master gratings to reduce the impact of the stray laser light. Spectral resolutions of 15-20 μeV can be achieved. In-situ tilting of the sample with respect to the direction of propagation of laser beams allows a transfer of a finite in-plane wavevector to the collective mode.

The impact of resonant light scattering in the investigation of intra- and inter-Landau level excitations of 2D electron states under quantizing magnetic field has been remarkable (Pinczuk, 1997; Hirjibehedin *et al.*, 2005; Pellegrini *et al.*, 1998; Luin *et al.*, 2005). These experimental and theoretical studies have uncovered some of the most intriguing many-body phenomena in contemporary physics. This is the regime where the integer and fractional quantum Hall effects were discovered (von Klitzing *et al.*, 1979; Tsui *et al.*, 1982; Laughlin, 1983). The multiplicity of quantum Hall states seen in superior quality systems at low temperatures can be regarded as sequences of compressible (metallic) and incompressible (insulating) states that occur as the external magnetic field changes. The distinct quantum Hall states are linked by quantum phase transitions (QPT) that emerge at constant temperature and are driven by changes in magnetic field (Sondhi *et al.*, 1997).

The purpose of this paper is to offer a review of our recent inelastic light-scattering work under the extreme conditions of temperature and magnetic field that prevail in the quantum Hall regimes. We also address the application of inelastic light scattering to study few-electron states in quantum dots. These experiments explore the collective excitation spectra of electron states and show

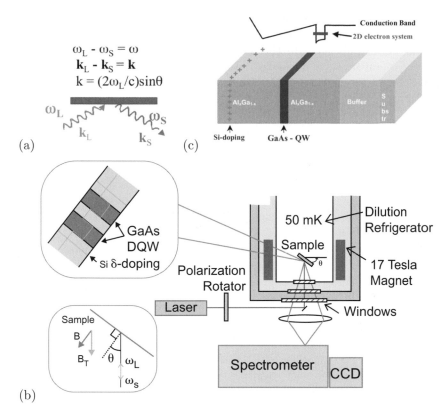

$$\omega_L - \omega_S = \omega$$
$$\mathbf{k}_L - \mathbf{k}_S = \mathbf{k}$$
$$k = (2\omega_L/c)\sin\theta$$

FIG. 12.1. (a) Kinematics of inelastic light scattering. (b) Typical set-up for inelastic light-scattering experiments at low temperature and high magnetic field. The sample (a coupled GaAs double quantum well – DQW – in the case shown in the diagram) is placed in a magnetocryostat. A double or triple spectrometer is used to collect the scattering light. A tunable single-mode dye, titanium–sapphire or diode laser is used as the source of the incident radiation. (c) Sequence of layers in a typical GaAs/AlGaAs modulation-doped single quantum well (QW) heterostructures grown for light-scattering experiments. The conduction-band profile along the growth axis is also shown. The high-mobility 2D electron system resides in the GaAs quantum well. Buffer and substrate layers are made of GaAs. Distance between Si doping and GaAs QW is around 100 nm

that their physics can be probed down to millikelvin temperatures with a sensitivity that goes far beyond what was imagined in 1978.

We focus on three different states in the quantum Hall regime. First we consider light-scattering results for fractional quantum Hall states at $\nu = 1/3, 2/5$ in single layers. We review next light-scattering research reported in the state at $\nu = 1$ in double layers. These results offer a glimpse of the remarkable behaviors of low-lying excitations that are due to Coulomb interaction terms in low-dimensional electron systems. In the last part of the paper we focus on light-scattering results obtained in the few-electron regime in nanofabricated semiconductor quantum dots. Prior to these reviews we introduce in the following section some of the fundamentals of light scattering by collective excitations in the quantum Hall regime.

12.3   Light scattering in the quantum Hall regimes

Figure 12.1(c) shows a typical layer sequence of a modulation-doped (Si-doping) single quantum well GaAs/Al$_x$Ga$_{1-x}$As sample for studies of quantum Hall fluids. The design of the structure allows fine control of parameters such as density, mobility and level splitting. In the quantum structures designed for optical experiments the aluminum composition is kept at $x = 0.10$ or lower to minimize the impact of layer-width fluctuations and other disorder (Pinczuk *et al.*, 1993). Low-temperature electron mobilities above $5 \times 10^6 \, \mathrm{cm}^2/\mathrm{Vs}$ are typical for these samples. A free electron density n below $10^9 \, \mathrm{cm}^{-2}$ was reported (Hirjibehedin and et al., 2002). In the case of double quantum well samples, low-temperature electron mobilities around $2 \times 10^6 \, \mathrm{cm}^2/\mathrm{Vs}$ and free electron densities close to $10^{11} \, \mathrm{cm}^{-2}$ are typically achieved.

In a magnetic field B perpendicular to the 2D plane the kinetic energy of the in-plane motion becomes quantized into discrete and highly degenerate Landau levels (LLs) separated by the cyclotron gap $\omega_C = (eB/m_e c)$, where m_e is the electron effective mass. The total external magnetic field B_T splits each the LL into spin-up and spin-down states.

For integer and "magic" fractional values of $\nu = n2\pi(l_B)^2$, where $(l_B)^2 = (\hbar c/eB)$ is the magnetic length, the longitudinal resistivity ρ_{xx} is vanishingly small and the Hall resistance R_H is quantized with extreme precision at values $h/\nu e^2$. Since the longitudinal resistivity in a perpendicular magnetic field is proportional to the longitudinal conductivity, in quantum Hall states the longitudinal conductivity also vanishes. The vanishing conductivity implies that there is a gap in spectra of low-lying excitations. The gaps in the integer ν states (IQHE) can be understood in terms of either the cyclotron gaps or the spin gaps (the gaps for spin reversal). In states with fractional ν (FQHE) the gaps are linked to the emergence of *incompressible* quantum liquids represented by a ground quantum state described by a single many-particle wavefunction (Laughlin, 1983). The quantum liquids emerge in response to the deceptively simple Coulomb repulsion in 2D and the Pauli principle that applies to fermions.

In recent years light-scattering spectroscopy has evolved into a primary experimental tool for the study of gap excitations in the quantum Hall regimes, and offers direct tests of key predictions of theories of the electron fluids. In studies of the quantum fluids of 2D electron systems in perpendicular magnetic fields light-scattering experiments play roles similar to neutron and X-ray scattering studies of elementary excitations in superfluid liquid helium.

Collective excitation modes of quantum Hall states are built from neutral particle–hole pair transitions. In these excitations a quasiparticle is promoted to an empty state and a quasihole is created in the ground state as shown schematically in Fig. 12.2(a). In excitations of the charge degree of freedom the spins of the particle and hole are identical. In spin excitations there is a spin-flip (SF). These pairs are the building blocks of the collective modes and can be regarded as magnetic-excitons with a dipole length x_o (Lerner and Lozovik, 1980; Kallin and Halperin, 1984), where x_o represents the displacement between the centers of the two cyclotron orbits (see Fig. 12.2a).

From a semiclassical point of view, the balance between the Lorentz and Coulomb forces leads to a center-of-mass velocity of the neutral pair that is inversely proportional to the separation between the particles. The excitonic pairs are thus characterized by the in-plane wavevector $\mathbf{q}$ which is a good quantum number for the excitation with $x_o = q \cdot l_B^2$. The coupling between particle and hole in the neutral pairs gives rise to many-body terms that have a strong $\mathbf{q}$-dependence (Sarma and Pinczuk, 1997; Lerner and Lozovik, 1980; Kallin and Halperin, 1984). The roton (or magneto-roton), a minimum of the dispersion at finite wavevector of the order of $1/l_B$, is the characteristic manifestation of these interactions. It is caused by excitonic terms (called vertex corrections) of the Coulomb interaction that binds particle–hole pairs and greatly reduces their energy.

Figure 12.2(b) highlights the salient features in the dispersion of charge excitation modes (CDE) predicted at $\nu = 1/3$ (Haldane and Rezayi, 1985; Girvin *et al.*, 1985). For the roton wavevector there is a peak in the structure factor of the quantum Hall liquid in analogy with that for excitations in superfluid ^{4}He. As the lowest energy excitation of the liquid, the magneto-rotons are barometers of its stability. In fact, roton instabilities have been predicted to trigger transitions from liquid to Wigner crystal states (Girvin *et al.*, 1985; Kamilla and Jain, 1997).

Different Coulomb interaction terms enter in the dispersions of spin and charge modes. In CDE, the direct terms of the Coulomb interaction (the depolarization shift) add to the excitonic term mentioned above and to differences in self-energy contributions. In spin excitations the depolarization shift is absent. In the large-wavevector limit both excitonic and direct terms of Coulomb interactions vanish and the mode energy is given by self-energy terms that add to single-particle contributions such as cyclotron spacings.

In the FQHE regime the 2DES condenses into a sequence of quantum fluids that manifest the peculiar way in which fundamental Coulomb interactions restructure the electron system into novel quantum ground states. Here the lowest CDEs are gapped intra-Landau level excitations, in which the opening of the gap reveals that the fluid is incompressible. It must be noted that the large-wavevector limit represents infinitely separated particle–hole pairs, as indicated in Fig. 12.2(a). The energies of the lowest charge excitations at large wavevectors are thus conceptually linked to the gaps measured in activated magnetotrans-

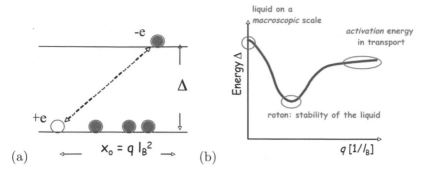

FIG. 12.2.   Schematic representation of the particle–hole pairs (mag-
neto-excitons) that enter in the construction of the collective excitations with
in-plane wavevector **q**. In the integer quantum Hall regime Δ can be the single-particle gap that separates Landau levels with different quantum numbers
and /or spin. In the fractional quantum Hall regime Δ is the many-body gap
that opens due to electron–electron interactions. The neutral pairs in the
collective mode in conduction-band states of GaAs/AlGaAs heterostructures
can be classified according to a change in the total angular momentum J.
Excitations with $\Delta J = 0$ are the singlets of charge-density modes (CDE).
Excitations with $\Delta J = 1$ are triplets in which the spin degree of freedom is
changed. Those are spin-density modes (SDE) without a fluctuating charge
density and SF excitations. The degeneracy of the triplet is removed at finite
magnetic field by the Zeeman energy. (b) Representative wavevector dispersion of CDE mode. The dispersion applies, for instance, to the intra-Landau
level excitations in the fractional quantum Hall state at filling factor $\nu = 1/3$.
The three critical points correspond to the long-wavelength limit (q = 0) conventionally probed in optical experiments, to the lowest energy excitations
at the roton minimum and to the long-wavevector limit ($q \gg 1/l_B$) relevant
in magnetotransport studies.

port.

A diverse class of spin excitation modes can occur in QH fluids. In spin wave
(SW) excitations only spin orientation changes. In SF excitations other quantum
numbers, such as the Landau level number or pseudospin, change too (see the
caption to Fig. 12.2). In these excitations the contributions due to Coulomb
interactions add to single-particle terms such as the Zeeman energy $E_Z = g\mu_B B_T$
where g is the Landé factor and μ_B is the Bohr magneton.

Collective excitations in nanofabricated semiconductor quantum dots (QDs)
are conveniently described within the parabolic approximation of single-particle
QD levels provided by Fock–Darwin (FD) orbitals. FD energies are given by
$\varepsilon_{nm} = \hbar\omega_0(2n + |m| + 1)$, where $n = 0, 1, \ldots,$ $m = 0, \pm1, \ldots$ are the radial and
azimuthal quantum numbers, respectively, and $\hbar\omega_0$ is the harmonic confinement

energy. The FD shells are defined by an integer value of $N_{\text{shell}} = 2n + |m|$ with well-defined atomic-like parity. QD states can be classified in terms of the z-component total angular momentum M, total spin S, and its z-component S_z. Selection rules in QDs dictate that the monopole transitions with $\Delta M = 0$ ($\Delta N_{\text{shell}} = 2, 4, \ldots$) and $\Delta S = 0, 1$ are the intershell charge and spin modes active in light-scattering experiments (Schuller *et al.*, 1998; Lockwood *et al.*, 1996).

12.4 Charge and spin excitations at $\nu = 1/3$

The large resonant enhancements in light-scattering experiments that probe the excitations of the quantum Hall fluids occur when ω_{L} is very close to the fundamental bandgap of the GaAs/AlGaAs heterostructure. This enhancement plays crucial roles because the measurements require extremely low incident power densities (of $\approx 10^{-4}\,W/cm^2$) that are compatible with the low temperatures encountered in dilution refrigerators.

At $\nu = 1/3$ the low-lying excitations are intra-Landau level excitations. Those include charge density modes (CDE) across the gap of the incompressible QH fluid with the dispersion shown in Fig. 12.2(b) and spin modes across the spin gap (or spin wave – SW) where the spin degree of freedom is modified. In the composite fermion Landau level (CF-LL) picture the $\nu = 1/3$ QH state corresponds to population of the lowest CF-LL $|0\rangle$ and the lowest energy excitations correspond to $|0\uparrow\rangle \rightarrow |1\uparrow\rangle$ and $|0\uparrow\rangle \rightarrow |0\downarrow\rangle$ CF-LL transitions for the CDE and SW, respectively.

The light-scattering spectra at $\nu = 1/3$ clearly display the major excitations associated with the critical points of the these CDE (obtained in a configuration with parallel incident and scattered light polarizations) and SW (with orthogonal incident and scattered polarizations) dispersions in the range 0.2–1 meV (Pinczuk *et al.*, 1993; Hirjibehedin *et al.*, 2003; Dujovne *et al.*, 2005). The observation of the long-wavelength $\mathbf{q} = \mathbf{0}$ CDE excitation is intriguing since the structure factor $S(q, \omega)$ for charge modes in the FQHE is predicted to vanishes as q^4 for small wavevectors. This observation thus highlights the impact of strong resonance enhancements. It also offers direct experimental tests for a possible new interpretation of the character of the long-wavelength CDE such as the one in terms of a two-roton bound state (Hirjibehedin *et al.*, 2005).

12.5 Spin-flip excitations at $\nu = 2/5$

At filling factor $\nu = 2/5$ two spin-up CF-LLs ($|0\uparrow\rangle$ and $|1\uparrow\rangle$) are populated so that a new low-lying SF excitation ($|1\uparrow\rangle \rightarrow |0\downarrow\rangle$) can occur. In this mode both the CF Landau level number and spin orientation change simultaneously.

SF excitations of composite fermions were predicted by Mandal and Jain (Mandal and Jain, 2001; Jain *et al.*, 2005). The new excitation modes unique to CF quasiparticles were introduced to interpret light-scattering spectra of low-lying excitations at $\nu > 1/3$ that were reported by Kang *et al.*, 2000.

Light-scattering observations of low-lying spin excitations at $\nu = 2/5$ were reported by Dujovne et al. (Dujovne et al., 2003b). These are the first experimental determinations of SF modes at $\nu = 2/5$. They offered direct evidence of a Landau level structure of CF quasiparticles and demonstrated remarkable applications of light-scattering methods to the study of the spin degree of freedom in the quantum liquid states (Dujovne et al., 2003a). Because of the simultaneous changes in orbital and spin quantum numbers, the study of SF excitations offer insights into the interplay of fundamental interactions that determine the state of spin polarization in the fluids of CF quasiparticles that emerge at $\nu > 1/3$ (Dujovne et al., 2005).

The quantum liquid at $\nu = 2/5$, with two CF Landau levels fully populated, has full spin-polarization when the SF energies are larger than the sequential CF level spacings. Park and Jain computed the spin phase diagram of FQH states and described it terms of an effective mass for spin polarization (Park and Jain, 1998). In agreement with these predictions, a loss of full spin polarization in the 2/5 state was reported in transport experiments and in luminescence measurements as function of density (Kang et al., 1997; Chen et al., 2003; Kukushkin et al., 1999). In light-scattering experiments the loss of full spin polarization appears as a collapse of SF excitation energies (Dujovne et al., 2005).

12.6   Spin excitations in double layers at $\nu = 1$

The quantum Hall states of the bilayer systems support unique realizations of highly-correlated phases that are driven by interplays between intra- and interlayer Coulomb interactions (Girvin and MacDonald, 1997). The remarkable behaviors of electron bilayers realized in GaAs/AlGaAs double quantum wells have been the focus of major experimental and theoretical research efforts (Yang and et al., 1994; Sarma et al., 1997; Pellegrini et al., 1997; Pellegrini et al., 1998).

The focus of recent research of electron bilayers has shifted to the study of peculiar electron–hole (e-h) quantum phases that could display spontaneous macroscopic coherence (Eisenstein, 2004). A paradigmatic configuration that illustrates how e-h excitonic pairs can spontaneously populate the ground state of the bilayers occurs when a tunneling gap Δ_{SAS} splits the single-particle levels of the two wells in their symmetric and antisymmetric linear combinations and B is such that electrons precisely fill the lowest spin-up symmetric Landau level (the total filling factor is $\nu = 1$). This single-particle-like configuration has full spin and pseudospin ferromagnetic order, where pseudospin $(T = \tau/2)$ is a quantum operator describing layer occupation: electrons in the left (right) quantum well have pseudospin along the $+z$ $(-z)$ direction normal to the plane, and the symmetric (antisymmetric) states have pseudospin aligned along $+x$ $(-x)$. Within mean-field theories, such as the time-dependent Hartree–Fock approximation (TDHFA), the pseudospin order parameter is $\langle \tau_x \rangle = 1$ (Fertig, 1989; MacDonald et al., 1990; Brey, 1990) (all electrons occupy the lowest spin-up symmetric LL). Within TDHFA the low-lying spin excitations are the long-wavelength SF mode across the tunneling gap and SW mode across the Zeeman gap with energy spitting given by $\delta E_\tau = E_{\mathrm{SF}} - E_{\mathrm{SW}} = \Delta_{\mathrm{SAS}}$. Correlation effects, however, reduce the pseudospin order parameter (Yoglekar and MacDonald, 2001). We have gone beyond the TDHFA to calculate the energies of the SF and SW excitations on a ferromagnetic, fully spin polarized, quantum Hall ground state with any degree of pseudospin polarization. We found that $\delta E_\tau = \langle \tau_x \rangle \cdot \Delta_{\mathrm{SAS}}$ (Luin *et al.*, 2005). This result highlights that the reduced pseudospin order parameter determines the tunneling SF energy. Recent measurements of SF and SW excitations by inelastic light scattering has offered direct probes of the correlation properties of the ground-state in the bilayers (Luin *et al.*, 2005). These experiments in fact demonstrated significant reductions of δE_τ with respect to the tunneling gap and allow the measurement of the pseudospin order parameter. We finally note that the pseudospin polarization can be written as $\langle \tau_x \rangle = (n_{\mathrm{S}} - n_{\mathrm{AS}})/(n_{\mathrm{S}} + n_{\mathrm{AS}})$, where n_{S} (n_{AS}) are the average value of electron density into the symmetric (antisymmetric) level showing that the new state with reduced pseudospin order is characterized by a high expectation value $n_{\mathrm{AS}}/(n_{\mathrm{S}} + n_{\mathrm{AS}})$. It is therefore tempting to describe this highly-correlated QH incompressible state in terms of bound electron–hole pairs across the tunneling gap making a particle–hole transformation in the lowest symmetric Landau level.

12.7 Overview of recent results

In this section we present representative examples of recent light-scattering determination of spin and charge excitations in QH fluids and in semiconductor quantum dots. These experiments demonstrate the capabilities of light scattering to probe fundamental interactions of electrons in low-dimensional systems. We begin presenting experiments in the fractional QH regime at $\nu = 1/3$ that have focussed on the dispersion of the long-wavelength CDE mode schematically shown in Fig. 12.2(b). For these investigations a tilt angle geometry like the one shown in Fig. 12.1(b) was employed. The remarkable observations are summarized in Fig. 12.3 that reports light-scattering spectra from long-wavelength CDE at $\nu = 1/3$ for various values of the tilt angle θ (Hirjibehedin *et al.*, 2005). Contrary to what is shown in Fig. 12.2(b), the spectra reported in Fig. 12.3 offer evidence for the existence of two distinct modes with energy splitting that increases at larger values of the in-plane wavevector. The two modes are interpreted as two branches of the long-wavelength CDE of the fractional QH liquid: one is the gap excitation as shown in Fig. 12.2(b) and the second branch corresponds to a two-roton bound state (Girvin *et al.*, 1986). The gap excitation is predicted to have a slight downward dispersion in the long-wavelength limit (see Fig. 12.2b) while the two-roton mode is expected to have an upward dispersion. This is in qualitative agreement with the increased energy splitting at large ql_B. It is also worth noting that the observed dispersion of the collective modes provides a measurement for the length-scale of the coherence in the QH fluid which has to be larger than $\approx 2\,\mu\mathrm{m}$.

An additional example of the application of light scattering to study interacting fluids in the QH regime and their spin properties is provided by the

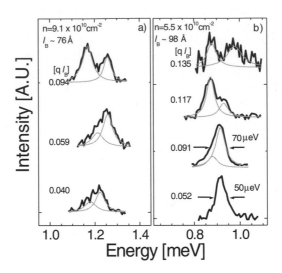

FIG. 12.3. Inelastic light-scattering spectra of long-wavelength charge modes at $\nu = 1/3$ at various angles (i.e., in-plane wavevector q) for two different asymmetrically-doped GaAs single quantum wells with well width and density of $25\,\mathrm{nm}$, $9.1 \times 10^{10}\,\mathrm{cm}^{-2}$ (a) and $33\,\mathrm{nm}$, $5.5 \times 10^{10}\,\mathrm{cm}^{-2}$ (b). Spectra are presented after conventional substraction of the background due to luminescence. The gray lines show fits with two Lorentzian line shapes. After Hirjibehedin *et al.* (2005). Reprinted figure with permission from [C. F. Hirjibehedin, I. Dujovne, A. Pinczuk, B. S. Dennis, L. N. Pfeiffer, and K. W. West, *Phys. Rev. Lett.* **95**, 066803 (2005)]. Copyright (2005) by the American Physical Society.

experiments reported in (Dujovne *et al.*, 2005). In these experiments, resonant inelastic light scattering was applied to study CF interaction effects that emerge when $\nu \to 1/2$. These experiments have probed low-lying SF modes in which electrons are excited from the highest occupied CF level to the lowest unoccupied CF level with opposite spin. At $\nu = 1/2$, CFs experience an effective zero magnetic field so that the energy spacing between CF LLs vanishes and the CF quasiparticles coalesce into a Fermi sea. In this framework the energy gap for the SF mode occurs only when the effective Zeeman energy for CFs (that includes the impact of residual CF interactions) is larger of the Fermi energy. Analysis of the energy of such SF modes can thus reveal unique information about the interaction effects at $\nu = 1/2$ and spin polarization of the Fermi sea of CFs. The results reported in (Dujovne *et al.*, 2005) showed the energy collapse of the SF mode as $\nu \to 1/2$. Collapse of this spin excitation offers direct evidence of a transition from spin-polarized to spin un-polarized CF states driven by the interplay between bare Zeeman term and Coulomb interaction energy.

Light scattering also has the sensitivity to study intershell spin excitations

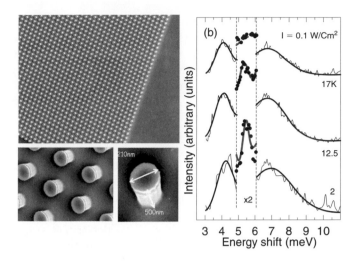

FIG. 12.4. Side panels: SEM pictures of the nanofabricated quantum dot arrays for the light-scattering experiments. (b) Inelastic light-scattering spectra of intershell spin excitations in semiconductor quantum dots as a function of temperature. The sharp spin mode in blue is attributed to the intershell (monopole) triplet-to-singlet excitations of the quantum-dot four-electron state. After García *et al.* (2005). Adapted figures with permission from [C. P. García, V. Pellegrini, A. Pinczuk, M. Rontani, G. Goldoni, E. Molinari, B. S. Dennis, L. N. Pfeiffer, and K. W. West, *Phys. Rev. Lett.* **95**, 266806 (2005)]. Copyright (2005) by the American Physical Society.

in semiconductor quantum dots with few electrons (Garcia *et al.*, 2005; Garcia *et al.*, 2006). Samples for these studies were fabricated from a 25 nm wide, one-side modulation-doped $Al_{0.1}Ga_{0.9}As$/GaAs quantum well with measured low-temperature electron density $n_e = 1.1 \times 10^{11}\,cm^{-2}$ and mobility of $2.7 \times 10^6\,cm^2$/Vs. QDs with different diameters were realized by inductive coupled plasma reactive ion etching. QD arrays (with sizes $100 \times 100\,\mu m$ containing 10^4 single QD replica) were defined by electron beam lithography. Deep etching (below the doping layer) was then achieved. Figure 12.4(b) shows representative results of spin excitations obtained on QDs having lateral lithographically-defined diameters of 210 nm (The SEM picture are shown in Fig. 12.4 side panels) (Garcia *et al.*, 2005). The spectra are detected with crossed polarizations between incident and scattered light. Two relatively-broad peaks are seen at energies close to 4 meV and 7–8 meV and interpreted as monopole spin excitations among the quantized energy shells of the QD with $\Delta S = 1$. A characteristic feature of these doublets is their significant linewidth that is attributed largely to inhomogeneous broadening due to the distribution of electron occupations of the dots. The spectra in Fig. 12.4(b) reveal at low temperature a much sharper spin excitation at 5.5 meV. To interpret the origin of this sharp spin mode we note that an additional triplet-to-singlet (TS) intershell spin mode with $\Delta S = -1$ can occur if the ground state is a triplet with $S = 1$ and the excited state is a $S = 0$ singlet state. On this basis we identify the sharp peak at 5.5 meV with the TS ($\Delta S = -1$) intershell spin excitation. According to Hund's rules a triplet ground state occurs only when two electrons are in a single partially populated shell as is the case of QDs with four electrons. This is consistent with theoretical calculations within a configuration-interaction approach that includes correlation effects (Garcia *et al.*, 2005). In this interpretation the narrow width is simply explained as due to the absence of inhomogeneous broadening from the distribution of the electron population of the QDs. The intensity of the TS spin excitation decreases significantly as temperature increases with an estimated activation gap of 0.7 ± 0.3 meV. At such low energy a possible thermally populated excited level is the singlet state without any change in orbital occupation. This energy thus provides an estimate of the low-lying intrashell singlet-triplet transition of the four-electron QDs and it compares well with CI estimate of 0.8 meV.

12.8   Concluding remarks

The experiments reviewed in this paper demonstrate the successful applications of resonant inelastic light-scattering methods to study low-energy collective spin and charge excitations of 2D electron liquids under extreme conditions of low temperature, high magnetic field and very low densities. These spectroscopic observations are remarkable in that they require the use of ultra-attenuated laser beams with power intensities of the order of 10^{-4} W/cm^2 or less. This review paper also shows that the sensitivity reached by light-scattering method also allows us to access the excitation spectra of few-electron states confined in nanofabricated semiconductor quantum dots. The method is particularly effective in revealing major features of spin states and of their quantum phase transitions that are not easily accessible by other experimental approaches.

More than 25 years have elapsed since the initial proposal and the first applications of inelastic light-scattering methods to the study of 2D electron systems. The unparalleled quality currently achieved in modulation doped GaAs quantum structures created by molecular beam epitaxy (MBE) and in nanofabricated semiconductor nanostructures is enabling the study of some of the most extraordinary behaviors due to fundamental electron interactions encountered in condensed matter. Light-scattering methods can offer unique insights on the novel quantum phases that emerge in these systems. It is our belief that light scattering will continue to play a pivotal role in the development of the new knowledge required to understand fundamental processes at the nanoscale.

Acknowledgments

The work presented here was supported by the Italian Ministry of Foreign Affairs, by the Italian Ministry of Research and by the European Community Human Potential Programme (Project HPRN-CT-2002-00291) by the National Science

Foundation under Award Number DMR-03-52738, by the Department of Energy under award DE-AIO2-04ER46133, by the Nanoscale Science and Engineering Initiative of the National Science Foundation under Award Number CHE-0117752, and by a research grant of the W. M. Keck Foundation. We gratefully acknowledge the contributions of Irene Dujovne, Cesar Pascual Garcia, Cyrus Hirjibehedin, Sokratis Kalliakos, and Stefano Luin. Finally we thank Brian S. Dennis for valuable technical help, L. N. Pfeiffer, and K. W. West for the growth of the semiconductor heterostructures and Massimo Rontani, Guido Goldoni and Elisa Molinari for the theoretical modeling in quantum dots.

References

Brey, L. (1990). *Phys. Rev. Lett.* **65**, 903.

Burstein, E., Pinczuk, A. and Buchner, S. (1979). *Physics of Semiconductors 1978*. Institute of Physics, London.

Chen, M., Kang, W. and Wegscheider, W. (2003). *Phys. Rev. Lett.* **91**, 116804.

Dujovne, I., Hirjibehedin, C. F., Luin, S., Pellegrini, V., Pinczuk, A., Dennis, B. S., Pfeiffer, L. N. and West, K. W. (2005). *Solid State Commun.* **135** 645.

Dujovne, I., Hirjibehedin, C. F., Pinczuk, A., Kang, M., Dennis, B. S., Pfeiffer, L. N. and West, K. W. (2003). *Solid State Commun.* **127**, 109.

Dujovne, I., Pinczuk, A., Kang, M., Dennis, B. S., Pfeiffer, L. N. and West, K. W. (2003). *Phys. Rev. Lett.* **90**, 036803.

Dujovne, I., Pinczuk, A., Kang, M., Dennis, B. S., Pfeiffer, L. N. and West, K. W. (2005). *Phys. Rev. Lett.* **95**, 056808.

Eisenstein, J.P. (2004). *Science* **305**, 950.

Fertig, H.A. (1989). *Phys. Rev.* B **40**, 1087.

Garcia, C. P., Pellegrini, V., Pinczuk, A., Dennis, B. S., Pfeiffer, L. N., West, K. W., Rontani, M., Goldoni, G. and Molinari, E. (2005). *Phys. Rev. Lett.* **95**, 266806.

Garcia, C. P., Pellegrini, V., Pinczuk, A., Dennis, B.S., Pfeiffer, L. N. and West, K. W. (2006). *Appl. Phys. Lett.* **88**, 113105.

Girvin, S. M. and MacDonald, A. H. (1997). *Perspectives in Quantum Hall effects*. Wiley, New York, pp. 161–224.

Girvin, S. M., MacDonald, A. H. and Platzman, P. M. (1986). *Phys. Rev.* B **33**, 2481.

Girvin, S. M., MacDonald, A. H. and Platzman, P. M. (1985). *Phys. Rev. Lett.* **54**, 581.

Haldane, F. D. M. and Rezayi, E. H. (1985). *Phys. Rev. Lett.* **54**, 237.

Hirjibehedin, C. F., Dujovne, Irene, Bar-Joseph, I., Pinczuk, A., Dennis, B. S., Pfeiffer, L. N. and West, K. W. (2003). *Solid State Commun.* **127**, 799.

Hirjibehedin, C. F., Dujovne, I., Pinczuk, A., Dennis, B. S., Pfeiffer, L. N. and West, K. W. (2005). *Phys. Rev. Lett.* **95**, 066803.

Hirjibehedin, C. F., Pinczuk, A., Dennis, B. S., Pfeiffer, L. N. and West, K. W. (2002). *Phys. Rev.* B **65**, 161309.

Jain, J. K., Park, K., Peterson, M. R. and Scarola, V. W. (2005). *Solid State Commun.* **135**, 602.

Kallin, C. and Halperin, B. I. (1984). *Phys. Rev.* B **30**, 5655.

Kamilla, R. K. and Jain, J. K. (1997). *Phys. Rev.* B **55**, 13417.

Kang, M., Pinczuk, A., Dennis, B. S., Eriksson, M. A., Pfeiffer, L. N. and West, K. W. (2000). *Phys. Rev. Lett.* **84**, 546.

Kang, W., Young, J. B., Hannahs, S. T., Palm, E., Campman, K. L. and Gossard, A. C. (1997). *Phys. Rev.* B **56**, 12776.

Kukushkin, I. V., von Klitzing, K. and Eberl, K. (1999). *Phys. Rev. Lett.* **82**, 3665.

Laughlin, R. B. (1983). *Phys. Rev. Lett.* **50**, 1395.

Lerner, I. V. and Lozovik, Yu. E. (1980). *Zh. Eksp. Teor. Fiz.* **78**, 1167. [*Sov. Phys. JETP* **51**, 588.]

Lockwood, D. J., Hawrylak, P., Wang, P. D., Sotomayor Torres, C. M., Pinczuk, A. and Dennis, B. S. (1996). *Phys. Rev. Lett.* **77**, 354.

Luin, S., Pellegrini, V., Pinczuk, A., Dennis, B. S., Pfeiffer, L. N. and West, K. W. (2005). *Phys. Rev. Lett.* **94**, 146804.

MacDonald, A. H., Platzman, P. M. and Boebinger, G. S. (1990). *Phys. Rev. Lett.* **65**, 775.

Mandal, Sudhansu S. and Jain, J. K. (2001). *Phys. Rev.* B **63**, 201310.

Park, K. and Jain, J. K. (1998). *Phys. Rev. Lett.* **80**, 4237.

Pellegrini, V., Pinczuk, A., Dennis, B. S., Plaut, A. S., Pfeiffer, L. N. and West, K. W. (1997). *Phys. Rev. Lett.* **78**, 310.

Pellegrini, V., Pinczuk, A., Dennis, B. S., Plaut, A. S., Pfeiffer, L. N. and West, K. W. (1998). *Science* **281**, 799.

Pinczuk, A. (1997). *Perspectives in Quantum Hall effects* (ed by S. Das Sarma and A. Pinczuk). Wiley, New York, pp. 307-341.

Pinczuk, A., Dennis, B. S., Pfeiffer, L. N. and West, K. W. (1993). *Phys. Rev. Lett.* **70**, 3983.

Das Sarma, S. and Pinczuk, A. (1997). *Perspectives in Quantum Hall effects.* (ed by S. Das Sarma and A. Pinczuk). Wiley, New York.

Das Sarma, S., Sachdev, S. and Zheng, L. (1997). *Phys. Rev. Lett.* **79**, 917.

Schuller, C., Keller, K., Biese, G., Ulrichs, E., Rolf, L., Steinebach, C. and Heitmann, D. (1998). *Phys. Rev. Lett.* **80**, 2673.

Sondhi, S. L., Girvin, S. M., Carini, J. P. and Shahar, D. (1997). *Rev. Mod. Phys.* **69**, 315.

Tsui, D. C., Stormer, H. L. and Gossard, A. C. (1982). *Phys. Rev. Lett.* **48**, 1559.

von Klitzing, K., Dorda, G. and Pepper, M. (1979). *Phys. Rev. Lett.* **45**, 494.

Yang, K., Moon, K., Zheng, L., MacDonald, A. H., Girvin, S. M., Yoshioka, D. and Zhang, Shou-Cheng. (1994). *Phys. Rev. Lett.* **72**, 732.

Yoglekar, Y. N. and MacDonald, A. H. (2001). *Phys. Rev.* B **65**, 235319.

13

REMARKS ON SURFACE-ATOM FORCES IN LONDON AND LIFSHITZ LIMITS

L. P. Pitaevskii

Dipartimento di Fisica, Università di Trento
and BEC Center, INFM-CNR I-38050 Povo, Trento, Italy;
Kapitza Institute for Physical Problems, ul. Kosygina 2, 119334
Moscow, Russia

Abstract

It is shown that the derivation of the Lifshitz equation for the energy of interaction between an atom and a solid body at large and small distances can be reduced to a solution of a simple electrostatic problem of the electric field distribution around a point charge near the body.

13.1 Introduction

The force acting on an atom near the surface of a dielectric body was for the first time calculated by E. M. Lifshitz for arbitrary distances and temperatures (Lifshitz, 1956). The Lifshitz theory is based on the calculation of the stress tensor of a fluctuating electromagnetic field between two bodies. The general equation for the atom-surface forces was obtained as a limiting case of interaction between a dielectric body 1 with dielectric permittivity $\varepsilon(\omega)$ and a body 2 in the limit when the second body can be considered as a dilute from the electrodynamic point of view, i.e., assuming that the dielectric function of the second body $\varepsilon_2(\omega) \approx 1$ and expanding with respect to $\varepsilon_2 - 1$. In general, both zero-point and thermal fluctuations contribute to the force. However, the zero point contribution is important at small distances l between atom and surface, with the condition:

$$l \ll \lambda_{\mathrm{T}} \equiv \hbar c / T. \tag{13.1}$$

Further, in this quantum case one can distinguish the short-distance London regime where the interaction energy decreases according to the $1/l^3$ law, correspondingly to the $1/r^6$ London law for the interaction energy between individual atoms. The Lifshitz result for the energy is

$$V(l) = -\frac{\hbar}{4\pi l^3} \int\limits_0^\infty \alpha(i\zeta) \frac{1 - \varepsilon(i\zeta)}{1 + \varepsilon(i\zeta)} d\zeta, \tag{13.2}$$

where $\alpha(i\zeta)$ and $\varepsilon(i\zeta)$ are, correspondingly, the electric polarizability of an atom and the dielectric permittivity of the body as functions of imaginary frequencies. This equation is valid for the condition:

$$l \ll \lambda_0 \equiv c/\omega_0 \,, \tag{13.3}$$

where ω_0 is a characteristic frequency in the absorption optical spectra of the atom and body. For larger distances, where an inequality opposite to equation (13.3) is satisfied, but equation (13.1) is still valid, the London-like law, equation (13.2), is changed to the Casimir–Polder law $V \propto 1/l^4$, which corresponds to the Casimir–Polder law of interaction between atoms. We will not consider this case in our note.

Lifshitz also discovered for the first time that, on the largest distances, with the condition opposite to equation (13.1)

$$l \gg \lambda_{\mathrm{T}} \equiv \hbar c/T \,, \tag{13.4}$$

the interaction energy decreases again according to the $1/l^3$ law:

$$V(l) = -\frac{T}{4l^3}\alpha(0)\frac{1-\varepsilon(0)}{1+\varepsilon(0)} \,. \tag{13.5}$$

One can notice an obvious similarity between equation (13.2) and equation (13.5). However, in the original Lifshitz derivation this similarity can be discovered in the final result only. In these remarks we give simple and transparent derivations of equations (13.2) and (13.5). The derivations reveal a deep physical reason for the similarity.

13.2   Atom-surface energy

As a first step we will derive an expression for the atom-surface interaction energy directly from the Green's function formalism of Dzyaloshinskii and Pitaevskii (1959) [below we follow the notation of Lifshitz and Pitaevskii (1984)]. The basis of this theory is the equation for the variation δF of the free energy due to a small change of the dielectric permittivity $\delta\varepsilon$:

$$\delta F = \frac{T}{4\pi}\sum_{s=0}^{\infty}\int \mathcal{D}_{ll}^{\mathrm{E}}(\zeta_s; \mathbf{r}, \mathbf{r})\,\delta\varepsilon(i\,|\zeta_s|\,, \mathbf{r})d^3x \,, \tag{13.6}$$

where $\mathcal{D}_{lk}^{\mathrm{E}}$ is the Matsubara Green's function of the electric field, $\zeta_s = 2s\pi T/\hbar$ and the $s = 0$ term is taken with the coefficient $1/2$. Notice now that the presence of an atom at the point $\mathbf{r}_a$ can be considered as a small change of the dielectric permittivity:

$$\delta\varepsilon(\omega, \mathbf{r}) = 4\pi\alpha(\omega)\,\delta(\mathbf{r} - \mathbf{r}_a) \,. \tag{13.7}$$

Substitution of equation (13.7) into equation (13.6) gives the final expression for the energy of the atom-surface interaction:

$$V(l) = T\sum_{s=0}^{\infty}\alpha(i\,|\zeta_s|)\left[\mathcal{D}_{ll}^{\mathrm{E}}(\zeta_s; \mathbf{r}, \mathbf{r}')\right]_{\mathbf{r}\to\mathbf{r}'\to\mathbf{r}_a} \,, \tag{13.8}$$

where $\mathcal{D}_{ik}^E(\zeta_s; \mathbf{r}, \mathbf{r}')$ is the Matsubara Green's function of the electric field for a dielectric half-space. The dielectric permittivity is equal to $\varepsilon(\omega)$ at $z < 0$ and is equal to 1 at $z > 0$. The function $\mathcal{D}_{ik}^{\mathrm{E}}$ satisfies the equation:

$$\left[\partial_i \partial_l - \delta_{il}\Delta + \frac{\zeta_s^2}{c^2}\varepsilon(i\,|\zeta_s|, \mathbf{r})\delta_{il}\right]\mathcal{D}_{lk}^{\mathrm{E}}(\zeta_s; \mathbf{r}, \mathbf{r}') = 4\pi\hbar\frac{\zeta_s^2}{c^2}\delta_{ik}\delta(\mathbf{r} - \mathbf{r}').\qquad(13.9)$$

Equation (13.8) allows us to obtain the general Lifshitz result for the atom-surface interaction in a much simpler way than in the original paper by Lifshitz (1956), because it is much easier to solve equation (13.9) for half-space than for the two-body geometry proposed by Lifshitz. We will show in this note that, using equation (13.8), one can obtain the limiting equations (13.2) and (13.5) in a unified and very simple way. The point is that relativistic retardation effects are not important in both limiting cases defined by equations (13.3) and (13.4). Indeed, the general condition of neglecting retardation, i.e., the condition of the quasistationarity of the field, is

$$\omega \ll c/l.\qquad(13.10)$$

It is obvious from equation (13.2) that in the London case the main contribution in the interaction is given by frequencies of the order of ω_0. Then, the inequality (13.3) ensures equation (13.10). This means that the atom is just in the "near-zone" of the fluctuating field (Landau and Lifshitz, 1975), where the retardation is not important. Analogously, for the "Lifshitz distances" determined by equation (13.4), only the $s = 0$ term in equation (13.8) is important. The problem becomes pure static and the retardation is not important again. So the problem is to calculate the Green's function neglecting the relativistic retardation effects. This problem will be solved in the next section.

13.3 Longitudinal Green's function

It is important for the derivation to separate the Green's function into its longitudinal and transverse parts, i.e., to present it in the form

$$\mathcal{D}_{ik}^{\mathrm{E}} = \mathcal{D}_{ik}^{\mathrm{E}(\mathrm{l})} + \mathcal{D}_{ik}^{\mathrm{E}(\mathrm{t})},\qquad(13.11)$$

where

$$(\mathrm{rot})_{il}\,\mathcal{D}_{lk}^{\mathrm{E}(\mathrm{l})} = 0 \quad\text{and}\quad \partial_l\mathcal{D}_{lk}^{\mathrm{E}(\mathrm{t})} = 0.\qquad(13.12)$$

The central point of the derivation is the statement that smallness of the retardation effects means that

$$\left|\mathcal{D}_{ik}^{\mathrm{E}(\mathrm{t})}\right| \ll \left|\mathcal{D}_{ik}^{\mathrm{E}(\mathrm{l})}\right|,\qquad(13.13)$$

i.e., that only fluctuations of the electrostatic nature are important. The meaning of equation (13.13) can be illustrated by calculations of $\mathcal{D}_{ik}^{\mathrm{E}}$ for a uniform dielectric, where one can easily calculate the coordinate Fourier transform of the function

$$\mathcal{D}_{ik}^{E}\left(\zeta_s; \mathbf{k}\right) \equiv \frac{k_i k_k}{k^2}\mathcal{D}^{E(l)} + \left(\delta_{ik} - \frac{k_i k_k}{k^2}\right)\mathcal{D}^{E(t)}. \tag{13.14}$$

It is easy to show that in this case:

$$\mathcal{D}^{E(l)}\left(\zeta_s; \mathbf{k}\right) = \frac{4\pi\hbar c^2}{\varepsilon(i\,|\zeta_s|)} \;,\quad \mathcal{D}^{E(t)} = \frac{4\pi\hbar\zeta_s^2 c^2}{\zeta_s^2\varepsilon(i\,|\zeta_s|)/c^2 + k^2}. \tag{13.15}$$

With the condition $\zeta_s^2 \ll k^2 c^2$, which in this case substitutes equation (13.13), we find

$$\mathcal{D}^{E(t)} \sim \left(\zeta_s/kc\right)^2\mathcal{D}^{E(l)} \ll \mathcal{D}^{E(l)}. \tag{13.16}$$

It is not difficult to prove that with the condition (13.10) the inequality (13.13) is valid also in the case of a half-space. Indeed, for this condition, the terms with space derivatives on the left-hand side of equation (13.9) are large in comparison with the last term. However, these terms actually combine in the operator $(\text{curlcurl})_{il}$ and disappear, acting on $\mathcal{D}_{ik}^{E(l)}\left(\zeta_s; \mathbf{r}, \mathbf{r}'\right)$, but not on $\mathcal{D}_{ik}^{E(t)}\left(\zeta_s; \mathbf{r}, \mathbf{r}'\right)$. An estimate of terms gives then equation (13.13). Thus, to obtain the asymptotic equations (13.2) and (13.5), we must calculate the longitudinal function $\mathcal{D}_{ik}^{E(l)}\left(\zeta_s; \mathbf{r}, \mathbf{r}'\right)$. To eliminate the transverse part of equation (13.8), let us apply the operator $\partial_i\partial_k'$ to both sides of equation (13.9). We obtain:

$$\partial_i\partial_k' D_{lk}^{E(l)} = -4\pi\hbar\Delta'\delta\left(\mathbf{r} - \mathbf{r}'\right). \tag{13.17}$$

The first equation (13.12) can be satisfied identically if we introduce a scalar function φ according to Volokitin and Persson (2001):

$$D_{ik}^{E} \approx D_{ik}^{E(l)} = \hbar\partial_i\partial_k'\varphi. \tag{13.18}$$

Substitution of equation (13.18) into equation (13.17) gives the equation for φ:

$$\Delta\varphi = -4\pi\delta\left(\mathbf{r} - \mathbf{r}'\right), \tag{13.19}$$

if the point $\mathbf{r}'$ is outside the dielectric. This is an equation for the potential of a unit charge, placed in point $\mathbf{r}'$ in vacuum near the surface of the dielectric. It is not difficult to check that, if φ satisfies the proper boundary conditions for this electrostatic problem,

$$\varphi_{z\to-0} = \varphi_{z\to+0}\;,\quad \varepsilon\left(\partial_z\varphi\right)_{z\to-0} = \left(\partial_z\varphi\right)_{z\to+0}, \tag{13.20}$$

and the boundary conditions for D_{ik}^{E} will be also satisfied. The electrostatic problem is solved in Problem 1 of § 7 in the book by Landau and Lifshitz (1984). The solution is:

$$\varphi = \frac{1}{|\mathbf{r} - \mathbf{r}'|} + \frac{1 - \varepsilon}{1 + \varepsilon}\frac{1}{\left[\left(z + z'\right)^2 + \left(\mathbf{x} - \mathbf{x}'\right)^2\right]^{1/2}}, \tag{13.21}$$

where $\mathbf{r} = \{z, \mathbf{x}\}$.

The first term in equation (13.21) does not depend on the presence of the dielectric and it must be omitted. Differentiating the second term to calculate $\mathcal{D}_{lk}^{E}(\zeta_{s}; \mathbf{r}, \mathbf{r}')$ and going to the limit $\mathbf{x} \to \mathbf{x}'$, $z \to z' \to l$, we find after simple calculations that

$$\mathcal{D}_{ii}^{E}(\zeta_{s}; \mathbf{r}, \mathbf{r}) = -\hbar \frac{1 - \varepsilon(i|\zeta_{s}|)}{1 + \varepsilon(i|\zeta_{s}|)} \frac{1}{2l^{3}} \, . \tag{13.22}$$

To obtain an equation for the London case $(T = 0)$, we must substitute equation (13.22) into equation (13.8) and change the summation to integration with respect to ζ:

$$\sum_{n} \to \int \dots dn = \int \dots \frac{\hbar}{2\pi T} d\zeta \, . \tag{13.23}$$

This gives immediately equation (13.2). The Lifshitz case, equation (13.5), is obtained if we substitute equation (13.22) into (equation 13.8) and take only the $n = 0$ term of the sum.

Notice that the results which we discussed are valid only in the state of complete thermodynamic equilibrium. In particular, equation (13.8) is valid only if the temperature of the body is equal to the temperature of the black-body radiations, which drops on the body from the $z > 0$ half-space. If these two temperatures are different, the interaction energy decays at $l \to \infty$ according to a slow law $1/l^{2}$ (Antezza *et al*, 2005).

13.4 Conclusion

This is a great pleasure for me to publish this short note in a book in honor of Leonid Keldysh. I wish him good health and permanent luck in his work. During the writing of this paper, I recollected when I saw Leonid for the first time.

It was about 1958. I was a Ph.D. student at the Institute for Physical Problems, at the Theoretical Department, where Landau was the Chair. Once I was walking along a corridor in the Institute and noticed Landau who was accompanied by a quite young person. Landau told me: "Hello. This fellow investigated the creation of electron–hole pairs in a semiconductor by an electric field. He claims that he calculated even the pre-exponential factor. Here he is wrong, of course. But to calculate the exponent is also not so bad." They entered the premises of the Department and I went to work. About an hour later Landau found me in a room for Ph.D. students. He said: "Can you imagine? This person, Keldysh, calculated the prefactor in a nice, completely correct way!"

References

Antezza, M., Pitaevskii, L. P. and Stringari, S. (2005). *Phys. Rev. Lett.* **95**, 113202.

Dzyaloshinskii, I. E. and Pitaevskii, L. P. (1959). *Zh. Eksp. Teor. Fiz.* **36**, 1797 [*Sov. Phys. JETP* **9**, 1282].

Landau, L. D. and Lifshitz, E. M. (1975). *The Classical Theory of Fields*. Pergamon Press, Oxford.

Landau, L. D. and Lifshitz, E. M. (1984). *Electrodynamics of Continuous Media*. Pergamon Press, Oxford.

Lifshitz, E. M. (1955). *Zh. Eksp. Teor. Fiz.* **29**, 94 [*Sov. Phys. JETP* **2**, 73 (1956)].

Lifshitz, E. M. and Pitaevskii, L. P. (1984). *Statistical Physics, Part II*. Pergamon Press, Oxford.

Volokitin, A. I. and Persson, B. N. J. (2001). *Phys. Rev.* B **63**, 205404.

14

MODERN TRENDS IN SEMICONDUCTOR SPINTRONICS

Emmanuel I. Rashba
Department of Physics, Harvard University,
Cambridge, Massachusetts 02138, USA

Abstract
Dedication: Dedicated to Leonid V. Keldysh on the occasion of his 75th birthday.

The fundamentals of semiconductor spintronics and some recent developments in this field are reviewed. Special attention is paid to electrical manipulation of electron spin, spin injection, and producing nonequilibrium spin populations via the spin–orbit interaction.

14.1   Introduction

Spin is the only internal degree of freedom of an electron. Traditionally, solid state physics relevant to electron charge and electron spin were only loosely related. With very few exceptions, semiconductor electronics and optoelectronics are based on employing the electron charge only. Currently, miniaturization of semiconductor devices approaches the quantum limits established by mesoscopic physics. At this spatial scale, in strongly confined systems, the basic laws of nature involve electron spin actively in most of electron phenomena. In physical processes in semiconductor micro- and nanostructures investigated during the last decade, electron charge and electron spin contribute on an equal footing. It is widely believed that the new physics based on the involvement of electron spin in the phenomena that were traditionally regarded as "charge related", will transform semiconductor electronics into spintronics. Combining electrical, magnetic, and optical phenomena for designing devices with new functionalities is the ambitious goal of semiconductor spintronics.

Many of the spectacular phenomena that are the subject of current research on low-dimensional semiconductor microstructures, like quantum wells, are based on physical mechanisms that bear similarity to the mechanisms discovered and investigated in the epoch when our world was still perceived as three-dimensional (3D) and artificial low-dimensional structures were considered as exotic. It was curiosity-driven research without any direct connection to applications. A considerable part of it was performed in the former Soviet Union. The underlying physics was widely discussed at public events, like conferences and seminars, and during private discussions inside the community where Leonid Keldysh has always been a distinguished member since publishing his very first paper in which the Franz–Keldysh effect was discovered. Therefore, I feel it is appropriate to review in this volume the essential basics and some of the more recent developments in spintronics, especially because its theoretical apparatus heavily relies on the Keldysh technique for nonequilibrium responses (Keldysh, 1965). This is true for analytical theories, calculational procedures and also for simulations of finite systems of various geometries.

14.2   Electrical spin operation

The final goal of spintronics is operating the electron spin at nanometer scales. Time-dependent magnetic fields $\boldsymbol{B}(t)$ usually have large spatial scale while the spatial scale for time-dependent electric fields $\boldsymbol{E}(t)$ applied through gates may be much less. Such fields couple to the electron spin through the spin–orbit interaction. For nonrelativistic electrons in vacuum it is described by the Thomas term

$$H_{\mathrm{so}}(\boldsymbol{r}) = \left(\frac{e\hbar}{4m_0 c^2} \right) \, \boldsymbol{\sigma} \cdot [\nabla V(\boldsymbol{r}) \times \boldsymbol{v}], \qquad (14.1)$$

where $\boldsymbol{v}$ is the electron velocity and $V(\boldsymbol{r})$ is the scalar potential. Because of the Dirac gap $2m_0 c^2$ in the denominator, in vacuum this interaction is very weak for a moderate and smooth external potential. As a result, electrical operation of the electron spin in vacuum is less effective than the magnetic operation through the Zeeman interaction by a factor $v/c \ll 1$. The situation changes drastically for an electron in a crystal because of the strong periodical potential of the crystal lattice and the large electron velocity $\boldsymbol{v}$ near nuclei. In this case, the Dirac gap is replaced by the forbidden gap $E_{\mathrm{G}} \sim 1$ eV, and the enhancement factor is typically as large as $m_0 c^2 / E_{\mathrm{G}} \sim 10^6$. Of course, there is an additional factor $\Delta_{\mathrm{SO}}/E_{\mathrm{G}}$, Δ_{SO} being the spin–orbit splitting of the bands. It is small for light elements, but of the order of unity for the compounds from the middle and lower parts of the periodic table. It is this enhancement factor that can make the electrical operation of electron spin in crystals highly efficient (Rashba, 1960). Electric dipole spin resonance (EDSR), driven by a field $\boldsymbol{E}(t)$ of frequency ω, which coincides with the spin flip frequency ω_{s}, has been observed in InSb by McCombe *et al.* (1967), and in a number of later studies.

An additional factor influencing spin–orbit coupling is the symmetry of the system. The effective Hamiltonian of spin–orbit coupling depends on the momentum $\boldsymbol{k}$, $H_{\mathrm{SO}} = H_{\mathrm{SO}}(\boldsymbol{k})$. This Hamiltonian is an invariant of the appropriate symmetry group (Bir and Pikus, 1974; Winkler, 2003). The lower the symmetry, the lower the powers of $\boldsymbol{k}$ appearing in $H_{\mathrm{SO}} = H_{\mathrm{SO}}(\boldsymbol{k})$, and the spin–orbit coupling is stronger. For electrons in the zinc blende modification of $A_3 B_5$ crystals, the expansion of $H_{\mathrm{SO}} = H_{\mathrm{SO}}(\boldsymbol{k})$ begins with the k^3 terms (Dresselhaus, 1955). However, for the wurtzite modification of the same crystals or for electrons confined in an asymmetric quantum well, the Hamiltonian is

$$H_\alpha = \alpha(\boldsymbol{\sigma} \times \boldsymbol{k}) \cdot \hat{\mathbf{z}}, \qquad (14.2)$$

where $\hat{\mathbf{z}}$ is a unit vector along the hexagonal axis or the confinement direction of a quantum well and α is a spin–orbit coupling constant. H_α is linear in $\mathbf{k}$, usually dominates in narrow gap systems, and is frequently used as a model Hamiltonian for 2D systems. Electrically driven spin dynamics in quantum wells strongly depends on the polarization of the field $\mathbf{E}(t)$. The EDSR is particularly strong with in-plane field $\mathbf{E}(t)$ while it is suppressed with $\mathbf{E}(t)\|\hat{\mathbf{z}}$ because orbital dynamics in the z direction is restricted by the confining potential (Rashba and Efros, 2004). Schulte *et al.* (2005) reported dominance of the EDSR over the traditional electron paramagnetic resonance (EPR) by a factor of about 10^4 for AlAs quantum wells, which is typical for weak spin–orbit coupling. The EDSR has been seen in in-plane polarization, in agreement with the above arguments. A theory of the shape of motionally-narrowed EDSR line has been developed by Duckheim and Loss (2006).

Besides spin–orbit coupling coming from orbital motion through the modified Thomas term, the spatial dependence of the Zeeman term provides an additional mechanism of spin–orbit coupling (Pekar and Rashba, 1964). Kato *et al.* (2003) took advantage of the small g-factor of $Al_x Ga_{1-x} As$ alloys that crosses zero at appropriate content x and used the dependence of the $\hat{g}$-tensor on the spatial coordinate z in the growth direction. Hence, the z-dependence of the Hamiltonian $H_Z = (\boldsymbol{\sigma}\hat{g}(z)\mathbf{B})$ was strong and efficient spin operation through the time-dependent gate voltage was achieved.

These examples illustrate the large variety of options that spin–orbit coupling provides for electric-field-driven spin dynamics in semiconductors, with different mechanisms and strength of this coupling.

14.3 Spin interference phenomena

The spin interference device proposed by Datta and Das (1990), usually termed as spin transistor, became one of the most influential concepts in the field of semiconductor spintronics. The basic idea of it was formulated in terms of the Hamiltonian H_α. The energy spectrum consists of two chiral branches with energies

$$\epsilon_\lambda(k) = \left(\frac{\hbar^2 k^2}{2m}\right) + \lambda\alpha k, \ \lambda = \pm 1. \tag{14.3}$$

For the eigenstates, the spin of electrons propagating in the $\mathbf{k}$ direction is polarized $\perp \mathbf{k}$; in mixed states, the spin precesses about an effective magnetic field $\mathbf{B}_\alpha = 2\alpha(\mathbf{k}\times\hat{\mathbf{z}})/g\mu_B$. This precession, described by a characteristic momentum $k_\alpha = m\alpha/\hbar^2$ and characteristic length $\ell_\alpha = \hbar^2/m\alpha$, was observed by Kalevich and Korenev (1990). If two ferromagnets with the magnetization F_1 and F_2 parallel to the propagation direction $\mathbf{k}$ serve as a source and drain, the conductivity of a straight wire of length L reaches a maximum when $Lk_\alpha = \pi n$ and a minimum when $Lk_\alpha = \pi(2n+1)/2$, with integer n. Actually, the same result follows when considering interference of electrons propagating with the same energy ϵ but with different momenta k_λ related by the equation $\epsilon_\lambda(k) = \epsilon$.

The basic ideas underlying the spin transistor are (i) injecting spin polarized electrons into a specimen with spin split bands, (ii) using and detecting spin-dependent electron phases, and (iii) electrical control of the spin–orbit coupling constant α by the gate voltage. Gate control of α has been achieved (Nitta *et al.*, 1997; Engels *et al.*, 1997), but a working device has never been reported. Nevertheless, the ideas underlying the spin transistor concept strongly influenced the research that followed.

The quantum phases acquired by electrons due to spin–orbit coupling are related to Berry phases (Aronov and Lyanda-Geller, 1993) and the role of singular points in electron trajectories have been clarified (Bercioux *et al.*, 2005). Extensive literature on the subject, including Berry phases in magnetic textures, is currently available. On the experimental side, there are several reports on observing these phases and gate control of them, of which the most recent are the papers by Koga *et al.* (2005) on loop arrays in InGaAs quantum wells and by Koenig *et al.* (2005) on HgTe quantum rings.

From the conceptual point of view, these recent experimental achievements support the spin transistor concept because they convincingly demonstrate the modulation of electrical resistance based on gate controlled spin interference. In these experiments quantum loops have been used instead of straight wires and, what is critically important, such setups do not require injection of spin polarized electrons which constitutes a problem by itself as will be explained in the next section.

14.4   Spin injection: Optical and electrical

Spin injection into semiconductors was achieved long ago through excitation by circularly polarized light (for a review see Meier and Zakharchenya, 1984). The discovery of the D'yakonov-Perel (1972) spin relaxation mechanism and prediction of the photogalvanic effect (Ivchenko and Pikus, 1978; Belinicher, 1978) strongly influenced the spin injection research. The former mechanism originates from spin precession in the field $\boldsymbol{B}_\alpha(\boldsymbol{k})$ interrupted by fast collisions and results in spin relaxation time $\tau_s^{-1} \approx \tau_p(2\alpha k_F/\hbar)^2$ and spin diffusion length-scale of ℓ_α; τ_p being the momentum relaxation time and k_F the Fermi momentum. The photogalvanic effect is a charge current induced by optically produced nonequilibrium spin polarization. It critically depends on the light polarization and allows detecting nonequilibrium spin populations by electrical measurements.

More recent research was focused on electrical spin injection that is of higher importance for device applications.

The discovery of spin injection from metallic ferromagnets into superconductors (Tedrow and Meservey, 1973) and paramagnetic metals (Johnson and Silsbee, 1985) suggested the use of similar sources for spin injection into a semiconductor spin transistor. However, early attempts at achieving this goal were unsuccessful, and the origin of the failure has been understood in terms of conductivity mismatch (Schmidt *et al.*, 2000). Namely, spin flow across a contact of two conductors, which are different in electron spin polarization degree, is controlled by the conductor with higher resistance. Hence, the spin injection co-efficient γ into a semiconductor from a ferromagnetic metal can be evaluated through the ratio of their conductivities, $\gamma \sim \sigma_S/\sigma_F \sim 10^{-5}$. This conductivity mismatch can be remedied by using resistive spin-selective contacts (tunnel or Schottky) between a metallic ferromagnet and a semiconductor, or by employing semimagnetic semiconductors or half-metals as spin sources. The last five years have witnessed an impressing increase in the spin injection coefficient, but the ultimate solution has not yet been found; for a review see Žutić *et al.* (2004). Another important problem regarding ferromagnetic spin emitters, irrespective of their nature, is the stray magnetic fields that perturb electron spin dynamics. Therefore, various concepts of spin emitters that do not include ferromagnetic elements and are based completely on spin–orbit coupling have been proposed.

One of the ideas was initiated by Voskoboynikov *et al.* (1999) and de Andrada e Silva and La Rocca (1999) and is based on the fact that electrons described by the Hamiltonian H_α are spin polarized by the effective field $\boldsymbol{B}_\alpha(\boldsymbol{k})$ either along this field or opposite to it. However, because spin–orbit coupling respects time inversion symmetry, the total spin magnetization averaged over the equilibrium distribution vanishes. Violating time reversal symmetry by inducing the in-plane electron drift and filtering the nonequilibrium part of spin populations by electron tunneling across a double or a triple barrier should produce spin polarized tunnel current. A prototype resonant-tunneling device fabricated with an AlSb/InAs/GaSb/AlSb heterostructure has been reported by Moon *et al.* (2004).

Spin injection can be achieved in a two-dimensional device without magnetic elements (Eto *et al.*, 2005; Silvestrov and Mishchenko, 2005). However, multi-terminal quantum rings were proposed as more efficient spin injectors, including injectors of pure spin currents, i.e., spin currents that are not accompanied by charge currents (Kiselev and Kim, 2003; Souma and Nikolić, 2005).

Furthermore, spin–orbit coupling allows creating beams of spin-polarized electrons either by means of magnetic focusing in a weak perpendicular field or even in the absence of it. Folk *et al.* (2003) achieved emission of spin-polarized electrons from an open quantum dot by applying an in-plane magnetic field. Rokhinson *et al.* (2004) employed spin–orbit coupling in a bulk crystal to spatially separate, in a magnetic focusing experiment, two spin components of a spin-unpolarized hole beam injected from a quantum point contact. Media with spin-split energy spectrum are similar to double-refractive crystals, hence, they allow spatial separation of two spin-polarized components of an originally unpolarized electron beam. Khodas *et al.* (2004) have established conditions under which such separation can be efficient. Spin polarization of electrons scattered from a lithographic barrier has already been achieved by Chen *et al.* (2005).

Generally, electron dynamics in media with spin–orbit coupling includes a number of exciting aspects. Spatially dependent α provides a solid-state analog of the Stern–Gerlach experiment that is free of complications related to the effect of the Lorentz force (Ohe *et al.*, 2005), and various types of spin–orbit coupling result in different versions of Zitterbewegung (Schliemann *et al.*, 2005; Zawadzki, 2005).

14.5   Transport in media with spin–orbit coupling

With the spin–orbit coupling playing the central role in semiconductor spintronics, developing a consistent theory of spin transport in systems with spin–orbit coupling is an important and also demanding goal. The eventful history of a related phenomenon, the anomalous Hall effect (AHE), indicates the existence of essential theoretical problems. As distinct from the regular Hall effect, the AHE is a transverse charge current due to electron magnetization M rather than due to an external magnetic field B; therefore, it is fundamentally based on spin–orbit coupling. The theory of AHE has been initiated by Karplus and Luttinger (1954), and 20 year long efforts resulted in the conclusion that the AHE is an extrinsic effect controlled by competing terms related to electron scattering. Extensive cancellations make calculations tricky (Nozières and Lewiner, 1973). Some of the extrinsic contributions to the anomalous Hall current do not depend on the scattering time, and recent research indicates existence of an intrinsic contribution to the anomalous Hall current that can be expressed in terms of the Berry curvature in k-space (Jungwirth *et al.*, 2002; Onoda and Nagaosa, 2002; Haldane, 2004). Experiment still does not provide any convincing evidence regarding the mechanisms involved in the AHE and this challenging problem remains a subject of current research.

The theory of spin transport meets even more fundamental difficulties. In the absence of spin–orbit coupling, spin transport in ferromagnets can be comfortably described in the framework of the Mott two-channel model whenever $\tau_\mathrm{p} \ll \tau_\mathrm{s}$. Indeed, under these conditions spin is conserved and spin transport theory is similar to the charge transport theory. Such an approach is routinely used in the theory of spin injection from ferromagnets into semiconductors, see Section 14.4. However, spin nonconservation stemming from spin–orbit coupling changes the situation drastically. The absence of the continuity equation for spin makes the very existence of a consistent definition of spin current problematic. The absence of a magnetization current in the Maxwellian equations implies a similar conclusion. Only the spin magnetization $M_\mathrm{s}(r)$ is an observable quantity, hence, it should play the central role in spin transport theory.

Under these considerations, two approaches have been applied. First, in an attempt to develop a theory of spin transport similarly to the theory of charge transport, the following definition of spin currents was applied (e.g., Murakami *et al.*, 2003; Sinova *et al.*, 2004):

$$j_j^i = \frac{1}{2} \langle \sigma_i v_j + v_j \sigma_i \rangle. \tag{14.4}$$

Here $\langle \ldots \rangle$ stands for the integration over the electron distribution function, and the anticommutator of v_j and σ_i is used because in media with spin–orbit coupling the velocity $v = \hbar^{-1} \partial H(k)/\partial k$ depends on the Pauli matrices $\boldsymbol{\sigma}$. This

definition is the simplest one and is widely used. However, because spin noncon-
servarion makes the physical meaning of j_j^i obscure, different definitions have
also been proposed; some of them result in the opposite sign of j_j^i. Second, because only spin magnetization is measured, Boltzmann and diffusive equations
have been derived for it (Mishchenko *et al.*, 2004; Burkov *et al.*, 2004; Tang *et
al.*, 2005; Liu and Lei, 2005; Khaetskii, 2006; Shytov *et al.*, 2006; Tse and Das
Sarma, 2006); such an approach does not rely on the spin current concept.

Because the energy spectrum of equation (14.3) comprises two branches, both
interbranch and intrabranch transitions contribute to j_j^i. Remarkably, when the
chemical potential satisfies $\mu > 0$, the interbranch contribution to the spin conductivity is universal in a perfect crystal, $j_x^z/E_y = e/4\pi\hbar$ (Sinova *et al.*, 2004). To
find the intrabranch contribution one has to violate momentum conservation by
impurities or by a magnetic field. Then, the inter- and intraband contributions
cancel (Inoue *et al.*, 2004), and this is an intrinsic property of the free electron
Hamiltonian of equation (14.3) (Dimitrova, 2005). For spin–orbit Hamiltonians
including higher powers of k, spin currents do not vanish, e.g., for heavy holes
because their spectrum includes k^3 terms.

14.6   Macro- and mesoscopic spin Hall effect

An in-plane electric field $\boldsymbol{E}$ induces homogeneous spin polarization of H_α electrons in the $(\boldsymbol{E} \times \hat{\mathbf{z}})$ direction (Edelstein, 1990; Aronov *et al.*, 1991). This was
observed by several groups (Silov *et al.*, 2004; Kato *et al.*, 2004a; Ganichev *et
al.*, 2006). A different related effect is the spin Hall effect (D'yakonov and Perel,
1971), that usually manifests itself as spin polarization $\boldsymbol{S}\|\hat{\mathbf{z}}$ near sample edges.
It has been observed in bulk *n*-GaAs by Kato *et al.* (2004) with distinct signa-
tures of the extrinsic mechanism. The magnitude of the effect is in agreement
with a theory, with no free parameters (Engel *et al.*, 2005; Tse and Das Sarma,
2006). On the contrary, a stronger effect observed in *p*-GaAs by Wunderlich *et
al.* (2005) was ascribed to the intrinsic spin Hall effect (Nomura *et al.*, 2005).

It is tempting to attribute spin accumulation near sample flanks to the bulk
current j_x^z and estimate it as $S_z \approx (\hbar/2)(j_x^z\ell_\alpha/D)$, D being the diffusion coefficient. Such an estimate is based on the assumption that bulk currents j_x^z have a
meaning of transport currents. However, several arguments indicate strict limitations on such an assumption. First, for H_α-electrons the bulk current vanishes,
$j_x^z = 0$. However, spins can accumulate at the flanks under the condition that
they can either cross them and move into $\alpha = 0$ regions (Adagideli and Bauer,
2005) or relax on them (Rashba, 2005); this seeming paradox is related to spin
nonconservation. Second, spin currents j_j^i are even with respect to time inversion,
hence, this tensor does not necessarily vanish in equilibrium while spin transport
and spin accumulation vanish.

Spin accumulation can be found only through solving the transport problem.
However, evaluation of the spatial dispersion of $j_j^i(q)$ suggests the following con-
jecture (Rashba, 2005). Fourier components of spin currents of equation (14.4)

at the wavevector k_{so} are scaled by the "universal" value of $eE/\hbar$ (Sinova *et al.*, 2004) and can be employed for estimating spin transport rate. Hence

$$j_{sH}^z(k_{so}) \sim \frac{eE}{\hbar}, \quad S^z(k_{so})/\hbar \sim \frac{1}{\hbar}k_{so}\tau eE, \quad (14.5)$$

where k_{so} is the spin precession momentum that reduces to k_α for H_α-electrons. Numerical coefficients in these equations depend on the specific form of H_{SO} and on the boundary conditions. The first of equations (14.5) indicates $e/\hbar$ as the "universal" scale for spin conductivities of two-dimensional systems.

The above discussion is applicable only to macroscopic samples of size $L \gtrsim \ell_\alpha$. For spintronic applications, mesoscopic three- or four-terminal devices with $L \lesssim \ell_\alpha$ seem to be of most importance. Applying the Keldysh technique in conjunction with the Landauer–Büttiker formalism to the mesoscopic spin Hall effect proved to be successful (Nikolić *et al.*, 2005). As far as numerical data and analytical estimates can be compared, numerical data seem to support the estimates of equations (14.5).

14.7 Related systems

A theory of quantized spin Hall conductivity through the edge states of insulating graphene was developed by Kane and Mele (2001) in the framework of the Haldane (1988) model, and the phase diagram for the dissipationless regime was found. These spin currents show remarkable robustness to a random potential (Sheng *et al.*, 2005).

For typical semiconductors, spin–orbit splitting is not strong and $k_\alpha \ll k_F$, hence, most theories have been developed for this limit. Recently, large spin–orbit splitting of surface states have been observed for metals and semimetals, of which the largest one is $\hbar^2 k_\alpha^2/2m \approx 0.2$ eV (Ast *et al.*, 2005). If such states can be used for interference devices, their sizes can be drastically reduced. There is current interest in a noncentrosymmetric superconductor Ce_2Pt_3Si where spin–orbit splittings at the Fermi level of 0.1 eV have been estimated (Mineev and Samokhin, 2005). Finally, spin–orbit effects in hopping conductors (Entin-Wohlman *et al.*, 2005) and optical lattices (Dudarev *et al.*, 2004) were also discussed.

The review does not cover a number of different avenues of semiconductor spintronics such as coherent operation with quantum dots in the context of quantum computing (Loss and DiVincenzo, 1998; Petta *et al.*, 2005).

Acknowledgments

I am very grateful to Dr. H.-A. Engel for reading the manuscript and making a number of valuable suggestions and to Prof. B. I. Halperin for inspiring discussions. Financial support from Harvard Center for Nanoscale Systems is gratefully acknowledged.

References

Adagideli, I. and Bauer, G.E.W. (2005). *Phys. Rev. Lett.* **95**, 256602.

Aronov, A. G. and Lyanda–Geller, Y. B. (1993). *Phys. Rev. Lett.* **70**, 343.

Aronov, A. G., Lyanda–Geller, Y. B. and Pikus, G. E. (1991). *Zh. Eksp. Teor. Fiz.* **100**, 973 [*Sov. Phys. JETP.* **73**, 537].

Ast, C. R., Pacilé, D., Falub, M., Moreschini, M., Papagno, M., Wittich, G., Wahl, P., Vogelgesang, R., Grioni, M. and Kern, K. (2005). *Cond-mat*/0509509, and references therein.

Belinicher, V. I. (1978). *Phys. Lett.* A **66**, 213.

Bercioux, D., Frustaglia, D. and Governale, M. (2005). *Phys. Rev.* B **72**, 113310.

Bir, G. L. and Pikus, G. E. (1974). *Symmetry and Strain-Induced Effects in Semiconductors.* Wiley, NY.

Burkov, A. A., Núñez, A. S. and MacDonald, A. H. (2004). *Phys. Rev.* B **70**, 155.

Datta, S. and Das, B. (1990). *Appl. Phys. Lett.* **56**, 665.

de Andrada e Silva, E. A. and La Rocca, G. C. (1999). *Phys. Rev.* B **59**, R15583.

Dimitrova, O. V. (2005). *Phys. Rev.* B **71**, 245327.

Dresselhaus, G. (1955). *Phys. Rev.* **100**, 580.

Duckheim, M. and Loss, D. (2006). *Nature Physics* **2**, 195.

Dudarev, A. M., Diener, R. B., Carusotto, I. and Niu, Q. (2004). *Phys. Rev. Lett.* **92**, 153005.

D'yakonov, M. I., and Perel', V. I. (1971). *Phys. Lett.* A **35**, 459.

D'yakonov, M. I. and Perel', V. I. (1972). *Fiz. Tv. Tela* **14**, 1452 [*Sov. Phys. Solid State.* **13**, 3023].

Edelstein, V. M. (1990). *Solid State Commun..* **73**, 233.

Engel H.-A., Halperin, B. I. and Rashba, E. I. (2005). *Phys. Rev. Lett.* **95**, 166605.

Engels, G., Lange, J., Schäpers, Th. and Lüth, H. (1997). *Phys. Rev..* B **55**, R1958.

Entin-Wohlman, O., Aharony, A., Galperin, Y. M., Kozub, V. I. and Vinokur V. (2005). *Phys. Rev. Lett.* **95**, 086603, and references therein.

Eto, M., Hayashi, T. and Kurotani M. (2005). *J. Phys. Soc. Jpn..* **74**, 1934.

Folk, J. A., Potok, R. M., Marcus, C. M. and Umansky, V. (2003). *Science* **299**, 679.

Ganichev, S. D., Danilov, S. N., Schneider, P., Bel'kov, V. V., Golub, L. E., Wegscheider, W., Weiss, D. and Prettl, W. (2006). *J. Magn. Magn. Mat..* **300**, 127.

Haldane, F. D. M. (1988). *Phys. Rev. Lett.* **61**, 2015.

Haldane, F.D.M. (2004). *Phys. Rev. Lett.* **93**, 206602.

Inoue, J., Bauer, G. E. W. and Molenkamp, L. W. (2004). *Phys. Rev..* B **70**, 041303(R).

Ivchenko, E. L. and Pikus, G. E. (1978). *Pisma Zh. Eksp. Teor. Fiz.* **27**, 640 [*Sov. Phys. JETP Lett.* **27**, 604].

Johnson, M. and Silsbee, R. H. (1985). *Phys. Rev. Lett.* **55**, 1790.

Jungwirth, T., Niu, Q. and MacDonald, A. H. (2002). *Phys. Rev. Lett.* **88**, 207208.

Kalevich, V. K. and Korenev, V. L. (1990). *Pisma Zh. Eksp. Teor. Fiz.* **52**, 859 [*Sov. Phys. JETP Lett.* **52**, 230].

Kane, C. L. and Mele, E. J. (2005). *Phys. Rev. Lett.* **95**, 226801.

Karplus, R. and Luttinger, J. M. (1954). *Phys. Rev.* **95**, 1154.

Kato, Y., Myers, R. C., Driscoll, D. C., Gossard, A. C., Levy, J. and Awschalom, D. D. (2003). *Science* **299**, 1201.

Kato, Y. K., Myers, R. C., Gossard, A. C. and Awschalom, D. D. (2004a). *Phys. Rev. Lett.* **93**, 176601.

Kato, Y. K., Myers, R. C., Gossard, A. C. and Awschalom, D. D. (2004b). *Science* **306**, 1910.

Keldysh, L. V. (1964). *Zh. Eksp. Teor. Fiz.* **47**, 1515 [(1965). *Sov. Phys. JETP* **20**, 1018].

Khaetskii, A. (2006). *Phys. Rev. Lett.* **96**, 056602.

Khodas, M., Shekhter, A. and Finkel'stein, A. M. (2004). *Phys. Rev. Lett.* **92**, 086602.

Kiselev, A. A. and Kim, K. W. (2003). *J. Appl. Phys.* **94**, 4001.

Koenig, M., Tschetschetkin, A., Hankiewicz, E., Sinova, J., Hock, V., Daumer, V., Schaefer, M., Becker, C., Buhmann, H. and Molenkamp, L. (2006). *Phys. Rev. Lett.* **96**, 076804.

Koga, T., Sekine, Y. and Nitta J. (2006). *Phys. Rev.* B **74**, 041302.

Liu, S. Y. and Lei, X. L. (2005). *Phys. Rev.* B **72**, 155314.

Loss, D. and DiVincenzo, D. P. (1998). *Phys. Rev.* A **57**, 120.

McCombe B. D., Bishop, S. G. and Kaplan, R. (1967). *Phys. Rev. Lett.* **18**, 748.

Meier, F. and Zakharchenya, B. P. eds (1984) *Optical Orientation.* North-Holland, Amsterdam.

Mineev, V. P. and Samokhin, K. V. (2005). *Phys. Rev.* B **72**, 212504.

Mishchenko, E. G., Shytov, A. V. and Halperin, B. I. (2004). *Phys. Rev. Lett.* **93**, 226602.

Moon, J. S., Chow, D. H., Schulman, J. N., Deelman, P., Zinck, J. J. and Ting, D. Z.- Y. (2004). *Appl. Phys. Lett.* **85**, 678.

Murakami, S., Nagaosa, N. and Zhang, S.-C. (2003). *Science* **301**, 1348.

Nikolić, B. K., Zârbo, L. P. and Souma, S. (2005). *Phys. Rev.* B **72**, 075361.

Nitta J., Akazaki, T., Takayanagi, H. and Enoki, T. (1997). *Phys. Rev. Lett.* **78**, 1335.

Nomura, K., Wunderlich, J., Sinova, J., Kaestner, B., MacDonald, A. H. and Jungwirth, T. (2005). *Phys. Rev.* B **72**, 245330.

Nozières, P. and Lewiner, C. (1973). *J. Phys. (Paris)* **34**, 901.

Ohe, J., Yamamoto, M., Ohtsuki, T. and Nitta, J. (2005). *Phys. Rev.* B **72**, 041308.

Onoda, M. and Nagaosa, N. (2002). *J. Phys. Soc. Jpn.* **71**, 19.

Pekar, S. I. and Rashba, E. I. (1964). *Zh. Eksp. Teor. Fiz.* **47**, 1927 [(1965) *Sov. Phys. JETP* **20**, 1295].

Petta, J. R., Johnson, A. C., Taylor, J. M., Laird, E. A., Yacoby, A., Lukin, M. D., Marcus, C. M., Hanson, M. P. and Gossard, A. C. (2005). *Science* **309**, 2180.

Rashba, E. I. (1960). *Fiz. Tv. Tela* **2**, 1224 [*Sov. Phys. Solid State* **2**, 1109].

Rashba, E. I. (2006). *Physica* E **34**, 31.

Rashba, E. I. and Efros, A. L. (2003). *Appl. Phys. Lett.* **83**, 5295.

Rokhinson, L. P., Larkina, V., Lyanda-Geller, Y. B., Pfeiffer, L. N. and West, K. W. (2004). *Phys. Rev. Lett.* **93**, 146601.

Schliemann, J., Loss, D. and Westervelt, R. M. (2005). *Phys. Rev. Lett.* **94**, 206801.

Schmidt, G., Ferrand, D., Molenkamp, L. W., Filip, A. T. and van Wees, B. J. (2000). *Phys. Rev.* B **62**, R4790.

Schulte M., Lok, J. G. S., Denninger, G. and Dietsche, W. (2005). *Phys. Rev. Lett.* **94**, 137601.

Sheng, L, Sheng, D. N., Ting, C. S. and Haldane, F. D. M. (2005). *Phys. Rev. Lett.* **95**, 136602.

Shytov, A. V., Mishchenko, E. G., Engel, H.-A. and Halperin, B. I. (2006). *Phys. Rev.* B **73**, 075316.

Silov A. Yu., Blajnov, P. A., Wolter, J. H., Hey, R., Ploog, K. H. and Averkiev, N. S. (2004). *Appl. Phys. Lett.* **85**, 5929.

Silvestrov, P. G. and Mishchenko, E. G. (2006). *Phys. Rev.* B **74**, 165301.

Sinova, J., Culcer, D., Niu, Q., Sinitsyn, N. A., Jungwirth, T. and MacDonald, A. H. (2004). *Phys. Rev. Lett.* **92**, 126603.

Souma, S. and Nikolić, B. (2005). *Phys. Rev. Lett.* **94**, 106602.

Tang, C. S., Mal'shukov, A. G. and Chao, K. A. (2005). *Phys. Rev.* B **71**, 195314.

Tedrow, P. M. and Meservey, R. (1973). *Phys. Rev.* B **7**, 318.

Tse, W.-K. and Das Sarma, S. (2006). *Phys. Rev. Lett.* **96**, 056601.

Voskoboynikov, A., Liu, S. S. and Lee, C. P. (1999). *Phys. Rev.* B **59**, 12514.

Winkler, R. (2003). *Spin-Orbit Coupling Effects in Two-Dimensional Electron and Hole System.* Springer, Berlin.

Wunderlich, J., Kaestner, B., Sinova, J. and Jungwirth, T. (2005). *Phys. Rev. Lett.* **94**, 047204.

Zawadzki, W. (2005). *Phys. Rev.* B **72**, 085217.

Žutić, I., Fabian, J. and Das Sarma, S. (2004). *Rev. Mod. Phys.* **76**, 323.

15

EXCITONIC INSULATORS, ELECTRON–HOLE LIQUIDS AND METAL–INSULATOR TRANSITIONS

T. Maurice Rice
Theoretische Physik, ETH-Zurich, 8093 Zurich, Switzerland

Abstract
The concepts of an excitonic insulator and of an electron–hole liquid go back many years to early work by Keldysh and collaborators. The term excitonic insulator was introduced to describe a Bose condensate of excitons which could possibly appear in a band insulator with a bandgap smaller than the excitonic binding energy. Later Keldysh proposed that a fluid of electrons and holes could have a metallic liquid ground state rather than a Bose condensate of excitons as an explanation of the luminescence spectrum of optically pumped Ge. The stability of a metallic electron–hole liquid in Ge and Si raised doubts that an excitonic insulator could be realized. Recently the necessary conditions of an electron–hole fluid with identical Fermi surfaces was realized in a quantum Hall bilayer system leading to an electron–hole condensate with spectacular properties. These developments have consequences for the nature of the metal to insulator transition when overlapping bands in a semimetal uncross to form a semiconductor and also may enter in the case of the Mott transition to a localized insulating state.

15.1 Introduction

The study of phase transitions between metallic and insulating phases has been a topic of continuing interest from the early days of solid state physics down to the present. The first paper that I am aware of on this topic was written by no less figures than Landau and Zeldovich (1943) during World War II at a time when life in Russia was very difficult, to put it mildly. Later came the key works by Mott (1949, 1956, 1961) on the role of the electron–electron interaction in driving a metal to an insulator as the lattice is expanded and by Anderson (1958) on the metal to insulator transition that occurs with increasing disorder. Of course Leonid Keldysh also contributed in important ways introducing to the subject the concepts of the excitonic insulator and the electron–hole liquid. So it is very appropriate to include a review of the development of these concepts and their relation to metal–insulator transitions in this volume dedicated to Leonid Keldysh. Metal–insulator transitions occur in a number of ways and this short review is restricted to metal–insulator transitions in ordered materials where these concepts are relevant.

The explanation of the difference between metallic and insulating solids was one of the earliest successes of the quantum theory of solids. Bloch energy bands, which are the energy levels of a single electron moving in the periodic potential of the crystal, can be exactly filled in a stoichiometric material with an even number of electrons per unit cell. This leads to a finite energy gap between the ground state and all excited states. Such a material will have insulating behavior. On the other hand, if the Bloch bands are partially filled, there is no energy gap and a current-carrying state can have an energy arbitrarily close to the ground state. This can happen with an even number of electrons per unit cell if Bloch bands overlap at the Fermi (i.e., the highest occupied) energy. It must be the case if the number of electrons per unit cell is odd, and then the single-electron theory predicts metallic behavior.

It was soon recognized that exceptions to this rule occurred. In particular de Boer and Verwey (1937) raised the case of NiO, which is an insulator despite the fact that the Ni-derived $3d$ bands are partially filled. The explanation was quickly provided in a remark by Peierls, who pointed out that the Coulomb interaction between electrons could destroy metallic behavior. This idea was extended and formalized by Mott (1949, 1956, 1961). In Mott insulators, electrons are localized to individual atoms by the Coulomb interaction, and a finite energy is required to move an electron from one atom to another. The localized electrons have magnetic moments from spin, and possibly orbital, degrees of freedom, and these moments usually undergo an ordering transition as the temperature is lowered. This magnetic behavior distinguishes Mott insulators from band insulators. There is no – or at least only a small – energy gap in the spin degrees of freedom, but there is a large energy gap for the charge degrees of freedom, and it is this charge gap that is responsible for the insulating behavior.

Metal–insulator transitions in ordered materials can be divided into two categories. The first occurs in systems with an even number of electrons per unit cell when the overlap between the valence and conduction bands is made to vanish. The second occurs in systems with an odd number of electrons per unit cell as the lattice is expanded leading to a transition into a Mott insulating state. These two categories will be reviewed in turn. The case of the Mott transition has generated strong interest over the years with a corresponding large literature which cannot be covered in this brief review. Fortunately an excellent and comprehensive review was written a few years ago by Imada, Fujimori and Tokura (1998).

15.2   Excitonic insulators, electron–hole liquids and metal–insulator transitions due to band crossing

As mentioned previously metallic or insulating states occur in a crystal depending on whether there is an overlap or an energy gap between the conduction and valence Bloch bands. Thus a transition between insulator and metal occurs if the bands which are initially separated by a bandgap are made to overlap by the application of an external perturbation such as pressure. Of course there are many materials where the two phases metal and insulator, have different crystal structures and then the transition is simply a form of crystallographic phase transition. The transition is more interesting without a crystallographic phase change. In this case, band theory predicts a simple continuous change as the bandgap changes to a band overlap. Initially Peierls (quoted by Landau and Zeldovich (1943)) and Mott (1949, 1956, 1961) argued that the fact that a finite density of electrons and holes would be needed to screen the long-range Coulomb force and suppress electron–hole binding, requires the metal–insulator transition to be first order. Later Keldysh and Kopaev (1964) pointed out that a continuous transition was in principle possible. Because of the finite exciton binding energy, the energy to create an exciton would go to zero before the bandgap, E_g, between conduction and valence bands went to zero. The result would be a condensation of excitons leading to a state that acquired the name excitonic insulator (for a review see Halperin and Rice, 1968a).

The case of an indirect band crossing is easiest to analyze. In the case of a direct bandgap, the nature of the bands enters as direct dipole valence – conduction band transition may be allowed leading to a divergence of the dielectric constant as the bandgap goes to zero (Halperin and Rice, 1968b). In the case of an indirect bandgap the low-energy electron–hole transitions do not contribute to the long-wavelength dielectric constant so that this can be taken as a constant independent of the bandgap, E_g. The exciton condensation leads to a symmetry breaking phase transition with an order parameter describing coherent hybridization of the Bloch states near the conduction band minima with those near the valence band maxima, i.e., $\langle a^+_{\mathbf{k},\sigma} b_{\mathbf{k}+Q,\sigma^1} \rangle$ where $a^+_{k,\sigma}$ $\left(b^+_{k+Q,\sigma^1} \right)$ are electron creation operators for conduction (valence) bands with extrema separated by a wavevector, Q. Such an order parameter describes a spin or charge density wave or possibly even an orbital current wave depending on the relative strengths and form of the residual interactions (Halperin and Rice, 1968a). The prediction then is that such modulated state should appear continuously as the indirect bandgap, E_g is lowered in a semiconductor.

On the semimetallic side, i.e., in the case of an overlap of the conduction and valence bands, $E_g < 0$, Keldysh and Kopaev (1964) pointed out that the electron–hole Coulombic attraction could still lead to a binding. The Hamiltonian in the effective mass approximation takes the form

$$H = -\frac{1}{2m_e} \sum_i \nabla_i^2 - \frac{1}{2m_h} \sum_j \nabla_j^2 + \frac{1}{2} \sum_{i \neq j} \frac{e^2}{\kappa |r_i^e - r_j^e|}$$

$$+ \frac{1}{2} \sum_{i \neq j} \frac{e^2}{\kappa |r_i^h - r_j^h|} - \sum_{i,j} \frac{e^2}{\kappa |r_i^e - r_j^h|} \tag{15.1}$$

with κ as the background dielectric constant. The characteristic energy scale is the binding energy of a single electron–hole pair which form an exciton, $E_x = (\mu / m\kappa^2) \, Ry$ where μ is the reduced mass $\left(\mu^{-1} = m_e^{-1} + m_h^{-1} \right)$. The corresponding length is the scaled Bohr radius, $a_x = (m\kappa/\mu) \, a_0$. Keldysh and Kopaev

(1964) pointed out that this Hamiltonian describes two Fermi fluids, electron and hole, with an attractive interaction between them and this will lead to a form of BCS pairing between them. Certainly this occurs in the ideal case with a single isotropic minimum in the conduction band and maximum in the valence band. Then the analogy to BCS theory is exact. The order parameter that results from this pairing is just the same as that which appears on the semiconducting side. In general however the band structure will deviate from the ideal form. This can lead to a critical value of the band overlap, $|E_g|$, and limit the existence of the excitonic insulator on the semimetallic side. The result would be that the broken symmetry excitonic insulator would be limited to a dome centered on $E_g = 0$ between the semimetallic and semiconducting regions.

To decide if this analysis is correct one needs to know the nature of the ground state of a fixed number of electrons and holes. At low densities, n, the energy per electron–hole pair will be slightly lower than E_x since two excitons can bind to form an excitonic molecule although this is only weakly bound for $m_e \sim m_h$. At high densities the system is two interpenetrating Fermi fluids of electrons and holes with an energy per electron–hole pair

$$E_M = \frac{3}{5}\left(E_F^e + E_F^h\right) + E_{xc}(n),\tag{15.2}$$

where E_F is the Fermi energy and E_{xc} the sum of exchange and correlation energies.

In 1968 Keldysh inspired by experiments on optically pumped Ge and Si, suggested that a high-density metallic state could have a lower energy with the result that the low-density gas of excitons would condense into a metallic high-density electron–hole liquid. This suggestion was soon confirmed by analysis of the line shape of the weak luminescence signal as the radiative recombination process requires emission of a phonon in these indirect gap semiconductors (for reviews of the electron–hole liquid see (Rice, 1977, Hensel *et al.*, 1977, Jeffries and Keldysh, 1983)). The slow recombination rate of the electrons and holes allows them to relax to the lattice temperature and creates ideal conditions to observe the electron–hole fluid at low temperatures. Microscopic calculations soon reached the same conclusion. In both Ge and Si there are many minima in the conduction band (4 and 6 respectively) and also the maximum in the valence band is doubly degenerate. The result is that E_F and so the kinetic energy term is greatly reduced while the interaction terms in E_{xc} are largely insensitive to these details of the band structure. The result is a deep minimum in $E_M(n)$ at a relatively high density, n_0 ($E_M(n_0) \approx -1.6E_x$). The study of the electron–hole liquid developed into an elegant subfield of semiconductor physics with excellent agreement between theory and experiment (Rice, 1977; Hensel *et al.*, 1977; Jeffries and Keldysh, 1983).

In their original work Keldysh and Kopaev (1964) considered the simplest case of a metallic liquid spherical and nondegenerate electron and hole band extrema. The Coulomb attraction leads to electron–hole pairing analogous to that of BCS theory and the resulting state is a neutral condensate of electron–hole pairs – the high-density analog of the excitonic condensate that appears when the exciton energy goes soft resulting in the excitonic insulator. This electron–hole condensate does not form in the electron–hole liquid in Ge and Si. The reason is simple. The degeneracies and anisotropies in the electron and hole bands lead to very different Fermi surfaces even though their densities are the same and so destroy the analogy with BCS theory which requires identical electron and hole Fermi surfaces. The result is that the normal Fermi liquid state is stable in the case of Ge and Si.

The enhanced stability of the high-density metallic state has also strong implications for the form of a insulator to metal transition in indirect bandgap semiconductors as pointed out by Brinkman and Rice (1973). The study of an excitonic instability at the semiconductor to semimetal transition reduces to the study of electrons and holes at a constant chemical potential rather than at a constant density. In the presence of a bandgap one adds a term $H' = (E_{\mathrm{g}}/2)\,(N_e + N_h)$ to equation (15.1) and the number of electrons (N_e) and holes (N_h) is allowed to vary. This term acts as a negative chemical potential. The proposition that excitonic phases occur near zero gap is equivalent to the proposition that in an electron–hole gas a Bose condensate phase of excitons is the stable phase at some value of the chemical potential. However, in the case of band degeneracies and anisotropies the state with minimum energy is the simple metallic liquid. Therefore, for such semiconductors if one were to reduce E_{g} continuously, a first-order transition would occur directly between the semiconducting and semimetallic phases when $E_{\mathrm{g}} = |E_{\mathrm{M}}(n_0)|$ bypassing all excitonic phases.

The idea of an exciton condensate is very appealing and inspired a lot of discussion down the years. The absence of a condensate in the cases of Ge and Si however was a disappointment but did not of course rule out the possibility completely. The stability of a metallic state is common in many bosonic systems, e.g., when one has a fluid of alkali atoms. Clearly the best way to achieve a bosonic condensate is to make the exciton–exciton interaction repulsive rather than the usual van der Waals attraction. Actually there is a simple way to make the exciton–exciton interaction repulsive. In a bilayer quantum well system in which the electrons and holes are confined to separate but nearby quantum wells, each exciton acquires an electric dipole moment. Since these dipole moments are parallel a repulsive interaction results rather than a van der Waals attraction (Lozovik and Yudson, 1976; Shevchenko, 1976). Further it also reduces the recombination rate which favors a low-temperature fluid. However, attempts to realize an exciton condensate in this way have not yet been successful.

An exciton condensate, however, has recently been realized and studied in an elegant series of experiments on a quantum Hall bilayer. A very readable review of these works has recently appeared written by two of the leading contributors, Eisenstein and MacDonald (2004). In this system two electron quantum wells with equal electron densities are separated by a barrier. An external magnetic field is tuned so that in each layer the lowest Landau level is half-filled. In this case each level has a Fermi liquid ground state. Applying an electron–hole transformation in one of the layers transforms the description of the system into an electron–hole bilayer system with equal numbers of electrons and holes in a nondegenerate ideal band structure. These are the ideal conditions for the Keldysh–Kopaev electron–hole paired state to form. By suitable adjustment of the barrier height and thickness it is possible to achieve conditions with a strong Coulomb attraction between the electrons and holes in the bilayer. Further the interlayer tunneling rate versus interlayer voltage can be monitored. When the layers are far apart there is a minimum tunneling rate at small interlayer voltages but there is dramatic transition to a tunneling spectrum with a strong narrow peak at zero voltage as the interlayer separation is narrowed. This narrow peak is a consequence of the formation of a coherent condensate of electron–hole pairs. In this condensate each electron lies opposite a hole in the other layer enabling hugely enhanced interlayer tunneling.

Another manifestation of electron–hole pairing appears when a parallel voltage is applied to one layer and an extremely large counterflow of electrons is observed in the other layer. The direct observation of the exciton condensate as a real phase different from just strong electron–hole correlations requires more subtle experiments as reviewed by Eisenstein and MacDonald (2004). They discuss the case of the Hall voltage in the bilayer system in the presence of counterflowing currents in the two layers. At a general magnetic field the Hall voltage in the two layers will be of opposite sign. However when the magnetic field is adjusted so that the lowest Landau level is just half-filled in each layer leading to the exciton condensate, then the Hall voltage vanishes since in this case the flowing objects are neutral electron–hole pairs which do not experience a Lorentz force. This vanishing of the Hall effect has been observed recently by two groups (Kellogg *et al.*, 2004; Tutuc *et al.*, 2004), confirming the existence of the exciton condensate.

One question of special interest is the existence of a small dissipation rather than true superflow in these experiments. The important issue here is whether the condensate order parameter, composed of a pair of fermions, breaks gauge symmetry or not. As discussed many years ago by Kohn and Sherrington (1970) this issue distinguishes the usual superconductor with electron–electron pairs from the excitonic condensate of electron–hole pairs. Some form of dissipation is to be expected in the latter case due to coupling of the order parameter to imperfections, etc. The exact nature of the dissipative processes that result is still an open topic (Eisenstein and MacDonald, 2004).

This evolution over many years from the original proposal of Keldysh and Kopaev to its final realization is a nice example that shows how an original and insightful theory eventually finds confirmation and realization.

15.3 The Mott transition

We turn now to the case of the Mott transition from a metal to a Coulomb localized insulator. This is a very active field (Imada *et al.*, 1998) and only aspects related to the topic discussed above will be covered in this brief review. In the simplest view, a Mott transition arises because of the competition between the intra-atomic Coulomb interaction U, which opposes charge fluctuation, and the bandwidth W, which represents the kinetic energy that can be gained by allowing charged excitations to propagate through the crystal. The criterion separating metallic from insulating behavior then is determined by the comparison $U \lesssim W$ or $U \gtrsim W$. Since U is an intra-atomic Coulomb interaction, it should not be sensitive to pressure, unlike bandwidth, W. Therefore, in principle it should be possible to convert all Mott insulators into metals by applying enough pressure. In practice this is not so easy since Mott insulators generally have large energy gaps for charge excitations; e.g., NiO, the classic Mott insulator, has an energy gap ≈ 5 eV. Thus very large pressures are usually required, which, in turn, limits detailed experimental study.

The simplest model that includes both the kinetic energy and intra-atomic Coulomb interaction is the one band Hubbard model at half-filling with one electron per site. A huge effort has gone into analysis of this model with only limited success. The two limiting cases are clear. When $U \ll W$ the system is simply metallic described by Landau Fermi liquid theory (leaving aside special cases with a perfectly nesting Fermi surface). In the opposite limit $U \gg W$ the system has a charge gap to form a separated pair of doubly occupied and empty sites with an approximate magnitude $U - W$. The physics in both limits is clear but very different and the challenge is to describe the transition between these two different states. In addition one must always keep in mind the effects of the long-range part of the Coulomb interaction as well as the intra-atomic interaction that drives the Mott transition.

Starting from the metallic side the first issue is how the interactions modify the properties of the metal as the transition to insulating behavior is approached. In Landau theory the interactions are treated perturbatively. The Fermi surface retains its full volume corresponding to a density of one electron per site but other properties such as the effective mass, Pauli spin susceptibility and charge compressibility are renormalized by the interactions. Brinkman and Rice (1970) calculated these renormalizations using the Gutzwiller approximation which introduces numerical renormalization factors to describe the effects of the reduced density of doubly occupied and empty sites as U increases. The key renormalizations are an enhanced effective mass, m^*, and spin susceptibility, χ, and a reduced compressibility, κ

$$\frac{m^*}{m} = \frac{1}{(1 - U/U_c)} \xrightarrow[U \to U_c]{} \infty, \tag{15.3}$$

$$\frac{\chi}{\chi_0} = \frac{m^*}{m} \left(1 + \frac{U}{U_c}\right)^2 \xrightarrow[U \to U_c]{} 4\left(\frac{m^*}{m}\right), \tag{15.4}$$

$$\frac{\kappa}{\kappa_0} = \frac{m^*}{m}\left(1 - \frac{U}{U_c}\right)^2 \xrightarrow[U \to U_c]{} 0. \qquad (15.5)$$

The Fermi surface volume remains intact but the enhanced effective mass means that the Drude weight in the dc conductivity which is proportional to n/m^* vanishes as U approaches the critical value U_c ($= 8|\bar\varepsilon|$ where $\bar\varepsilon$ is the average kinetic energy in the half-filled band). Note that the reduction in the compressibility leads to a vanishing Fermi-Thomas screening wavevector as $U \to U_c$ and the metal–insulator transition is approached.

Important limitations on this approach are that it does not describe the insulating state that occurs for $U > U_c$ and also does not describe fully the spectral properties of the single-particle Green's function, only those of the coherent quasiparticles near the Fermi energy. In addition it ignores that magnetic interactions which lead to antiferromagnetic order in the insulating state and which can cause precursor ordered phases in the metallic region as U nears U_c. The first two limitations have been rectified by the recently developed dynamic mean field theory (DMFT) (Georges *et al.*, 1996). In this approach the problem is mapped to an effective single impurity Anderson model which is then solved nonperturbatively. The low-energy properties in the metallic state agree well with the Gutzwiller approximation but the final transition to the metallic state is first order, not continuous. The approach however does not remedy the deficiencies in regard to the role of antiferromagnetism at the transition.

The greatest success of the Brinkman–Rice theory is the description of the Landau Fermi liquid properties of the Fermi liquid ^{3}He as the crystalline solid phase is approached under pressure (Vollhardt, 1984). In this case the antiferromagnetic (AF) interactions are absent since the real value of U here is effectively infinite. In particular the ratio of the enhancements of χ and m^* approaches with increasing pressure the predicted value of 4. This is a nontrivial result since a mapping of the liquid phase with infinite U to a half-filled lattice Hubbard model with $U \lesssim U_c$ is not at all obvious. It would seem that this factor of 4 between the enhancements of the spin susceptibility and effective mass has a deeper origin. A number of transition metal oxides show similar behavior in their metallic phase as the Mott insulating state is approached (de Boer and Verwey, 1937). The best known of these is V_2O_3 which inspired the original work. The metallic phase shows Landau Fermi liquid properties with enhanced m^* and χ. The Mott transition in V_2O_3 is first order between a AF ordered insulator and a metal which does not have a precursor AF ordered metallic state. This system is generally considered to be the best experimental example of a Mott transition under pressure. However closer examination has shown that V_2O_3 is a material with a complex band and magnetic structure and this inhibits a detailed comparison between theories based on simple Hubbard models and experiment (Shiina *et al.*, 2001). Recent progress on the theory side allows a more realistic treatment of the full electronic structure within the DMFT approach and gives hope that better theoretical descriptions will be available in the future (Keller *et al.*, 2004).

The biggest deficiency of these approaches is the omission of the antiferromagnetic interactions which are present in Hubbard models at finite U and in real Mott insulators through superexchange interactions. In the Hubbard model for $U \gg W$ virtual hopping processes generate an AF Heisenberg spin–spin coupling, $J = 4t^2/U$, where t is the nearest neighbor hopping element. In general the resulting Heisenberg $S = 1/2$ AF model has an AF ordered ground state. On the metallic side the onset of AF order depends sensitively on the electronic band structure. For example if one takes a simple cubic lattice and only nearest neighbor hopping processes the resulting Fermi surface is perfectly nested at half-filling in one, two or three dimensions. This leads to a simple AF ordered ground state for arbitrarily small U which opens an energy gap over the whole Fermi surface. Since this state is a simple commensurate ordered state the Umklapp processes pin the AF ordered state leading to an insulating ground state. In this case there is never a metallic ground state. However, if longer range hopping processes are introduced, the Fermi surface is no longer perfectly nested and a finite value of U is required to stabilize a magnetic state. Further the magnetic order may be an incommensurate spin density wave (SDW). This will only partially truncate the Fermi surface in general. The volume enclosed by the Fermi surface will be finite but reduced from the paramagnet. A further increase in U will lead in general to a modification of the **Q**-vector of the SDW and a further reduction in the Fermi surface. The evolution to a complete truncation of the Fermi surface and a commensurate value of **Q** will depend sensitively on the details of the electronic band structure. While there is no nice example of a metal–insulator transition proceeding in this way, the case of Cr and its alloys (Fawcett *et al.*, 1994) offers a nice experimental example. Pure Cr has an incommensurate SDW ground state and a partially truncated Fermi surface. Changing the electron-atom ratio leads to an evolution in the **Q**-vector of the SDW state. Reducing the electron-atom ratio by alloying V leads to a rapid suppression of the SDW state and a paramagnetic metal with a full Fermi surface. Increasing the electron-atom ratio by alloying with Mn quickly leads to a simply commensurate AF ordered state with a stronger truncation ($\sim 50\%$) of the paramagnetic Fermi surface. A similar evolution should occur for a general electronic band structure in the half-filled Hubbard model through a series of incommensurate SDW states with partially truncated Fermi surfaces. Finally the Umklapp processes at larger values of the ratio U/W stabilize an AF commensurately ordered and insulating state beyond a critical value of U/W. In this case the Fermi surface is progressively truncated to zero, possibly by a series of transitions that change the **Q**-vector with increasing U/W. The nature of the final transition to the insulator will be influenced by the long-range part of the Coulomb interaction as discussed earlier.

The discussion on the metal–insulator transition in transition metal oxides has been very strongly influenced by the intense study of the cuprates in the past two decades triggered by the discovery of high-temperature superconductivity. Their simple planar square lattice structure with only a single band near the Fermi level is an additional attraction. When stoichiometric, the cuprates are

AF ordered Mott insulators with an energy gap of order 2 eV. This large value of the energy gap in the stoichiometric insulators prevents a Mott transition into a metallic state under laboratory pressures. Of course the behavior of the cuprates upon even light doping is very fascinating. In particular the anomalous pseudogap phase of the underdoped cuprates is a challenge to our understanding of Fermi liquids (Timusk and Statt, 1999). On the one hand the deviations from the perturbative Landau theory of Fermi liquids are striking, e.g., thermally activated spin susceptibility, Fermi arcs not a closed Fermi surface in angle resolved photoemission (ARPES) experiments, etc. However, there is no evidence that these anomalous properties are the result of a phase transition even in very well ordered underdoped materials, such as $YBa_2Cu_4O_8$. So the question is how to describe this anomalous phase. The most promising proposal was made early on by Anderson (1987) (also Anderson *et al.*, 2004) namely that this is a lightly doped resonant valence bond (RVB) state. A RVB state arises in an antiferromagnet if the quantum fluctuations are so strong as to prevent long-range AF order and stabilize instead a spin liquid with only short-range spin correlations. The case of $S = 1/2$ moments that occur in the cuprates is most favored in view of the large energy gain of the singlet state for two spins. The RVB concept is that the spin liquid should be viewed as a quantum mechanical superposition of many singlet pair configurations. The square lattice however does not have an RVB ground state, at least in the absence of strongly frustrating interactions. But the RVB proposal for the cuprates requires that only weak doping is sufficient to stabilize a doped RVB spin liquid and suppress the AF order.

All this makes it especially interesting to study the case of a Mott transition in the case where the Mott insulator is a spin liquid rather than an AF ordered insulator. However there are not many models where the insulator has strictly short-range order. One such example is the case of two leg ladders at half-filling. This system has been studied in great detail within the Hubbard model both at weak coupling, using RG and bosonization techniques, and strong coupling where the system reduces to a two-leg AF Heisenberg ladder (for a review see Dagotto and Rice, 1996). Actually the properties in both limits are similar and there is a continuous crossover as the strength of the Hubbard U is varied. The system is insulating for all values of U, with both charge and spin gaps. The charge gap is driven by Umklapp processes in the particle–hole and particle–particle channels. At small values of U there are enhanced correlations in the commensurate AF and "d-wave" pairing channels but in both cases the correlation functions decay exponentially due to the presence of both spin and charge gaps. In the weak coupling limit the two-leg Hubbard ladder at low energies is described by an effective field theory with dynamically generated SO(8) symmetry. The single electron Green's function was derived by Essler and Konik, (2005) and takes an interesting form

$$G_a^{\mathrm{L}}(k_x, \omega) = \frac{z_a(\omega + \varepsilon_a(k_x))}{\omega^2 - \varepsilon_a^2(k_x) - \triangle^2} + G_{\mathrm{inc}}^{\mathrm{L}}. \tag{15.6}$$

Here $\varepsilon_a\,(k_x)$ is bare dispersion of the bonding (antibonding) bands that cross the Fermi energy at zero energy and $\triangle$ is the single-particle energy gap. The first term on the right-hand side of equation (15.6) represents fully coherent quasi-particle bands with infinite lifetime since there are no allowed decay processes due to the finite energy gaps for charge and spin excitations. It is interesting to note that the coherent part has the BCS form but unlike BCS theory there is no anomalous part of the Green's function.

Recently Konik, Rice and Tsvelik 2005 (KRT) studied a higher dimensional model with long-range interladder hopping processes. By choosing a form for the interladder hopping $t_a(\mathbf{k}_\perp)$ that is sharply peaked in $\mathbf{k}_\perp$ space ($\mathbf{k}_\perp \perp k_x$ – the ladder direction) KRT could justify treating it within a random phase approximation (RPA). This led KRT to a form for the coherent part of the Green's function,

$$G_a^{\mathrm{RPA}}\,(\mathbf{k},\omega) \;=\; \frac{1}{G_a^{\mathrm{L}}\,(k_x,\omega)^{-1} - t_a\,(\mathbf{k}_\perp)}\,. \qquad (15.7)$$

The quasiparticle spectrum is given by solutions of

$$\omega - \varepsilon_a\,(k_x) - \frac{\triangle^2}{\omega + \varepsilon_a\,(k_x)} - t\,(\mathbf{k}_\perp) \;=\; 0 \qquad (15.8)$$

leading to a dispersion relation for the coherent quasiparticle bands $E_a\,(\mathbf{k})$

$$E_{a,\pm}\,(\mathbf{k}) \;=\; \frac{1}{2}t\,(\mathbf{k}_\perp) \pm \sqrt{\left[\varepsilon_a\,(k_x) + \frac{t\,(\mathbf{k}_\perp)}{2}\right]^2 + \triangle^2} \;=\; 0\,. \qquad (15.9)$$

Particle–hole symmetry is maintained when $t(\mathbf{k}_\perp)$ is chosen to have sharp peaks with equal magnitude but opposite signs at $\mathbf{k}_\perp = 0$ and $\mathbf{k}_\perp = \mathbf{G}/2$. The system remains insulating until the quasiparticle bands cross the Fermi level. This occurs when $\max |t(k_\perp)| > 2\triangle$. At this point a set of Fermi pockets appears consisting of hole and electron ellipsoids on either side of lines with k_x fixed at one of the Fermi points of the noninteracting two-leg ladder.

KRT pointed out that the Luttinger Sum Rule relating the area where the zero frequency Green's function is positive, i.e., $G(\mathbf{k},0) > 0$, to the total density, is satisfied in a novel way in this doped spin liquid. As $|k_x|$ moves away from zero, there are sign changes in $G^{\mathrm{RPA}}(\mathbf{k},0)$ at 4 lines where $\varepsilon_a(k_x) = 0$, i.e., at the 4 Fermi points of isolated noninteracting ladders. There are also sign changes on the Fermi surface of the electron and hole pockets and the total electron density remains at one electron per site since the electron and hole pockets enclose equal areas.

In this KRT model the metal–insulator transition occurs through a shrinking of the Fermi pockets to zero as U and therefore the gap, $\triangle$, is increased. The form of the metal–insulator transition therefore is completely analogous to the case of a band crossing from semimetal to semiconductor. As discussed above this should be generally a first-order transition. There is however a question of the stability of the semimetallic state with respect to antiferromagnetism or '*d*-wave' superconductivity. An isolated two-leg ladder at half-filling has enhanced polarizabilities in these channels which will lead to enhanced residual interactions between the quasiparticle in the electron and hole pockets (Konik *et al.*, 2005) As a result for certain parameter choices there may be magnetic order on the semimetallic rather than in the insulating region at larger values of U.

15.4 Conclusions

Metal–insulator transitions is a many facetted subject and this brief review covers only a small part. The ideas that Leonid Keldysh introduced of a possible Bose condensed excitonic insulating phase and of a stable metallic electron–hole liquid that could occur in the vicinity of the metal–insulator transition, have been beautifully realized but in other contexts. Nonetheless these ideas have important consequences for the metal–insulator transition not only in the case of band crossing but also for the precise form that the metal to insulator transition takes in the Mott transition.

Acknowledgement

I am especially grateful to Alexei Tsvelik and Robert Konik at Brookhaven National Laboratory for their hospitality and scientific collaboration. Support from the MANEP program of the Swiss Nationalfonds is also gratefully acknowledged.

References

Anderson, P. W. (1958). *Phys. Rev.* **109**, 1492.

Anderson, P. W. (1987). *Science* **237**, 1196.

Anderson, P. W., Lee, P. A., Randeria, M., Rice, T. M., Trivedi, N. and Zhang, F. C. (2004). *J. Phys. C: Condensed Matter* **16**, R755.

de Boer, J. H. and Verwey, E. J. W. (1937). *Proc. Phys. Soc. London* A **49**, 59.

Brinkman, W. F. and Rice, T. M. (1970). *Phys. Rev.* B **2**, 4302.

Brinkman, W. F. and Rice, T. M. (1973). *Phys. Rev.* B **7**, 1508.

Dagotto, E. and Rice, T. M. (1996). *Science* **271**, 618.

Eisenstein, J. P. and MacDonald, A. H. (2004). *Nature* **432**, 691.

Essler, F. H. L. and Konik R. M. (2005). *From Fields to Strings: Circumnavigating Theoretical Physics* (ed by M. Shifman, A. Vainshtein and J. Wheater). World Scientific Singapore (see also *cond-mat*/0412421).

Fawcett, E., Alberts, H. L., Galkin, V. Yu., Noakes, D. R. and Yakhai, J. V. (1994). *Rev. Mod. Phys.* **66**, 25.

Georges, A., Kotliar, G., Krauth, W. and Rozenberg, M. J. (1996). *Rev. Mod. Phys.* **68**, 13.

Halperin, B. I. and Rice, T. M. (1968a). *Solid State Physics* **21**, 115.

Halperin, B. I. and Rice, T. M. (1968b). *Rev. Mod. Phys.* **40**, 755.

Hensel, J. C., Phillips, T. G. and Thomas, G. A. (1977). *Solid State Physics* **32**, 88.

Imada, M., Fujimori, A. and Tokura, Y. (1998). *Rev. Mod. Phys.* **70**, 1039.

Jeffries, C. D. and Keldysh, L. V., ed (1983). *Electron-Hole Droplets in Semiconductors*. North-Holland, Amsterdam.

Keldysh, L. V. (1968). *Proceedings of the Ninth International Conference on Semiconductors* Moscow (ed by S. M. Ryvkin and V. V. Shmastev). Nauka Leningrad, p. 1303.

Keldysh, L. V. and Kopaev, Yu. V. (1964). *Fiz. Tverd. Tela* **6**, 279 [*Sov. Phys., Solid State* **6**, 2219 (1965)].

Keller, G., Held, K., Eyert, V., Vollhardt, D. and Anisimov, V. I. (2004). *Phys. Rev. B* **70**, 205116.

Kellogg, M., Eisenstein, J. P., Pfeiffer, L. N. and West, K. W. (2004). *Phys. Rev. Lett.* **93**, 036801.

Kohn, W. and Sherrington, D. (1970). *Rev. Mod. Phys.* **42**, 1.

Konik, R. M., Rice, T. M. and Tsvelik, A. (2006). *Phys. Rev. Lett.* **96**, 086407. (see also *cond-mat*/0511268).

Landau, L. D. and Zeldovich, G. (1943). *Acta Phys. Chem. (USSR)* **18**, 194 (see *Collected Papers of L. D. Landau* (ed by D. ter Haar). Gordon and Breach, New York).

Lozovik, Yu. E. and Yudson, V. I. (1976). *Zh. Eksp. Teor. Fiz.* **71**, 738 [*Sov. Phys. JETP* **44**, 389 (1976)].

Mott, N. F. (1949). *Proc. Phys. Soc.* A **62**, 416.

Mott, N. F. (1956). *Can. J. Phys.* **34**, 1356.

Mott, N. F. (1961). *Phil. Mag.* **6**, 287.

Rice, T. M. (1977). *Solid State Physics* **32**, 1.

Shevchenko, S. I. (1976). *Fiz. Nizk. Temp.* **2**, 505 [*Sov. J. Low Temp. Phys.* **44**, 389].

Shiina, R., Mila, F., Zhang, F. C. and Rice, T. M. (2001). *Phys. Rev. B* **63**, 144422.

Timusk, T. and Statt, B. (1999). *Rep. Prog. Phys.* **62**, 61.

Tutuc, E., Shayegan, M. and Huse, D. (2004). *Phys. Rev. Lett.* **93**, 036802.

Vollhardt, D. (1984). *Rev. Mod. Phys.* **56**, 99.

16

ELECTRON–HOLE LIQUID IN SEMICONDUCTORS

N. N. Sibeldin

Lebedev Physical Institute, Russian Academy of Sciences, Moscow, Russia

Abstract

Condensation of excitons in semiconductors into an electron–hole liquid (EHL) is a beautiful and multifaceted phenomenon. In this paper, we survey its main features and summarize basic characteristics of the EHL, relying mainly on experimental studies of optically excited germanium and silicon. We consider microscopic properties of the EHL, internal and external factors that affect the stability of the liquid state, phase diagrams of the exciton gas–EHL system, kinetics of the condensation and recombination in this nonequilibrium system, processes that determine the size and concentration of the EHL droplets (EHDs) in the crystal, motion of the droplets in static and dynamic strain fields, giant EHDs, phonon-wind effects, spatial structure and dynamics of the EHD cloud.

16.1   Introduction

In his closing speech at the 9th International Conference on the Physics of Semiconductors held in Moscow, L. V. Keldysh (1968) proposed that excitons in a semiconductor may condense into a metallic electron–hole liquid (EHL) with electrons and holes (e-h) bound collectively by internal interaction forces. Such a phase transition is similar to the condensation of an alkali-metal atomic vapor into liquid metal. Like conventional gas–liquid transitions, exciton condensation is a first-order phase transition occurring, e.g., at temperatures lower than the critical one if the exciton gas density exceeds the "saturated-vapor density".

Not later than the next year, the EHL was discovered experimentally in germanium (Pokrovskii and Svistunova, 1969; Bagaev *et al.*, 1969; Vavilov *et al.*, 1969). Although less clearly, the existence of the liquid phase was also confirmed by Asnin and Rogachev (1969).

After these first results an intensive research of the exciton condensation started around the world. This considerable attention was caused by the beauty and unusual nature of the phenomenon itself, where a gas–liquid transition takes place in the nonequilibrium e-h system in a semiconductor, and by the diversity and uniqueness of the properties of EHL droplets. In addition, excitons in a semiconductor represent an excellent model system where the behavior of matter can be studied experimentally under conditions that are extremely difficult to implement in the laboratory for substances consisting of conventional atoms. For instance, a magnetic field as low as $\sim 0.1\,\mathrm{T}$ has the same effect upon the excitons in certain semiconductors as a field of $\sim 10^5$ T, unattainable in the laboratory, upon the hydrogen atom. In this sense, the EHL is a unique condensed-matter object suitable for laboratory experiments in ultrahigh magnetic fields, equivalent to astrophysical-scale fields acting on matter within stars.

By now it can be stated that the main properties of the exciton gas–EHL system are fairly well studied and, in general, understood for a number of semiconductors. The EHL was observed in Ge, Si, GaP, GaAs, CdS and many other semiconductors. In most detail, the EHL properties are investigated in germanium and silicon, where the material parameters and characteristics of excitons are rather favorable for the EHL observation.

Because of the space constraints we shall try, without going into much detail, to summarize the main features of the exciton condensation, primarily discussing experimental results related to bulk Ge and Si. The comprehensive information can be found in review articles and monographs (Keldysh, 1970; Keldysh, 1971; Keldysh, 1986; Pokrovskii, 1972; Bagaev, 1975; Jeffries, 1975; Voos and Benoît à la Guillaume, 1976; Rice, 1977; Hensel et al., 1977; Jeffries and Keldysh, 1983; Tikhodeev, 1985; Keldysh and Sibeldin, 1986)[10] and references therein.

16.2   Microscopic properties and main thermodynamic parameters of the EHL

A large-radius or Wannier–Mott exciton in semiconductor is an electron and a hole bound together by the Coulomb force. Exciton of this kind may be considered as a free neutral quasi-atom similar to the hydrogen, or, because the effective masses of electrons and holes are close, to the positronium. It should be stressed that, like the positronium, excitons have a finite lifetime. The ground-state energy and the effective Bohr radius of Wannier–Mott excitons are

$$E_{\mathrm{ex}} = -\frac{1}{2}\frac{m^* e^4}{\kappa^2 \hbar^2}, \quad a_{\mathrm{ex}} = \frac{\kappa \hbar^2}{m^* e^2}, \tag{16.1}$$

where κ is the static permittivity of the crystal, $m^* = (m_e^{-1} + m_h^{-1})^{-1}$, m_e and m_h are, respectively, the reduced, electron and hole effective masses.[11]

Coulomb interaction in a crystal is reduced due to the medium polarization and, as a rule, the e-h effective masses are smaller than that of free electron. Thus, the exciton binding energies are 2–4 orders of magnitude smaller than typical atomic energies, and their size considerably exceeds the interatomic distance in the crystal.

Usually, the exciton binding energy $|E_{\mathrm{ex}}| \ll E_{\mathrm{g}}$, where E_{g} is the bandgap. At temperatures sufficiently low, $k_{\mathrm{B}} T < |E_{\mathrm{ex}}|$, there are virtually no excitons at equilibrium, and excitation of the crystal is required for their generation. In most cases, a sample is pumped by light with the photon energy exceeding E_{g}. Electrons and holes produced by the optical excitation thermalize rapidly and

---

[10]See also contributions of Kuwata-Gonokami and Rice in this volume.

[11]For a detailed description of excitons see (Rashba and Sturge, 1982).

bind together into excitons. At low excitation levels, an insulating ideal Bose gas of excitons with a relatively low concentration is formed in the crystal.

In many respects, excitonic systems behave like ordinary atomic systems. As their density increases, interaction between the excitons can lead to the formation of biexcitons (exciton molecules) and condensation into a liquid phase, a bound state of macroscopically large number of particles. The effective e-h masses are usually comparable, and heavy particles like atomic nuclei are absent in this system. Consequently, quantum effects such as delocalization of particles are prominent. By this reason, the amplitude of zero-point oscillations in the excitonic molecule is large, and its dissociation energy is quite small (below $0.1|E_{\mathrm{ex}}|$)).[12] Zero-point oscillations also bring about instability (melting) of the excitonic or e-h crystal, and, in the EHL state, they cause delocalization of electrons and holes over the entire liquid volume. In other words, rather than being excitonic or biexcitonic, the EHL is a metallic state with collectivized electrons and holes (Keldysh, 1968).

The main EHL thermodynamic parameters are its density n_0 and binding energy $\varphi_0 = E_{\mathrm{ex}} - E_0$ (the exciton work function in the EHL). Here E_0 is the e-h pair energy in the liquid ground state (at $T = 0$). At high concentrations ($na_{\mathrm{ex}}^3 \sim 1$), the e-h system is a dense plasma. Its potential energy is of the order of the kinetic energy. The total energy per e-h pair in the plasma can be written as a function of density as[13]

$$E(n) = (3/5)(E_{\mathrm{F}}^e + E_{\mathrm{F}}^h) + E_{\mathrm{exch}} + E_{\mathrm{corr}} \,. \qquad (16.2)$$

The first term here is the average e-h pair kinetic energy in the plasma. At low temperatures ($k_{\mathrm{B}}T \ll E_{\mathrm{F}}^{e,h}$), the electrons and holes are degenerate. For isotropic parabolic nondegenerate bands, the Fermi energies and wavevector are

$$E_{\mathrm{F}}^{e,h} = \hbar^2 k_{\mathrm{F}}^2/2m_{e,h}, \qquad k_{\mathrm{F}} = \pi(3n/\pi)^{1/3} \,, \qquad (16.3)$$

where spin-related degeneracy is taken into account.

The last two terms in equation (16.2) account for Coulomb-interaction potential energy. Although plasma as a whole is locally neutral, there are spatial correlations in the mutual arrangement of particles that lead to a decrease in its total energy. Correlations due to Pauli principle, which prohibits two identical particles with parallel spins from occupying the same space region, reduce the Coulomb-interaction energy by the exchange energy $E_{\mathrm{exch}} \propto n^{1/3}$. Correlations of dynamic origin, related to the repulsion between similarly charged particles and attraction between oppositely charged ones, are responsible for the correlation energy E_{corr}. Like the exchange energy, E_{corr} is negative in accordance with its physical meaning and its absolute value increases with the plasma density.

---

[12]In what follows, we disregard biexcitons. Their role is surveyed in (Rice, 1977; Keldysh, 1983; Keldysh, 1986; Kulakovskii and Timofeev, 1983; Ivanov *et al.*, 1998).

[13]A detailed discussion of the EHL theory can be found in review articles (Rice, 1977; Vashishta *et al.*, 1983; Kulakovskii and Timofeev, 1983; Tikhodeev, 1985).

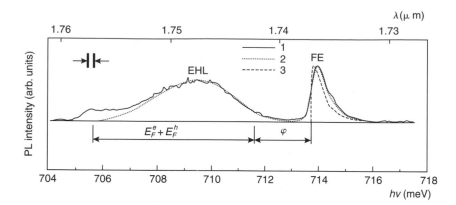

FIG. 16.1. LA components of the EHL and FE luminescence spectrum of bulk
Ge recorded at $T = 3.04\,\mathrm{K}$ (curve 1) and fitted with and without (curves
2 and 3) allowance for a finite spectrometer resolution. From (Lo, 1974).
Reprinted figure with permission from Elsevier Ltd: [T. K. Lo, *Solid State Commun.* **15**,
1231 (1974)], copyright (1974).

The total e-h pair energy as a function of density, equation (16.2), exhibits
a minimum at the equilibrium density n_0. This means that dense plasma will
experience self-compression or self-expansion until it fills the volume corresponding to the equilibrium density. Thus, in contrast to the gas or ordinary plasma,
which tend to spread over the entire available space, the system under study
behaves as a liquid. The e-h liquid is a neutral quantum two-component (in the
simplest case) Fermi liquid, where the internal pressure of the degenerate plasma
is balanced by the Coulomb attraction forces. The energy per one e-h pair in the
ground state is $E_0 \equiv E(n_0)$.

As it was already explained, at low density the e-h system is a gas of excitons.
If the energy E_0 (note that $E_0 < 0$) exceeds the exciton ground-state energy given
by equation (16.1), then, even for $T \to 0$, the lowest-energy state of the system
is a gas of excitons, and condensation into a liquid is impossible. In the opposite
case $\varphi_0 = E_{\mathrm{ex}} - E_0 > 0$ the liquid state is thermodynamically stable and, for
any value of the average e-h pair density, the EHL will be the ground state of
the system.

We turn now to the experimental results. Among the techniques used to
measure the EHL binding energy and density, the one based on the analysis of
the luminescence spectrum of the crystal stands out for its information capacity
(see reviews Pokrovskii 1972; Hensel *et al.* 1977; Kulakovskii and Timofeev 1983).
For the first time, analysis of the shape of the emission spectrum of Ge in this
context was carried out by Pokrovskii and Svistunova (1970), and of Si, by
Kaminskii *et al.* (1970). Later, this method was widely used to determine the
EHL binding energies and densities in a number of semiconductors.

The luminescence spectrum of the crystal shows the free exciton (FE) and EHL emission. The shape of the latter reflects the e-h energy spectrum and distribution in the liquid. Bulk Ge and Si are indirect semiconductors, and the luminescence spectrum consists of several phonon replicas. In Ge the most intense FE and EHL luminescence lines, shown in Fig. 16.1, are related to the radiative recombination processes with the emission of longitudinal acoustic (LA) phonons.

The EHL density n_0 is determined from the best fit of the calculated lineshape to the experimental one, see Fig. 16.1. Then, we can obtain the e-h Fermi energies, equation (16.3), and the EHL binding energy φ (as shown in Fig. 16.1) for a given temperature. Since the Fermi energies and, thus, the binding energy depend upon the temperature, in order to find φ_0 we need to measure their temperature dependencies and to extrapolate to $T = 0$ (Thomas *et al.*, 1973). Note that, in order to obtain the energy E_0 per e-h pair in the ground state of the liquid, it is necessary to measure independently E_{ex}. The values of n_0, φ_0, E_0 and E_{ex} for Ge and Si from the review by Hensel *et al.* (1977) are listed in Table 16.1.

Using the e-h Fermi distribution for the lineshape fitting, we actually assume the metallic EHL nature. A good agreement between the theoretical and experimental EHL luminescence spectra in Ge and Si validates this assumption.

The metallic EHL properties manifest themselves in a number of other experiments, e.g., on the electric and electromagnetic properties, for details see reviews (Pokrovskii, 1983; Markiewicz and Timusk, 1983). Here, we only mention the far-infrared (IR) resonance absorption of the EHL droplets (Vavilov *et al.*, 1969). Studying the IR transmission spectra of photoexcited Ge, these authors observed at temperatures below 2 K a broad absorption band peaked at $\hbar\omega \approx 9\,\mathrm{meV}$, originating from the excitation of plasma oscillations in the EHL droplets. The plasma resonance absorption appears only if the particles are much smaller than the electromagnetic radiation wavelength. For spherical particles,

TABLE 16.1. Ground-state energies of exciton E_{ex} and EHL E_0, EHL binding energy φ_0, density n_0, critical density n_c and temperature T_c for unstrained and highly uniaxially strained Ge and Si.

	Ge	Si	Ge	Si
			$P \parallel \langle 111 \rangle$	$P \parallel \langle 100 \rangle$
E_{ex} (meV)	-4.17	-14.7 ± 0.4	-2.33	-12.8
φ_0 (meV)	1.8 ± 0.2	8.2 ± 0.1	0.65	2.0
E_0 (meV)	-6.0 ± 0.2	-22.9 ± 0.5	-2.98	-14.8
n_0 ($\times 10^{17}\,\mathrm{cm}^{-3}$)	2.38 ± 0.05	33.3 ± 0.5	0.274	4.5
T_c (K)	6.5 ± 0.1	25 ± 5	3.5	13
n_c ($\times 10^{17}\,\mathrm{cm}^{-3}$)	0.8 ± 0.2	12 ± 5	0.077	1.5

absorption peaks at frequency $\omega = \omega_{\text{pl}}/\sqrt{3}$, where $\omega_{\text{pl}} = \left[(4\pi e^2 n_0)/(\kappa m^*)\right]^{1/2}$ is the plasma frequency. The latter allows to find the EHL density from the measured resonance frequency. In addition, the results of these experiments indicate that the size of metallic EHL droplets is of the order of several micrometers.

Another indication of the metallic Fermi-liquid EHL nature is the observation of oscillations, periodic in reciprocal magnetic field, of the EHL luminescence intensity (Bagaev *et al.*, 1972) and of the far-IR resonance absorption by the EHL droplets (Murzin *et al.*, 1973) in Ge, similar to de Haas–van Alphen and Shubnikov–de Haas oscillations. The intensity oscillations occur as, with an increase of the magnetic field, the Fermi level of electrons bound into the EHL is crossed by successive Landau levels. They originate from the EHL density oscillations, as shown by Keldysh and Silin (1973). Their calculations were confirmed experimentally by Karuzskii *et al.* (1975), who observed the corresponding oscillations of the EHL luminescence linewidth (roughly the sum of the electron and hole Fermi energies, see Fig. 16.1) and of the EHL lifetime (see Section 16.7). In more detail, the magnetic EHL properties are considered in reviews (Rice, 1977; Hensel *et al.*, 1977; Pokrovskii, 1983; Markiewicz and Timusk, 1983; Keldysh and Sibeldin, 1986) and, especially, in (Silin, 1983).

It should be mentioned that in early 1970s some authors adhered to an alternative view on the nature of the EHL luminescence line in Ge, which they attributed to the biexciton emission (Rogachev, 1975). At that time, the issue provoked a fairly heated discussion only of a historical interest now.

16.3   Stability of the EHL

The EHL exists if it is stable against the dissociation into free excitons, i.e., if $E_{\text{ex}} > E_0$. Let us consider the factors stabilizing the condensed EHL state.[14]

According to theoretical estimates, the EHL is unstable in covalent direct-bandgap semiconductors with a simple band structure (i.e., with isotropic, parabolic and nondegenerate bands); however, already a low degree of chemical bonds ionicity in the crystal is enough to make the EHL stable, due to the interaction of electrons and holes with longitudinal optical phonons (Keldysh and Silin, 1975).

The most significant factors for the EHL stability are the electronic spectrum anisotropy and degeneracy. If, for a given density of the liquid, the electron or hole density of states is increased, the kinetic energy (cf. equations 16.2, 16.3) decreases, while the potential energy, determined by the EHL density, remains virtually unchanged. The balance between the attraction and repulsion forces will be affected, and the liquid will experience self-compression until the balance is restored for some higher density and absolute value of the ground-state energy per one e-h pair. In semiconductors with degenerate and/or multivalley bands, the density of states is noticeably higher than in a semiconductor with simple bands. At the same time, E_{ex} is independent of the number of valleys. Thus, the

---

[14]See reviews (Rice, 1977; Vashishta *et al.*, 1983; Keldysh, 1983; Keldysh, 1986; Kulakovskii and Timofeev, 1983).

multivalleys lead to a considerable enhancement of the EHL stability (Bagaev *et al.*, 1969). It can be shown analogously (see Keldysh 1983) that the electronic spectrum anisotropy is a stabilizing factor as well. In highly anisotropic e-h systems (e.g., quasi-one- and quasi-two-dimensional), an "ultradense" EHL ($n_0 \gg a_{\mathrm{ex}}^{-3}$) with the ground-state energy exceeding considerably (by its absolute value) the exciton binding energy is predicted (Andryushyn *et al.*, 1976).

External perturbations that modify the electronic spectrum of the crystal can either stabilize or destabilize the EHL. A uniaxial strain applied to the crystal results in lifting the degeneracy at the valence-band maximum and, in multivalley semiconductors, in a relative energy shift of the conduction-band minima. Thus, the density of states near the band edges decreases, and so do n_0, $|E_0|$ and φ (Bagaev *et al.*, 1970; Benoît à la Guillaume *et al.*, 1972).

The EHL main parameters in highly strained Ge (compressed along $\langle 111 \rangle$ axis) and Si (along $\langle 100 \rangle$ axis), taken from the review article of Kulakovskii and Timofeev (1983), are listed in Table 16.1. "Highly strained" means here that electrons occupy only one lowest conduction band minimum in Ge or two minima in Si, and holes occupy only one subband of the deformation-split off valence band. A reduction in the exciton binding energy under these conditions (see Table 16.1) is related to the change of equal-energy surfaces of holes that accompanies the valence band splitting off. A more comprehensive information on the EF–EHL system under the strain fields can be found in the reviews cited above and in (Hensel *et al.*, 1977), (Bagaev *et al.*, 1983) and (25).

Unlike the strain field, a strong magnetic field stabilizes the EHL. In a magnetic field, the electronic spectrum is quasi-one-dimensional. In ultrahigh magnetic fields, where the Landau-level spacing $\hbar\omega_{\mathrm{c}} \gg |E_{\mathrm{ex}}|$ (here, $\omega_{\mathrm{c}}^{e,h} = eB/m_{e,h}$ is the cyclotron frequency), "ultradense" EHL is predicted (Keldysh and Onishchenko, 1976), with $n_0 \gg (a_{\mathrm{ex}}a_B^2)^{-1}$, where $a_B = \sqrt{\hbar/eB}$. The density of liquid increases with the magnetic field as $n_0 \propto B^{8/7}$ and $|E_0| \propto B^{2/7}$. Note that the exciton binding energy increases with the magnetic field at a lower rate ($\propto \ln B^2$). Thus, in a sufficiently strong field, the EHL is anticipated to become stable even if it is unstable without the field. The theory of the EHL in ultrahigh magnetic fields is reviewed by Silin (1983). This and related issues are also discussed in reviews by Keldysh (1983, 1986).

A magnetic-field stabilized EHL was found by Kavetskaya *et al.* (1982, 1997) in InSb,[15] where EHL appears only for $B > 2\,\mathrm{T}$; with the field increasing from 2.3 to 5.5 T, n_0 grows from 3.2×10^{15} to $6.7 \times 10^{15}\,\mathrm{cm}^{-3}$ and φ – from 0.4 to 1.2 meV. Similar results were obtained by Chernenko and Timofeev (1997) in highly strained Ge: under compression, the EHL emission line disappears without the magnetic field and reappears if $B > 4\,\mathrm{T}$, n_0 and φ grow with a further increase of B. In unstrained Ge, the liquid density also increases with the magnetic field. However, fields up to $B = 19\,\mathrm{T}$ do not affect the stability of the liquid state (Störmer and Martin, 1979).

---

[15] See also in (Sibeldin, 2003).

Completing the discussion of the EHL stability, let us recall that, commonly, the density of various electronic liquids is characterized by the dimensionless parameter r_s. In case of EHL $r_s = [(4/3)\pi a_{ex}^3 n_0]^{-1/3}$. For EHL in Ge and Si, $r_s = 0.5$ and 0.86, respectively. For electronic Fermi liquids in metals, $r_s \geq 1.88$ (Kittel, 1971). Thus, EHL in Ge and Si is characterized by the highest density among "earthly" liquids. Moreover, of all quantum liquids, EHL is the "most quantum" one (Hensel *et al.*, 1977).

16.4 Phase diagrams

Phase equilibrium in the FE–EHL system can be described graphically by diagrams in the same way as for conventional liquid–vapor systems. In the (n, T) plane, the region of the liquid–gas coexistence is bounded by two lines meeting at the critical point (n_c, T_c). On the liquid side, the boundary line follows the temperature dependence of the liquid density. On the gas side, the boundary line for conventional liquids follows the temperature dependence of the saturated vapor density; in the EHL case it is given by

$$n_{0T} = \nu_d \left(\frac{M_d k_B T}{2\pi \hbar^2} \right)^{3/2} \exp(-\varphi/k_B T) \qquad (16.4)$$

for flat liquid–gas boundary (where ν_d and M_d are the exciton ground-state degeneracy factor and density-of-states effective mass). As we shall see below, at low temperatures $(T \ll T_c)$ nonequilibrium nature of the FE–EHL system leads to deviation of the gas branch from that defined by equation (16.4) (see Section 16.6).

Thermodynamics of the FE–EHL system in the critical region is described fairly well theoretically: the calculated critical parameters (density n_c and temperature T_c) are close to the experimentally obtained ones in Ge and Si. Theoretical studies are reviewed in (Rice, 1977; Vashishta *et al.*, 1983; Kulakovskii and Timofeev, 1983); here, we address the experimental results only.

We have already discussed in Section 16.2 the spectroscopic methods to determine the EHL density n_0; its temperature dependence is the liquid branch of the phase diagram. In order to measure the free exciton density temperature dependence (the gas branch of the phase diagram), threshold or spectroscopic techniques were used. Let us briefly dwell on more illustrative threshold measurements, see details in (Hensel *et al.*, 1977; Kulakovskii and Timofeev, 1983).

As established in the very first experiments, the EHL lines in the luminescence spectra (Pokrovskii and Svistunova 1969, 1970) and the far IR resonance-absorption feature (Vavilov *et al.*, 1969) exhibit a threshold behavior with raising excitation level at a given temperature or with reducing temperature at a given excitation. The exciton density, measured at the EHL formation threshold as a function of temperature refers to the gas branch of the phase diagram. The binding energy found then from equation (16.4) appears to be lower than that measured spectroscopically (see Section 16.2). The explanation is that some degree of the exciton vapor supersaturation is required to form a nuclei of the

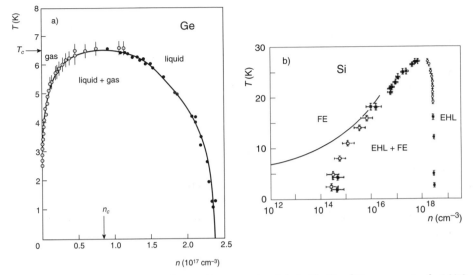

FIG. 16.2. The FE–EHL phase diagram in (a) bulk Ge (Thomas *et al.* 1973, 1974) and (b) bulk Si (Dite *et al.*, 1977). Reprinted figures with permission from [G. A. Thomas, T. M. Rice, and J. C. Hensel, *Phys. Rev. Lett.* **33**, 219 (1974)], copyright (1974) by the American Physical Society and from [A. F. Dite, V. D. Kulakovskii, and V. B. Timofeev, *Zh. Eksp. Teor. Fiz.* **72**, 1156 (1977)], copyright (1977) by Zhurnal Teoreticheskoi i Eksperimentalnoi Fiziki.

liquid phase, which, similar to conventional liquids, increases with lowering the temperature (Section 16.6). At the same time, equation (16.4) describes the gas branch of the phase diagram under the equilibrium conditions.

Figure 16.2 shows phase diagrams of bulk Ge and Si obtained as outlined above. At low temperatures, the gas branch deviates from equation (16.4), which can be seen most clearly in the case of Si (compare the solid line in Fig. 16.2b with the experimental dots for $T < 20\,\mathrm{K}$). The EHL critical parameters in these semiconductors (including that for highly strained Ge and Si) taken from (Hensel *et al.*, 1977) and (Kulakovskii and Timofeev, 1983) are listed in Table 16.1.

We have discussed the simplest gas–liquid phase diagrams. More complicated phase diagrams are possible if, apart from the gas–liquid transition, other phase transitions may take place in the liquid or gas phase. Different possibilities are discussed by Keldysh (1983, 1986) and in reviews cited in this section.

16.5 Macroscopic structure of the EHL

At low temperatures $(T < T_{\mathrm{c}})$, the EHL density is considerably higher than the exciton gas density at equilibrium with liquid and, in the range of optical excitation levels commonly used, the liquid occupies a small fraction $(10^{-3}-10^{-2})$ of the excited volume in the crystal. The very first experiments on the far IR resonance absorption (Vavilov *et al.*, 1969) had shown that the EHL in Ge exists

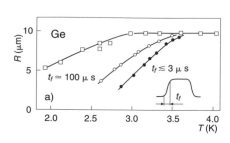 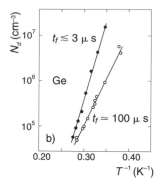

FIG. 16.3. The EHD radius (a) and concentration of droplets (b) in Ge under quasi-cw excitation by long-front and short-front pulses as functions of temperature. Squares in panel (a) correspond to a higher excitation level, and inset shows the excitation pulse shape; the pulse duration was $t_\mathrm{p} = 500\,\mu s$. The values of R on the two lower curves in panel (a) were used to calculate the concentration in panel (b). From (Bagaev *et al.*, 1976*a*). Reprinted figure with permission from [V. S. Bagaev, N. V. Zamkovets, L. V. Keldysh, N. N. Sibel'din, and V. A. Tsvetkov, *Zh. Eksp. Teor. Fiz.* **70**, 1501 (1976)], copyright (1976) by Zhurnal Teoreticheskoi i Eksperimentalnoi Fiziki.

in form of droplets several micrometers in size (see Section 16.2). Existence of the EHL droplets, or e-h droplets (EHDs), also manifests itself in the observation of intense photocurrent bursts on a p–n junction (Asnin *et al.*, 1970), whose magnitude corresponds to 10^7–10^9 e-h pairs per droplet.

The most direct and convincing evidence of the EHDs existence as well as detailed information on their size and concentration under different conditions was obtained in light-scattering experiments (Pokrovskii and Svistunova, 1971; Sibeldin *et al.*, 1973; Bagaev *et al.*, 1973; Worlock *et al.*, 1974). In these studies the excited region of the sample is probed by a laser beam (usually, at a wavelength of $3.39\,\mu m$) for which the crystal is transparent. The radius of scattering particles (EHDs) is obtained from the angular distribution of the scattered light intensity (scattering indicatrix). After that, the EHD concentration can be found from the scattered light intensity at certain angle with respect to the incident probe beam (or the probe beam absorption, which is recorded simultaneously with the scattering).

Experiments on light scattering demonstrated that both size and concentration of droplets depend upon quite a number of factors: experiment configuration, excitation regime (pulsed or cw, bulk or surface), etc. Here, we briefly consider the results obtained under cw excitation, stressing some general features. Under typical experimental conditions, the EHD radius in Ge is 1–10 μm, depending on the temperature and the excitation level. Figure 16.3 shows the temperature dependencies of the EHD radius and concentration (Bagaev *et al.*, 1976*a*). The size of the droplets increases and their concentration decreases at higher temperatures. However, with increasing temperature the EHD radius grows up to $R \approx 10\,\mu\text{m}$, and then does not depend neither on T (Fig. 16.3a) nor on the excitation level. At lower temperatures, the EHD radius may either increase (under quasi-cw excitation by pulses with a slowly rising front) or decrease (pulses with a steep front) with growing excitation. Such a difference is due to the different behavior of the EHD concentration in this temperature range in the two cases: with increasing excitation level, it increases fairly weakly in the former case but superlinearly in the latter case. It will be shown in the next section that the size and concentration of the EHDs are determined by the exciton condensation kinetics. In particular, at low temperatures, the size of droplets is determined by their concentration and, consequently, by the rate of the liquid-phase nucleation. Dependence of the EHD concentration upon the front duration of the excitation pulses appears due to a relationship between the nucleation rate and the exciton-vapor supersaturation.

Data on the size and concentration of EHDs in Ge are reviewed in much more detail in (Hensel *et al.*, 1977; Tikhodeev, 1985; Keldysh and Sibeldin, 1986). Unfortunately, there are virtually no data on EHDs in Si. From the magnitude of the photocurrent bursts in a *p–n* junction, Capizzi *et al.* (1975) estimated the EHD radius in Si to be $R = 0.75\,\mu\text{m}$ at $T \approx 2\,\text{K}$. Other indirect estimates yield even smaller values of R. Thus, EHDs in Si are considerably smaller than in Ge.

16.6 Kinetics of exciton condensation

Condensation of any vapor begins from a liquid-phase nuclei capable of further growth (critical-radius nuclei). Growth of the nuclei proceeds until a vapor–liquid equilibrium is established. However, e-h pairs in a semiconductor have a finite lifetime, and this gives rise to a range of novel properties – both in the condensation kinetics and in the steady-state conditions – that conventional liquids do not possess. In particular, the nonequilibrium nature of the e-h system causes qualitative changes in the vapor branch of the phase diagram.

The kinetics of exciton condensation was the subject of quite a number of studies, see reviews (Rice, 1977; Hensel *et al.*, 1977; Westervelt, 1983; Tikhodeev, 1985; Keldysh and Sibeldin, 1986). Here, following (Bagaev *et al.*, 1976a), we briefly consider the most important features.

The balance equation for the total number of e-h pairs in a spherical EHD of radius R is (Pokrovskii and Svistunova, 1970; Bagaev *et al.*, 1976a):

$$\frac{d}{dt}\left(\frac{4}{3}\pi n_0 R^3\right) = 4\pi R^2 \left[n(R) - n_{0\text{T}} \exp\left(\frac{2\sigma}{n_0 R k_\text{B} T}\right)\right] v_\text{T} - \frac{4}{3}\pi R^3 \frac{n_0}{\tau_0}, \quad (16.5)$$

where $n(R)$ is the exciton density near the droplet surface (which, in general, differs from the average exciton-gas density n in the sample bulk but can be expressed through the latter), the saturated-vapor density $n_{0\text{T}}$ is given by equation (16.4), v_T is the mean thermal exciton velocity, σ is the EHL surface tension coefficient, and τ_0 is the e-h pair lifetime in EHL. The first term on the right-hand side is the difference between the rates of exciton capture and evaporation from the EHD, and the second term accounts for e-h recombination in the droplet.

A steady-state ($dR/dt = 0$) solution to equation (16.5) represents the dependence of the exciton-gas density n on the radius of an EHD in equilibrium with the gas. Such a dependence has two branches. The branch descending with increasing R is unstable (for $\tau_0 \to \infty$, it is described by the second term in square brackets in equation (16.5), like in ordinary liquids) and corresponds to the critical nuclei. The branch ascending with increasing R is stable; its existence is a characteristic feature of phase equilibrium in a system with finite particle lifetime and is related to the last term on the right-hand side of equation (16.5). Recombination has to be compensated by an increase in the exciton density with increasing R, so that the exciton flux onto the surface of the droplet increases and condensation counterbalances recombination. Due to the presence of this branch, there is a minimum in the dependence of n on R. Importantly, the lowest density $n_{\min}$ of the exciton vapor for which the EHDs (of radius $R_{\min}$) in a stable equilibrium with the exciton gas still may exist, exceeds the value n_{0T} corresponding to thermodynamic equilibrium. The relative difference $(n_{\min} - n_{0T})/n_{0T}$ increases with decreasing temperature. Hence, the gas branch of the phase diagram is not correctly described by equation (16.4) in the low-temperature region, as mentioned in Section 16.4.

In order to determine n and R under given temperature and excitation level, we also have to take into account a condition of the e-h pairs number conservation, or of a balance between the generation and recombination (both in gas and liquid phases),

$$g = \frac{n}{\tau} + \frac{4\pi R^3 n_0 N_d}{3\tau_0} \,, \tag{16.6}$$

where g is the e-h generation rate, τ is the exciton lifetime, and N_d is the concentration of EHDs. The critical-nucleus radius and the steady-state EHD radius can be found for a given exciton density from equations (16.5) and (16.6). At low temperatures and high excitation levels the recombination proceeds mainly in the liquid phase and the second term in (16.6) is much greater than the first one. In this case, $R \approx \left[(3g\tau_0) / (4\pi n_0 N_d)\right]^{1/3}$. In the opposite case (high temperatures, n only slightly exceeds $n_{\min}$), $n \approx g\tau$ and the droplet radius is independent of N_d.

The final link comes from the classical kinetic theory of condensation: the concentration of EHDs is governed by the rate of critical nuclei formation (nucleation rate)

$$\frac{dN_d}{dt} \propto n^2 \exp\left(-\frac{4\pi\sigma R_{cr}^2}{3k_B T}\right). \tag{16.7}$$

Here, R_{cr} is the critical radius. In the case of sufficiently large supersaturation $\Delta n = n - n_{0T}$ (and $n > n_{\min}$), $R_{cr} \approx 2\sigma/n_0 k_B T \ln(n/n_{0T})$, which is valid near the thermodynamic equilibrium and neglecting the finite lifetime effects. One can

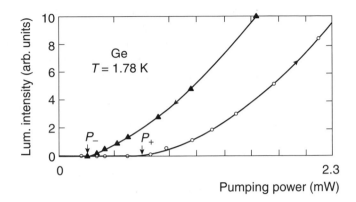

FIG. 16.4. Hysteresis of EHL luminescence in Ge. The pump is uniformly increased from zero to 7.5 mW (off scale) and than back to zero in a total time 192 s. From (Lo *et al.*, 1973). Adapted figure with permission from [T. K. Lo, B. J. Feldman, and C. D. Jeffries, *Phys. Rev. Lett.* **31**, 224 (1973)]. Copyright (1973) by the American Physical Society.

see from the latter and equation (16.7) that the nucleation rate increases considerably with the growth of the exciton-vapor supersaturation. In particular, for a given generation rate, the degree of supersaturation and, thus, the concentration of the droplets increase as the temperature is lowered, in agreement with the experimental data in Fig. 16.3. These formulas also show that relative supersaturation $\Delta n/n_{0T}$ needed to sustain a given nucleation rate increases as the temperature decreases. For low temperatures, this leads to discrepancy between the measured gas branch of the phase diagram and that defined by equation (16.4) discussed in Section 16.4.

Supersaturation effects are most pronounced in the experiments on optical hysteresis (Lo *et al.*, 1973), see a detailed discussion in (Westervelt, 1983). When the excitation power P is raised slowly, EHL luminescence appears above a threshold value P_+ (see Fig. 16.4). A further increase of P results primarily in the formation of new EHDs and, to less extent, in the growth of the existing ones; thus, the EHL luminescence intensity increases. If, after reaching some maximum value, P is started to be gradually reduced, the concentration of EHDs does not change, while their radii start to decrease. The EHL luminescence disappears at $P = P_- < P_+$, when the droplet radii decrease down to nearly R_{min}.

Dependence of the EHD size and concentration on the excitation pulse front duration (Section 16.5) is another manifestation of the hysteresis. Hysteresis phenomena, when the system remembers its previous history, are not unusual for the first-order phase transitions. However, uniqueness of the EHL lies in the fact that billions of e-h pairs generations, created long after the EHDs have been formed (several hours, in some experiments on the hysteresis in Ge), still remember the initial nucleation conditions.

The nucleation rate, eqn (16.7), depends crucially on the EHL surface tension σ. In Ge, the surface tension was measured in (Alekseev *et al.*, 1974; Bagaev *et al.*, 1975; Etienne *et al.*, 1976; Westervelt, 1976) and by other authors. At $T = 0$, $\sigma = (2.6 \pm 0.3) \times 10^{-4} \, \mathrm{erg/cm^2}$ (see the review by Westervelt 1983), which is in a satisfactory agreement with the theoretical estimates; see (Rice, 1977; Vashishta *et al.*, 1983), and references therein.

It should be added that, at sufficiently low temperatures ($T \leq 1.4 \, \mathrm{K}$ in Ge) the recombination losses inside the under-critical size droplets exceed the surface evaporation, and the condensation goes without supersaturation effects and hysteresis (Silver, 1975), see also in reviews (Keldysh, 1983; Westervelt, 1983). This behavior is quite general (Tikhodeev, 1983): the kinetics of the first-order transition in a finite lifetime system may become analogous to that of the second order. Most likely in this case, at least in Si, the EHD nucleation is triggered by multiexciton–impurity complexes (Kaminskii *et al.*, 1970; Pokrovskii, 1972).

The picture of the condensation kinetics outlined in this section explains almost all experimental results except for the limitation of the droplet radii below $R \approx 10 \, \mu\mathrm{m}$ (see Section 16.5). The explanation for the latter is, apparently, related to the phonon-wind effects (Section 16.10).

16.7   Recombination kinetics

In this section, we consider the recombination kinetics of the exciton – liquid two-phase system after the pump is turned off. The recombination kinetics is described by equation (16.5) and the analogous one for the exciton concentration. Several characteristic cases of the recombination process will be discussed, on example of Ge (when e-h pairs in EHL live longer than FE, $\tau_0 \gg \tau$).

At low temperatures the exciton evaporation from droplets may be neglected. The droplet radius and, thus, the liquid phase volume ($\propto R^3 N_\mathrm{d}$) decrease primarily due to the e-h recombination in the liquid. Then equation (16.5) gives

$$R(t) \approx R(0)\exp(-t/3\tau_0)\,, \qquad R^3(t)N_\mathrm{d} \approx R^3(0)N_\mathrm{d}\exp(-t/\tau_0)\,, \qquad (16.8)$$

where $R(0)$ is the initial radius of the EHD.

At higher temperatures, the evaporation rate becomes comparable to the e-h pair recombination rate in the liquid. Neglecting the backward flux of the excitons into the droplets, we obtain from equation (16.5) that

$$R(t) \approx R(0)\exp(-t/3\tau_0) - 3\frac{n_{0T}}{n_0}v_\mathrm{T}\tau_0\big[1 - \exp(-t/3\tau_0)\big]\,. \qquad (16.9)$$

The EHD radius drops nonexponentially down to zero at some moment t_c (Hensel *et al.*, 1973). Thus, $R(0)$ can be determined from the measurements of t_c.

The kinetics of EHD luminescence in Ge is shown in Fig. 16.5 (Westervelt *et al.*, 1974). The EHL lifetime $\tau_0 = 36 \, \mu\mathrm{s}$ can be found from the line slope at $T = 1.8 \, \mathrm{K}$. In a qualitative agreement with equation (16.9), the total time of EHD luminescence t_c decreases with increasing temperature. However, when it

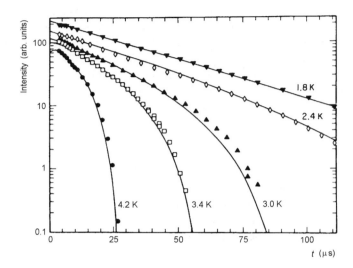

FIG. 16.5. Kinetics of EHD luminescence in ultrapure germanium excited by
GaAs-laser pulses (pulse duration $t_{\mathrm{p}} = 80\,\mathrm{ns}$). From (Westervelt *et al.*, 1974).
Reprinted figure with permission from [R. M. Westervelt, T. K. Lo, J. L. Staehli, and
C. D. Jeffries, *Phys. Rev. Lett.* **32**, 1051 (1974)]. Copyright (1974) by the American
Physical Society.

comes to the evaluation of $R(0)$ from equation (16.9), certain complications arise
(see Hensel *et al.* 1977; Keldysh and Sibeldin 1986). This method appears to be
valid only for sufficiently low pulsed excitation levels.

As the pump intensity is increased (as well as the supersaturation), the concentration of EHDs grows steeply, according to eqn (16.7). Then, the exciton
capture rate by droplets can exceed the exciton recombination rate. In this
collective-relaxation regime the measurement of t_{c} does not allow us to find $R(0)$.
Instead, the total number of e-h pairs in liquid phase and thus its volume (immediately after the excitation is switched off) can be extracted from t_{c} (Ashkinadze
and Fishman, 1977). The data shown in Fig. 16.5 were obtained under these experimental conditions. In this case, excitons are in a quasi-equilibrium with the
EHDs, and, in the course of the decay, the exciton concentration $n \approx n_{0\mathrm{T}}$ remains nearly constant for a fairly long time $\approx (2-3)\tau_0$ (Westervelt *et al.*, 1974;
Manenkov *et al.*, 1976).

The EHL lifetime in Ge is $\tau_0 \approx 40\,\mu\mathrm{s}$. Typical exciton lifetimes in different
samples are $\tau = 2-9\,\mu\mathrm{s}$ (Pokrovskii, 1972; Hensel *et al.*, 1977). The rate of par-
ticle recombination in the liquid is $\tau_0^{-1} = \tau_{\mathrm{r}}^{-1} + \tau_{\mathrm{A}}^{-1}$, with $\tau_{\mathrm{r}}^{-1} \propto n_0$ the radiative
recombination rate and $\tau_{\mathrm{A}}^{-1} \propto n_0^2$ the nonradiative Auger recombination rate.
The quantum efficiency for EHL luminescence in Ge, $Q = \tau_0/\tau_{\mathrm{r}} \approx 25\%$ (Betzler
et al., 1975).

In Si the EHL density n_0 exceeds that in Ge by more than an order of magnitude (see Table 16.1). Therefore, for the EHL in Si, the Auger recombination rate $\tau_A^{-1} \gg \tau_r^{-1}$, and the quantum efficiency $Q \sim 5 \times 10^{-4}$ is very low (Cuthbert, 1970). The EHL lifetime in Si is $\tau_0 \approx 0.15\,\mu s$ (Cuthbert, 1970), and the exciton lifetime is $\tau = 2 - 5\,\mu s$ (Ashkinadze *et al.*, 1971; Cuthbert, 1970). Thus, unlike Ge, in Si we have $\tau_0 \ll \tau$, and the EHDs decay faster than excitons. Hence, the pattern of the recombination kinetics in the FE–EHL system in Si differs qualitatively from the one outlined above for Ge, see further details in reviews (Hensel *et al.*, 1977; Westervelt, 1983).

16.8   Motion of the EHDs under external forces

Since the charge carriers in the EHDs are Fermi degenerate, their scattering by phonons and defects is strongly suppressed: only electrons and holes in narrow energy intervals $\sim k_B T \ll E_F^{e,h}$ around the Fermi levels participate in the scattering. Thus, the mobility of EHDs is fairly high at low temperatures, and they can easily be accelerated by external forces up to the sound velocity in a crystal and transferred on macroscopic distances. High mobility represents one of the main properties of the EHDs, which, to large extent, makes possible most of the phenomena discussed in this and the next sections.

The dominant mechanism of the EHD momentum loss in sufficiently pure crystals is scattering of the constituent electrons and holes by acoustic phonons, and, at low velocities $v \ll s$ the friction force is proportional to v (Keldysh, 1971). Thus, under an external force F, a drift motion establishes after a short time of the order of momentum relaxation time τ_p, with a drift velocity

$$\mathbf{v} = (\tau_p/MN)\,\mathbf{F} = (\tau_p/M)\,\mathbf{f}\,, \qquad (16.10)$$

where N is the total number of e-h pairs in the droplet, M is the e-h pair effective mass, and $\mathbf{f} = \mathbf{F}/N$ is the external force per e-h pair. The coefficient in front of $\mathbf{F}$ in eqn (16.10) is the mobility of EHDs. As a rule, $F \propto N$, and the drift velocity is independent of the droplet size. Thus when speaking about "high mobility" of EHDs, τ_p/M ratio is implied.

The EHD momentum relaxation time τ_p was calculated by Keldysh (1971) for a model semiconductor with idealized band structure; later, a number of authors carried out calculations taking into account the multivalley structure of the conduction band, screening of the deformation potential by the EHD-bound charge carriers and other factors.[16] The EHD momentum relaxation time considerably increases with decreasing temperature. At low temperatures ($T < 1\,K$ for Ge), $\tau_p \propto T^{-5}$, i.e., similarly to the mobility of electrons in metals is limited by acoustic-phonon scattering.

In this section, we consider the EHDs motion in nonuniform static (Bagaev *et al.*, 1969) and dynamic (ultrasound waves (Alekseev *et al.* 1975, 1980) and

---

[16]See reviews (Bagaev *et al.*, 1983; Keldysh and Sibeldin, 1986) for the whole range of issues related to the motion of the EHDs. See also reviews (Hensel *et al.*, 1977; 25; Tikhodeev, 1985).

strain pulses (Bagaev *et al.*, 1980)) strain fields. Effects related to the phonon wind (Bagaev *et al.*, 1976*b*) will be discussed in Sections 16.10 and 16.11.

In a nonuniform strain field the force per each e-h pair is

$$\mathbf{f} = -\nabla(E_{\mathrm{g}} + E_0) = -\sum_{i,k} \mathcal{D}_{ik} \nabla \varepsilon_{ik} , \qquad (16.11)$$

where ε_{ik} is the strain tensor and $\mathcal{D}_{ik} = \partial(E_{\mathrm{g}} + E_0)/\partial\varepsilon_{ik}$ is the total deformation potential for an e-h pair in the liquid.

A planar longitudinal ultrasound wave with wavelength $\lambda_{\mathrm{s}} \gg R$ in a cubic semiconductor ($\mathcal{D}_{ik} = \mathcal{D}\delta_{ik}$) pushes e-h pairs with a force $\mathbf{f} = \mathbf{q}\mathcal{D}\varepsilon \sin(\omega t - \mathbf{qr})$, where $\varepsilon = \mathrm{Tr}(\varepsilon_{ik})$ is the volumetric strain amplitude of the wave, $\omega = sq$ and $\mathbf{q}$ ($q = 2\pi/\lambda_{\mathrm{s}}$) are its frequency and wavevector, and s is the longitudinal sound velocity. Under the force $\mathbf{f}$, the droplet oscillates as a whole and is dragged along $\mathbf{q}$. If the sound wave intensity is sufficiently low, the droplet drift velocity $v \ll s$, and the ultrasound absorption coefficient has a maximum as a function of τ_p at $\omega\tau_p = 1$ (Keldysh and Tikhodeev, 1975). In high-intensity sound waves, the droplets can be "captured" by the wave, being dragged at sonic speed.[17]

Ultrasound absorption upon propagation of sound through an EHD cloud in Ge was observed by Alekseev *et al.* (1975), the sound frequency was 160 MHz, $\lambda_{\mathrm{s}} \approx 30 \, \mu\mathrm{m}$. The temperature dependence of the absorption coefficient of sound waves exhibited a maximum at $T = 2.4 \, \mathrm{K}$. Then the momentum relaxation time at 2.4 K is $\tau_p = \omega^{-1} \approx 1 \, \mathrm{ns}$. This technique provides the most accurate method for measuring τ_p. The EHDs drag by ultrasound was demonstrated by Alekseev *et al.* (1980): the EHD cloud became elongated along the sound wave propagation direction.

An experimental study of the EHD drag by pulses of longitudinal and transverse (shear) strain was carried out by Bagaev *et al.* (1980). Such pulses were generated in a Ge sample upon irradiation of a thin metallic film, deposited onto one of the side facets of the sample, by intense laser pulses. Propagating through the EHD cloud, strain pulses pushed the droplets with the speed of longitudinal and transverse sound at distances as long as 4 mm.

Motion of the EHDs in nonuniform static-strain fields is studied most comprehensively. Here, we discuss briefly experiments by Alekseev *et al.* (1974) carried out using specially tailored Ge samples with the cross-section varying along the compression axis (hourglass-like shape). The distribution of elastic stress, the strain gradient in the sample – and, thus, the force $\mathbf{f}$ – can easily be calculated. Therefore, one can determine the position of the droplets along the axis monitoring the spectral position of the EHL luminescence line, which varies with the magnitude of the strain. It was found that the EHDs created near the top or bottom of the sample move towards the middle section, where the strain is highest. For time-resolved measurements, the strained sample was excited by short

---

[17]The EHD motion at near- and supersonic velocities (and, in particular, their damping and shape) is a separate interesting problem. See reviews cited in the previous footnote.

laser pulses ($t_\mathrm{p} \approx 0.2\,\mu$s) and the evolution of the luminescence spectra was registered. It was found that the line corresponding to the EHD emission from the top section of the sample, where the droplets were created, fades away from the spectrum as the time passes, and the line corresponding to the EHD emission from the middle section appears. Under given experimental conditions the time necessary for this transformation amounted to $\approx 1\,\mu$s, which corresponds to the average velocity of the EHD motion $v \approx 3 \times 10^5$ cm/s.

In conclusion, let us summarize briefly the data on the EHD momentum relaxation time τ_p. In Ge different measurements differ significantly (see footnote 16 above). For example, converting the discussed above value to temperature $T = 2\,$K (assuming $\tau_p \propto T^{-5}$) we obtain $\tau_p \approx 2\,$ns. Experiments of Tamor and Wolfe (1981), see the next section, give $\tau_p = 0.53\,$ns for the same $T = 2\,$K. The EHD momentum relaxation time in Si determined by similar measurements is $\tau_p \approx 0.6\,$ns at $T = 2\,$K.

16.9   Giant EHDs

Qualitatively new properties of the EHL arise when the e-h system is confined by a nonuniform strain field applied to the sample. A potential well (or wells) for electrons and holes may appear in a certain region (or regions) in the sample due to a compressive force applied to it by a screw or a tip, see details in reviews (25; Jeffries, 1975; Hensel *et al.*, 1977; Pokrovskii, 1983; Markiewicz and Timusk, 1983). Here we discuss only the experiments with the large, or giant droplets (called γ-droplets[18]).

As the strained sample is photoexcited, EHDs, free excitons, electrons and holes that appear in the excitation region move in the nonuniform strain field towards the potential well and accumulate there. The structure of the resulting liquid phase in the well proves to be completely different in Ge and Si.

Figure 16.6 shows spatially-resolved photoluminescence from a nonuniformly stressed Ge sample obtained by raster scanning an image of the sample across the entrance aperture of the spectrometer tuned to the EHL line (Tamor and Wolfe, 1981). The stress was applied to the center of the upper sample face, and a giant EHD (egg-shaped) was formed in the well. A "jet" streaming from the excited region of the sample (near the laser focus) is a cloud of small-size EHDs moving in the nonuniform strain field. In this field, the EHDs are pushed first along the x-axis into the sample bulk and then, after passing about 1 mm, move towards the potential well, where they merge into the giant droplet.

Giant EHDs in nonuniformly strained Ge were discovered by Wolfe *et al.* (1975). The radius of a giant EHD can be determined from its photographic image (Fig. 16.6). However, such an image does not allow us to distinguish a cloud of small-size droplets from a monodroplet (cf. their images in Fig. 16.6). The fact that the giant droplet represents a single whole was proved by the

---

[18]In contrast to conventional EHDs, also called α-droplets. Sometimes, giant EHDs are also called monodroplets.

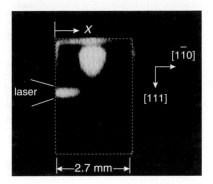

FIG. 16.6. Spatially-resolved image of the photoluminescence signal from a nonuniformly strained Ge sample under cw excitation. Egg-like spot is a giant EHD, a smaller elongated spot is a stream of small-size droplets. The sample dimensions are $3.5 \times 3.3 \times 2.7 \, \text{mm}^3$, $T = 2 \, \text{K}$. From (Tamor and Wolfe, 1981). Adapted figure with permission from [M. A. Tamor and J. P. Wolfe, *Phys. Rev. B* **24**, 3596 (1981)]. Copyright (1981) by the American Physical Society.

appearance of Alfvén dimensional resonances (Wolfe *et al.*, 1975) and by the absence of the IR light scattering, which is characteristic of a cloud of α droplets (Pokrovskii and Svistunova, 1976).

The quantitative characteristics of giant EHDs depend on the orientation and magnitude of the stress. The radius of a γ droplet grows with the excitation level, since the total number of e-h pairs in the potential well increases, and can be as large as $\approx 0.5 \, \text{mm}$. Since the giant droplet is located in the region of highest sample deformation, its density is considerably lower (by a factor of 3–5 in most experiments) than the EHL density in unstrained Ge (see Table 16.1). Because of the low density, the lifetime of giant EHDs $\tau_0 \sim 500 \, \mu\text{s}$ is more than an order of magnitude longer than the lifetime of small EHDs (see Section 16.7).

A different scenario is realized in Si. Small EHDs in a potential well do not merge into a single droplet, see review (25); the formation of a giant EHD is prevented by the forces of mutual repulsion between the droplets due to the phonon wind (Section 16.10), for details see review (Keldysh and Sibeldin, 1986).

16.10 Phonon wind

Nearly the whole energy spent on the formation of e-h pairs finally dissipates into heat, i.e., is transferred eventually into acoustic phonons. Thus, appearance of intense flows of nonequilibrium acoustic phonons is inevitable. Being absorbed by excitons and EHDs, nonequilibrium phonons transfer their momentum and, thus, exert a force upon excitons and EHDs. As we shall see below, this force, called the phonon-wind force (Keldysh, 1976; Bagaev *et al.*, 1976*b*), is of the key importance for many properties of the gas-liquid e-h system, discussed in this and the next sections (see reviews Keldysh and Sibeldin 1986; Bagaev *et al.*

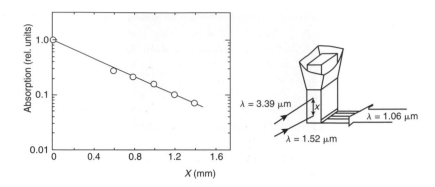

FIG. 16.7. Absorption of the probe radiation by the EHL droplets as a function of the distance between the center of the EHD generation region and the probe beam ($P_{\mathrm{pump}} \approx 40\,\mathrm{mW}$, $T = 1.95\,\mathrm{K}$); right-hand side panel shows the experiment layout. From (Bagaev *et al.*, 1976*b*). Adapted figure with permission from [V. S. Bagaev, L. V. Keldysh, N. N. Sibel'din, and V. A. Tsvetkov, *Zh. Eksp. Teor. Fiz.* **70**, 702 (1976)], copyright (1976) by Zhurnal Teoreticheskoi i Eksperimentalnoi Fiziki.

1983; Hensel *et al.* 1977; Tikhodeev 1985).

In an elastically isotropic semiconductor with simple bands, the phonon-wind force acting on a pair of particles in the liquid at point $\mathbf{r}$ can be estimated as

$$\mathbf{f}(\mathbf{r}) = \frac{m^2 \mathcal{D}^2}{2\pi \hbar^3 \rho s^2 n_0} \, \overline{|\mathbf{q}|}_{\mathbf{q} \lesssim 2k_{\mathrm{F}}} \, \mathbf{J}(\mathbf{r}) \,, \tag{16.12}$$

where ρ is the crystal density, m is the carrier effective mass, $\mathbf{J}(\mathbf{r})$ is the phonon energy flow, and $\overline{|\mathbf{q}|}_{\mathbf{q} \lesssim 2k_{\mathrm{F}}}$ is the average quasimomentum of absorbed phonons. Due to the energy and quasimomentum conservation, only relatively long-wavelength phonons with $q \lesssim 2k_{\mathrm{F}}$ can be absorbed in a degenerate EHL.

In a cubic semiconductor with simple band structure charge carriers interact only with longitudinal phonons. In multivalley semiconductors absorption of transverse phonons by electrons in the EHL is also possible. In this case the electron–phonon interaction is anisotropic, i.e., for each conduction-band valley $\mathcal{D}$, m and k_{F} in equation (16.12) depend on the phonon propagation direction and polarization. Due to the valence band degeneracy, holes interact with longitudinal and transverse phonons too. However, the main contribution to the absorption of phonons by the EHL comes from electrons.

Under force eqn (16.12), EHDs acquire a drift velocity $\mathbf{v}$ given by eqn (16.10). Phenomena considered in this and the next sections originate from this effect of the EHD drag by the phonon wind, predicted by Keldysh (Bagaev *et al.*, 1976*b*).

The EHDs drag by phonon wind in Ge was first observed in the experiment explained in Fig. 16.7. The EHDs cloud was generated by the cw $1.52\,\mu$m laser, EHDs absorption was probed by light of $\lambda = 3.39\,\mu$m. Nonequilibrium phonons were generated in the bottom part of the sample by the cw $1.06\,\mu$m laser. EHDs were pushed by phonon wind towards the probe beam. In these experiments, EHDs were dragged by the phonon wind over distances of more than 2 mm. The measured spatial dependence of probe absorption (straight line in Fig. 16.7) corresponds to a spatial distribution of averaged e-h pairs concentration in EHDs $\bar{n} \propto \exp(-x/L_d)$, typical for a drift motion, where $L_d = v\tau_0$ is the drift length. The slope of dependence in Fig. 16.7 gives $L_d \approx 0.5\,\text{mm}$. For $\tau_0 = 40\,\mu\text{s}$ we get the EHD drift velocity $v \approx 1.3 \times 10^3$ cm/s. In agreement with equations (16.10) and (16.12), the measured EHD drift velocity increased linearly with $1.06\,\mu\text{m}$ pump intensity.

We discuss now very briefly the mechanisms of the phonon wind generation, i.e., of relatively long-wavelength acoustic phonons with $q \lesssim 2k_F$, upon optical excitation of semiconductor crystals. Firstly, such phonons (so-called primary phonons) are emitted at the final stage of thermalization of photoexcited charge carriers. Secondly, they can be generated in the decay of optical phonons into the high-frequency acoustic ones and the relaxation of the latter, because at the initial thermalization stages the carrier excess energy is released mostly into short-lifetime optical phonons. Modes of relaxation and propagation of nonequilibrium acoustic phonons can be rather diverse (Levinson, 1986).[19] Without dwelling into details, the latter component of the phonon wind will be referred to as "hot spot" phonons.[20]

The third contribution to the phonon wind comes from the EHDs themselves. The Auger process (Section 16.7), being the main channel of EHL recombination, produces hot particles. Some part β of their energy $\approx E_g$ goes eventually into longwavelength acoustic phonons, via EHL overheating (Hensel *et al.*, 1977). Then, e.g., a droplet of radius R emits a phonon flux

$$\mathbf{J(r)} = \frac{1}{3}\beta E_g \frac{n_0}{\tau_0} \frac{R^3}{r^3}\mathbf{r}\,. \tag{16.13}$$

This phonon flux pushes away the excitons, the gas concentration near the droplet and the droplet growth rate decrease. As far as $J(R) \propto R$, the growth of the droplet slows down abruptly as R increases. This may explain the EHD radius limitation observed experimentally, see Section 16.5 (Bagaev *et al.*, 1976*b*).

Another more effective mechanism of the EHD size limitation is due to "electrostatic"-like instability of large liquid volumes (Keldysh, 1976). Some of the nonequilibrium phonons emitted by the EHL are absorbed in the liquid. Thus, two EHL volumes δV and $\delta V'$ separated by a distance r repel each other with a force

$$\mathbf{F} = \frac{\beta E_g}{8\pi^2} \frac{n_0}{\tau_0} \frac{m^2}{\hbar^3} \frac{\mathcal{D}^2}{\rho s^2} \overline{|\mathbf{q}|} \frac{\delta V \delta V'}{r^3}\mathbf{r} = \rho_{\text{ph}}^2 \frac{\delta V \delta V'}{r^3}\mathbf{r}\,. \tag{16.14}$$

Thus, EHL behaves like uniformly charged liquid with the charge density ρ_{ph}. As a result, e.g., droplets with $R > R_b = (15\sigma/2\pi\rho_{\text{ph}}^2)^{1/3}$ become unstable

---

[19]See also (Bagaev *et al.*, 1983; Keldysh and Sibeldin, 1986) and references therein.

[20]This term was introduced by Hensel and Dynes (1977).

(analogously to large nuclei). An estimation for Ge ($\rho_{\mathrm{ph}} \approx 300\,\mathrm{g}^{1/2}\mathrm{cm}^{-3/2}\mathrm{s}^{-1}$) gives $R_{\mathrm{b}} \approx 20\,\mu\mathrm{m}$, in a reasonable agreement with the available experimental data, see review (Keldysh and Sibeldin, 1986).

Certainly, these repulsion forces act between all EHDs in the droplet cloud (Keldysh, 1976). Durandin *et al.* (1977) observed expansion of the EHD cloud due to these mutual repulsion forces over time spans from 10 to 100 μs after bulk excitation of Ge by short light pulses, i.e., in the situation where the formation of an EHD layer and the relaxation of the hot spot have completed (see Section 16.11). The EHD layer in this experiment was ring-shaped, and its radius grew with time with the average velocity $\approx 4 \times 10^3$ cm/s allowing to obtain the cited above value of ρ_{ph}.

A direct experimental proof of repulsion forces between EHDs was obtained by Doehler and Worlock (1978): the droplets drift velocity in Ge (see Section 16.11) was increased when the EHD cloud was heated by the IR radiation.

There are no direct experimental estimates of ρ_{ph} in Si. An estimate according to equation (16.14) yields $\rho_{\mathrm{ph}} \approx 2 \times 10^4\,\mathrm{g}^{1/2}\mathrm{cm}^{-3/2}\mathrm{s}^{-1}$ (Keldysh and Sibeldin, 1986). This gives $R_{\mathrm{b}} \approx 2\,\mu\mathrm{m}$. Under strong compression of Si (see Table 16.1 and Section 16.7), ρ_{ph} decreases by nearly one order of magnitude. However, even in highly compressed Si, ρ_{ph} exceeds that in unstrained Ge almost by a factor of 10. Strong repulsive interaction between the EHDs prevents formation of giant droplets in a potential well in nonuniformly strained Si (Section 16.9).

It may be added here that the EHDs can be considered as reservoirs for the crystal excitation energy, and their motion under external forces – as directional excitation energy transfer. Energy stored in the EHDs may be transformed into other forms of energy (Tsvetkov *et al.*, 1985; Zamkovets *et al.*, 1994).

16.11   Spatial structure and dynamics of the EHD cloud

For a long time, before the possibility of nondiffusive drag of EHDs on macroscopic distances by phonon wind was predicted by Keldysh and confirmed experimentally (Keldysh, 1976; Bagaev *et al.*, 1976b), there was a mystery why the "diffusion coefficient" of EHDs differs in different experiments by up to six orders of magnitude (Pokrovskii 1972; Hensel *et al.* 1977; Bagaev *et al.* 1983; Keldysh and Sibeldin 1986, and references therein). The phonon wind effects appear to be of decisive importance for the dynamics of the droplet cloud, they determine the cloud shape and size.

If the excitation light is focused into a point, in an isotropic approximation a hemispherical EHD cloud should be formed. Under steady-state excitation the droplets move from the excitation region along the radius of the hemisphere with a velocity depending on the distance r from its center,

$$v(r) = \frac{1}{2\pi\bar{n}_{\mathrm{max}}\tau_0 r^2} \left(\Gamma G \tau_0 + 2\pi \int_0^r \bar{n}(r) r^2 dr \right). \tag{16.15}$$

Here, G is the total e-h pair generation rate, $\bar{n}(r)$ is the average EHL density, $\Gamma = \Delta E/\beta E_g$, ΔE is the energy (per one e-h pair) transferred to the phonon wind as a result of carrier thermalization in the excitation region, and

$$\bar{n}_{\max} = \frac{Mn_0^2}{4\pi\rho_{\mathrm{ph}}^2\tau_p\tau_0}. \tag{16.16}$$

In equation (16.15), the term $\propto \Gamma$ describes the phonon wind from the excitation region (primary and hot spot phonons, see Section 16.10). The integral term is due to the repulsion between the droplets.

It can be seen from eqn(16.15) that EHDs drift velocity, drift length, and thus the size of the cloud increase with decreasing temperature ($v \propto \tau_p$; see Section 16.8) and increasing excitation level. At high excitation levels $\Gamma g\tau_0 \gg \bar{n}_{\max}/\Gamma$ (where g is the generation rate per unit volume in the excitation region) and low temperatures, so that evaporation of the droplets during their motion may be neglected, the EHD drift length can be estimated as

$$L_{\mathrm{eff}} = \left[\frac{3G\tau_0}{2\pi\bar{n}_{\max}}(1+\Gamma)\right]^{1/3}. \tag{16.17}$$

As far as the total number of e-h pairs in the droplets is $\propto G$ and the volume of the droplet cloud is $\propto L_{\mathrm{eff}}^3$, the average concentration of e-h pairs in the cloud is independent of the generation rate at high excitation levels. In the excitation region $\bar{n} \approx \bar{n}_{\max}/\Gamma$, and it is the highest possible concentration in the conditions of steady-state nonbulk excitation.

The fact that the volume of the EHD cloud increases proportionally to the excitation intensity and its average concentration saturates at $\bar{n} \approx 10^{15}\,\mathrm{cm}^{-3}$ at high excitation levels was established in a number of studies, discussed in detail in review articles cited above. The maximum concentration $\bar{n}_{\max}$ can be estimated by using eqn (16.16) with the value of ρ_{ph} cited in Section 16.10; this yields $\bar{n}_{\max} \approx 1.8 \times 10^{14}\,\mathrm{cm}^{-3}$ and $\Gamma \approx \bar{n}_{\max}/\bar{n} \approx 0.2$.

A direct confirmation of the dynamic nature of the EHD cloud was obtained by Doehler *et al.* (1977), where the spatial distribution of the EHD drift velocities at different excitation levels (Fig. 16.8) was measured by means of the Doppler frequency shift of the scattered probe radiation.

It should be noted that the exciton condensation kinetics discussed in Section 16.6 cannot be directly applied to the dynamic EHDs cloud. Here the memory of the initial excitation conditions is absent. The main macroscopic parameters characterizing the cloud depend significantly on the EHD drift velocity and thus on the phonon-wind intensity (Zamkovets *et al.*, 1985).

The most beautiful and convincing manifestation of the EHD drag by the phonon wind is the anisotropic spatial structure of the droplet cloud in Ge (Greenstein and Wolfe, 1978), appearing upon sharply focused surface excitation and caused by the anisotropy of the phonon flows, as well as anisotropy of the electronic spectrum of the crystal and the electron–phonon interaction, as mentioned in Section 16.10.

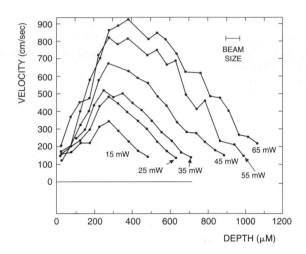

FIG. 16.8. Dependences of the EHD drift velocities in Ge ($T = 2\,\mathrm{K}$) on the distance between the probe laser beam and the excited surface at different excitation levels. From (Doehler *et al.*, 1977). Adapted figure with permission from [J. Doehler, J. C. V. Mattos, and J. M. Worlock, *Phys. Rev. Lett.* **38**, 726 (1977)]. Copyright (1977) by the American Physical Society.

Because of the anisotropy of the elastic properties in crystals, acoustic phonons propagate predominantly (focus) along certain crystallographic directions.[21] In Ge, longitudinal acoustic phonons are focused along $\langle 111 \rangle$. T_1 transverse phonons (fast transverse mode along $\langle 110 \rangle$) propagate close to the planes $\{100\}$. T_2 phonons (slow transverse mode) – close to the $\langle 100 \rangle$ and $\langle 111 \rangle$ axes. Correspondingly, the EHD cloud has wide lobes along the $\langle 111 \rangle$ crystallographic directions and narrow flares stretched along the $\langle 100 \rangle$ directions. Photographs of the anisotropic EHD cloud are shown in Fig. 16.9.

Experimental data for Si is not abundant. Since the EHL lifetime in Si is considerably shorter than in Ge (see Section 16.7), the size of the droplet cloud is small. The average concentration of the e-h pairs in the liquid saturates at $\bar{n} \approx 1.5 \times 10^{16}\,\mathrm{cm}^{-3}$ as the excitation level increases (Tamor and Wolfe, 1980). An estimation by equation (16.16) and using ρ_{ph} cited in Section 16.10 yields $\bar{n}_{\max} \approx 1.9 \times 10^{16}\,\mathrm{cm}^{-3}$. These values of the average concentration correspond to $\Gamma \approx 1.3$. Anisotropic structure of the EHD cloud in Si is only weakly pronounced, mainly because the droplet cloud is nearly of the same size as the excitation region.

To this point, we considered the properties of a steady-state droplet cloud. Measurements with both spatial and temporal resolution under pulsed excitation reveal effects originating from different components of the phonon wind (see

---

[21]Phonon focusing is considered in detail in a review by Maris (1986). See also (Bagaev *et al.*, 1983; Keldysh and Sibeldin, 1986) and references therein.

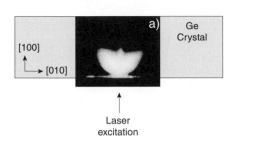

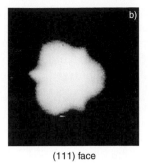

FIG. 16.9. Images of an EHD cloud in Ge taken through the (001) (a) and (111) (b) facets of the sample. In panel (b) the excitation is focused on the opposite surface of the sample. From (Greenstein and Wolfe, 1978). Reprinted figures with permission from [M. Greenstein and J. P. Wolfe, *Phys. Rev. Lett.* **41**, 715 (1978)]. Copyright (1978) by the American Physical Society.

Section 16.10) because they are separated in time.

The dynamics of the EHD cloud in Ge under pulsed excitation was studied first by Damen and Worlock (1976), who carried out spatially and temporally resolved measurements of the probe radiation scattering by the droplets. They observed that, under sufficiently intense excitation, a layer of EHDs moving from the excited surface into the bulk of the crystal is formed.

In a series of similar experiments (see (Sibeldin *et al.*, 1983) and references therein) it was established that the dynamics of the droplet cloud passes consecutively through three stages according to the three mechanisms of the phonon wind generation (see also (Keldysh and Sibeldin, 1986) and references therein): (i) Near-sonic expansion of the cloud and formation of the layer of EHDs under the primary phonon wind. The duration of this stage is $\lesssim 0.5\,\mu s$. (ii) Approximately $\sim 8\,\mu s$-long expansion of the EHD layer with a gradually decreased velocity under the hot-spot phonons. (iii) A slower expansion due to mutual repulsion forces between the EHDs, $v \sim 10^3$ cm/s and duration of this final stage of the cloud dynamics exceeds $100\,\mu s$.

Similar to a steady-state droplet cloud, a cloud under pulsed excitation has anisotropic spatial structure (Tamor *et al.*, 1983).

Finally, we note that the model considered by Sibeldin *et al.* (1983) to describe the first stage of the EHD cloud dynamics is also fairly well applicable to exciton transport in bulk semiconductors (Bulatov and Tikhodeev, 1992; Tikhodeev *et al.*, 2000) and quantum wells (Monte *et al.*, 2000) at high levels of optical excitation.

16.12 Conclusion

In this paper, we surveyed a wide range of issues related to the exciton condensation and properties of the nonequilibrium FE–EHL system. At the same time, due to inevitable restrictions on the length of this review, a significant part of the problems under discussion was dealt with very briefly and many important and interesting topics were left aside without consideration (multicomponent EHL, condensation of excitons in doped semiconductors, EHL in materials other than Ge and Si, EHL in thin-film structures and a large number of others). Many of these issues are analyzed in detail in review publications cited in the Introduction and in other places throughout this Chapter, although the results obtained in recent years were not covered in these papers.

Among the recent accomplishments, one should certainly mention the discovery of the EHL in diamond, characterized by exceedingly high parameters ($n_0 \approx 10^{20}$ cm^{-3}, $T_c = 165$ K) (Shimano *et al.*, 2002; Nagai *et al.*, 2003; Jiang *et al.*, 2005) and of metastable EHL in CuCl (Nagai *et al.*, 2001). Critical parameters of the EHL and liquid alkali metals were compared by Leys *et al.* (2002). Observation of the EHL in thin Si layers of silicon-on-insulator structures (Tajima and Ibuka, 1998), Si quantum wells in Si/SiO$_2$ structures (Pauc *et al.* 2004, 2005) and alloy layers of Si/SiGe nanostructures (Burbaev *et al.*, 2006) opens prospects for studying the properties of a quasi-two-dimensional EHL in Si- and Ge-based heterostructures, which can be controlled through the structure design.

Studying the possibilities of the EHL formation in low-dimensional structures represents a separate exciting problem. Much interest is currently attracted to the investigation of Bose condensation in such systems (see contributions by Butov, Timofeev and Zimmermann to this volume). Studies of the real-space condensation are fairly scarce (see, e.g., Kalt *et al.* 1992; Ding and Wang 2005). Moreover, conditions for the EHL formation in such systems are not really clear. However, work is going on and further progress in this direction is anticipated.

Acknowledgments

In conclusion, I would like to express my deep appreciation to L. V. Keldysh; long-lasting personal contact and scientific collaboration with him, which largely determined my own interests in science. I am sincerely grateful to V. S. Bagaev, V. A. Tsvetkov, S. G. Tikhodeev and N. V. Zamkovets for many years of joint work and numerous discussions of the issues considered in this paper. I thank M. L. Skorikov for his help in preparing the English text of this paper and useful discussions that helped its improvement.

References

Alekseev, A. S., Astemirov, T. A., Bagaev, V. S., Galkina, T. I., Penin, N. A., Sibeldin, N. N. and Tsvetkov, V. A. (1974). In *Proceedings of the 12th International Conference on Physics of Semiconductors* (ed. M. Pilkuhn), Teubner, Stuttgart, p. 91.

Alekseev, A. S., Galkina, T. I., Maslennikov, V. N., Khakimov, R. G. and Schebnev, E. P. (1975). *Pisma Zh. Eksp. Teor. Fiz.* **21**, 578 [*Sov. Phys. JETP Lett.* **21**, 291].

Alekseev, A. S., Galkina, T. I., Maslennikov, V. N. and Tikhodeev, S. G. (1980). *Zh. Eksp. Teor. Fiz.* **79**, 216. [*Sov. Phys. JETP* **52**, 109].

Andryushyn, E. A., Babichenko, V. S., Keldysh, L. V., Onishchenko, T. A. and Silin, A. P. (1976). *Pisma Zh. Eksp. Teor. Fiz.* **24**, 210 [*Sov. Phys. JETP Lett.* **24**, 185].

Ashkinadze, B. M. and Fishman, I. M. (1977). *Fiz. Tekh. Poluprovodn.* **11**, 408 [*Sov. Phys. Semicond.* **11**, 235].

Ashkinadze, B. M., Kretsu, I. P., Patrin, A. A. and Yaroshetskii, I. D. (1971). *Phys. Status Solidi (b)* **46**, 495.

Asnin, V. M. and Rogachev, A. A. (1969). *Pisma Zh. Eksp. Teor. Fiz.* **9**, 415 [*Sov. Phys. JETP Lett.* **9**, 248].

Asnin, V. M., Rogachev, A. A. and Sablina, N. I. (1970). *Pisma Zh. Eksp. Teor. Fiz.* **11**, 162 [*Sov. Phys. JETP Lett.* **11**, 99].

Bagaev, V. S. (1975). In *Excitons at High Density* (ed. H. Haken and S. Nikitine), Volume 73 of *Springer Tracts Mod. Phys.*, Chapter 5, p. 72.

Bagaev, V. S., Galkina, T. I., Gogolin, O. V. and Keldysh, L. V. (1969). *Pisma Zh. Eksp. Teor. Fiz.* **10**, 309 [*Sov. Phys. JETP Lett.* **10**, 195].

Bagaev, V. S., Galkina, T. I. and Gogolin, O. V. (1970). *Kratkie soobscheniya po fizike FIAN, No. 2*, p. 42 (in Russian).

Bagaev, V. S., Galkina, T. I., Penin, N. A., Stopachinskii, V. B. and Churaeva, M. N. (1972). *Pisma Zh. Eksp. Teor. Fiz.* **16**, 120 [*Sov. Phys. JETP Lett.* **16**, 83].

Bagaev, V. S., Penin, N. A., Sibeldin, N. N. and Tsvetkov, V. A. (1973). *Fiz. Tverd. Tela* **15**, 3269 [*Sov. Phys. Solid State* **15**, 2179].

Bagaev, V. S., Sibeldin, N. N. and Tsvetkov, V. A. (1975). *Pisma Zh. Eksp. Teor. Fiz.* **21**, 180 [*Sov. Phys. JETP Lett.* **21**, 80].

Bagaev, V. S., Zamkovets, N. V., Keldysh, L. V., Sibeldin, N. N. and Tsvetkov, V. A. (1976*a*). *Zh. Eksp. Teor. Fiz.* **70**, 1501 [*Sov. Phys. JETP* **43**, 783].

Bagaev, V. S., Keldysh, L. V., Sibeldin, N. N. and Tsvetkov, V. A. (1976*b*). *Zh. Eksp. Teor. Fiz.* **70**, 702 [*Sov. Phys. JETP* **43**, 362].

Bagaev, V. S., Bonch-Osmolovskii, M. M., Galkina, T. I., Keldysh, L. V. and Poyarkov, A. G. (1980). *Pisma Zh. Eksp. Teor. Fiz.* **32**, 356 [*Sov. Phys. JETP Lett.* **32**, 341].

Bagaev, V. S., Galkina, T. I. and Sibeldin, N. N. (1983). In (Jeffries and Keldysh, 1983), Chapter 4, p. 267.

Benoît à la Guillaume, C., Voos, M. and Salvan, F. (1972). *Phys. Rev.* B **5**, 3079.

Betzler, K., Zhurkin, B. G. and Karuzskii, A. L. (1975). *Solid State Commun.* **17**, 577.

Bulatov, A. E. and Tikhodeev, S. G. (1992). *Phys. Rev.* B **46**, 15058.

Burbaev, T. M., Kurbatov, V. A., Rzaev, M. M., Sibeldin, N. N., Tsvetkov, V. A. and Schäffler, F. (2006). In *Proceedings of the 14th International Symposium "Nanostructures: Physics and Technology"*, Ioffe Institute, St. Petersburg, p. 132.

Capizzi, M., Voos, M., Benoît à la Guillaume, C. and McGroddy, J. C. (1975). *Solid State Commun.* **16**, 709.

Chernenko, A. V. and Timofeev, V. B. (1997). *Zh. Eksp. Teor. Fiz.* **112**, 1091 [*JETP* **85**, 593].

Cuthbert, J. D. (1970). *Phys. Rev.* B **1**, 1552.

Damen, T. C. and Worlock, J. M. (1976). In *Proceedings of the 3rd International Confnerence on Light Scattering in Solids*, Flammarion, Paris, p. 183.

Ding, C. R. and Wang, H. Z. (2005). *Phys. Rev.* B **71**, 085304.

Dite, A. F., Kulakovskii, V. D. and Timofeev, V. B. (1977). *Zh. Eksp. Teor. Fiz.* **72**, 1156 [*Sov. Phys. JETP* **45**, 604].

Doehler, J., Mattos, J. C. V. and Worlock, J. M. (1977). *Phys. Rev. Lett.* **38**, 726.

Doehler, J. and Worlock, J. M. (1978). *Phys. Rev. Lett.* **41**, 980.

Durandin, A. D., Sibeldin, N. N., Stopachinskii, V. B. and Tsvetkov, V. A. (1977). *Pisma Zh. Eksp. Teor. Fiz.* **26**, 395 [*Sov. Phys. JETP Lett.* **26**, 272].

Etienne, B., Benoît à la Guillaume, C. and Voos, M. (1976). *Phys. Rev.* B **14**, 712.

Greenstein, M. and Wolfe, J. P. (1978). *Phys. Rev. Lett.* **41**, 715.

Hensel, J. C. and Dynes, R. C. (1977). *Phys. Rev. Lett.* **39**, 969.

Hensel, J. C., Phillips, T. G. and Rice, T. M. (1973). *Phys. Rev. Lett.* **30**, 227.

Hensel, J. C., Phillips, T. G. and Thomas, G. A. (1977). *Solid State Physics* **32** (ed. H. Ehrenreich, F. Seitz, and D. Turnbull), Academic Press, New York, p. 88.

Ivanov, A. L., Haug, H. and Keldysh, L. V. (1998). *Phys. Rep.* **296**, 237.

Jeffries, C. D. (1975). *Science* **89**, 955.

Jeffries, C. D. and Keldysh, L. V. (ed.) (1983). *Electron–Hole Droplets in Semiconductors*, Volume 6 of *Modern Problems in Condensed Matter Sciences* (ed. of the Series: V. M. Agranovich and A.A. Maradudin). North-Holland, Amsterdam.

Jiang, J. H., Wu, M. W., Nagai, M. and Kuwata-Gonokami, M. (2005). *Phys. Rev.* B **71**, 035215.

Kalt, H, Nötzel, R. and Ploog, K. (1992). *Solid State Commun.* **83**, 285.

Kaminskii, A. S., Pokrovskii, Ya. E. and Alkeev, N. V. (1970). *Zh. Eksp. Teor. Fiz.* **59**, 1937 [*Sov. Phys. JETP* **32**, 1048].

Karuzskii, A. L., Betzler, K. V., Zhurkin, B. G. and Balter, B. M. (1975). *Zh. Eksp. Teor. Fiz.* **69**, 1088 [*Sov. Phys. JETP* **42**, 554].

Kavetskaya, I. V., Kost', Ya. Ya., Sibeldin, N. N. and Tsvetkov, V. A. (1982). *Pisma Zh. Eksp. Teor. Fiz.* **36**, 254 [*Sov. Phys. JETP Lett.* **36**, 311].

Kavetskaya, I. V., Zamkovets, N. V., Sibeldin, N. N. and Tsvetkov, V. A. (1997). *Zh. Eksp. Teor. Fiz.* **111**, 737 [*JETP* **84**, 406].

Keldysh, L. V. (1968). In *Proceedings of the 9th International Conference on Physics of Semiconductors* (ed. S.M. Ryvkin and V.V. Shmastev), Nauka, Leningrad, p. 1303.

Keldysh, L. V. (1970). *Usp. Fiz. Nauk* **100**, 514 [*Sov. Phys. Usp.* **13**, 292].

Keldysh, L. V. (1971). In *Eksitony v poluprovodnikakh (Excitons in Semicon-ductors)* (ed. B. M. Vul), Nauka, Moscow (in Russian), p. 5.

Keldysh, L. V. (1976). *Pisma Zh. Eksp. Teor. Fiz.* **23**, 100 [*Sov. Phys. JETP Lett.* **23**, 86].

Keldysh, L. V. (1983). In (Jeffries and Keldysh, 1983), p. xi.

Keldysh, L.V. (1986). *Contemp. Phys.* **27**, 395.

Keldysh, L. V. and Silin, A. P. (1973). *Fiz. Tverd. Tela* **15**, 1532 [*Sov. Phys. Solid State* **15**, 1026].

Keldysh, L. V. and Silin, A. P. (1975). *Zh. Eksp. Teor. Fiz.* **69**, 1053 [*Sov. Phys. JETP* **42**, 535].

Keldysh, L. V. and Tikhodeev, S. G. (1975). *Pisma Zh. Eksp. Teor. Fiz.* **21**, 582 [*Sov. Phys. JETP Lett.* **21**, 273].

Keldysh, L. V. and Onishchenko, T. A. (1976). *Pisma Zh. Eksp. Teor. Fiz.* **24**, 70 [*Sov. Phys. JETP Lett.* **24**, 59].

Keldysh, L. V. and Sibeldin, N. N. (1986). In *Nonequilibrium Phonons in Nonmetallic Crystals* (ed. W. Eisenmenger and A. Kaplyanskii), Volume 16 of *Modern Problems in Condensed Matter Sciences*, North-Holland, Amsterdam, Chapter 9, p. 455.

Kittel, C. (1971). *Introduction to Solid State Physics* (4th edn). Wiley, New York.

Kulakovskii, V. D. and Timofeev, V. B. (1983). In (Jeffries and Keldysh, 1983), Chapter 2, p. 95.

Levinson, Y. B. (1986). In *Nonequilibrium Phonons in Nonmetallic Crystals* (ed. W. Eisenmenger and A. Kaplyanskii), Volume 16 of *Modern Problems in Condensed Matter Sciences,* North-Holland, Amsterdam, Chapter 3, p. 91.

Leys, F. E., March, N. H., Angilella, G. G. N. and Zhang, M.-L. (2002). *Phys. Rev.* B **66**, 073314.

Lo, T. K. (1974). *Solid State Commun.* **15**, 1231.

Lo, T. K., Feldman, B. J. and Jeffries, C. D. (1973). *Phys. Rev. Lett.* **31**, 224.

Manenkov, A. A., Milyaev, V. A., Mikhailova, G. N., Sanina, V. A. and Seferov, A. S. (1976). *Zh. Eksp. Teor. Fiz.* **70**, 695 [*Sov. Phys. JETP* **43**, 359].

Maris, H. J. (1986). In *Nonequilibrium Phonons in Nonmetallic Crystals* (ed. W. Eisenmenger and A. A.A. Kaplyanskii), Volume 16 of *Modern Problems in Condensed Matter Sciences,* North-Holland, Amsterdam, Chapter 2, p. 51.

Markiewicz, R. S. and Timusk, T. (1983). In (Jeffries and Keldysh, 1983), Chapter 7, p. 543.

Monte, A. F. G., da Silva, S. W., Cruz, J. M. R., Morais, P. C. and Chaves, A. S. (2000). *Phys. Rev.* B **62**, 6924.

Murzin, V. N., Zayats, V. A. and Kononenko, V. L. (1973). *Fiz. Tverd. Tela* **15**, 3634 [*Sov. Phys. Solid State* **15**, 2421].

Nagai, M., Shimano, R. and Kuwata-Gonokami, M. (2001). *Phys. Rev. Lett.* **86**, 5795.

Nagai, M., Shimano, R., Horiuchi, K. and Kuwata-Gonokami, M. (2003). *Phys. Rev.* B **68**, 081202.

Pauc, N., Calvo, V., Eymery, J., Fournel, F. and Magnea, N. (2004). *Phys. Rev. Lett.* **92**, 236802.

Pauc, N., Calvo, V., Eymery, J., Fournel, F. and Magnea, N. (2005). *Phys. Rev.* B **72**, 205324.

Pokrovskii, Ya. E. (1972). *Phys. Status Solidi (a)* **11**, 385.

Pokrovskii, Ya. E. (1983). In (Jeffries and Keldysh, 1983), Chapter 6, p. 509.

Pokrovskii, Ya. E. and Svistunova, K. I. (1969). *Pisma Zh. Eksp. Teor. Fiz.* **9**, 435 [*Sov. Phys. JETP Lett.* **9**, 261].

Pokrovskii, Ya. E. and Svistunova, K. I. (1970). *Fiz. Tekh. Poluprovodn.* **4**, 491 [*Sov. Phys. Semicond.* **4**, 409].

Pokrovskii, Ya. E. and Svistunova, K. I. (1971). *Pisma Zh. Eksp. Teor. Fiz.* **13**, 297 [*Sov. Phys. JETP Lett.* **13**, 212].

Pokrovskii, Ya. E. and Svistunova, K. I. (1976). *Pisma Zh. Eksp. Teor. Fiz.* **23**, 110 [*Sov. Phys. JETP Lett.* **23**, 95].

Rashba, E. I. and Sturge, M. D. (ed.) (1982). *Excitons*, Volume 2 of *Modern Problems in Condensed Matter Sciences* (ed. of the Series: V. M. Agranovich and A. A. Maradudin), North-Holland, Amsterdam.

Rice, T.M. (1977). *Solid State Physics* **32** (ed. H. Ehrenreich, F. Seitz, and D. Turnbull), Academic Press, New York, p. 1.

Rogachev, A. A. (1975). In *Excitons at High Density* (ed. H. Haken and S. Nikitine), Volume 73 of *Springer Tracts Mod. Phys.*, Chapter 8, p. 127.

Shimano, R., Nagai, M., Horiuch, K. and Kuwata-Gonokami, M. (2002). *Phys. Rev. Lett.* **88**, 057404.

Sibeldin, N. N. (2003). *Usp. Fiz. Nauk*, **173**, 999 [*Phys. Usp.* **46**, 971].

Sibeldin, N. N., Bagaev, V. S., Tsvetkov, V. A. and Penin, N. A. (1973). *Fiz. Tverd. Tela* **15**, 177 [*Sov. Phys. Solid State* **15**, 121].

Sibeldin, N. N., Stopachinskii, V. B., Tikhodeev, S. G. and Tsvetkov, V. A. (1983). *Pisma Zh. Eksp. Teor. Fiz.* **38**, 177 [*Sov. Phys. JETP Lett.* **38**, 207].

Silin, A. P. (1983). In (Jeffries and Keldysh, 1983), Chapter 8, p. 619.

Silver, R. N. (1975). *Phys. Rev.* B **11**, 1569; B **12**, 5689.

Störmer, H. L. and Martin, R. W. (1979). *Phys. Rev.* B **20**, 4213.

Tajima, M. and Ibuka, S. (1998). *J. Appl. Phys.* **84**, 2224.

Tamor, M. A. and Wolfe, J. P. (1980). *Phys. Rev.* B **21**, 739.

Tamor, M. A. and Wolfe, J. P. (1981). *Phys. Rev.* B **24**, 3596.

Tamor, M. A., Greenstein, M. and Wolfe, J. P. (1983). *Phys. Rev.* B **27**, 7353.

Thomas, G. A., Phillips, T. G., Rice, T. M. and Hensel, J. C. (1973). *Phys. Rev. Lett.* **31**, 386.

Thomas, G. A., Rice, T. M. and Hensel, J. C. (1974). *Phys. Rev. Lett.* **33**, 219.

Tikhodeev, S. G. (1983). *Pisma Zh. Eksp. Teor. Fiz.* **37**, 215 [*Sov. Phys. JETP Lett.* **37**, 255].

Tikhodeev, S. G. (1985). *Usp. Fiz. Nauk*, **145**, 3 [*Sov. Phys. Usp.* **28**, 1].

Tikhodeev, S. G. (1997). *Phys. Rev. Lett.* **78**, 3225.

Tikhodeev, S. G., Gippius, N. A. and Kopelevich, G. A. (2000). *Phys. stat. sol. (a)* **178**, 63.

Tsvetkov, V. A., Alekseev, A. S., Bonch-Osmolovskii, M. M., Galkina, T. I., Zamkovets, N. V. and Sibeldin, N. N. (1985). *Pisma Zh. Eksp. Teor. Fiz.* **42**, 272 [*Sov. Phys. JETP Lett.* **42**, 335].

Vashishta, P., Kalia, R. K. and Singwi, K. S. (1983). In (Jeffries and Keldysh, 1983), Chapter 1, p. 1.

Vavilov, V. S., Zayats, V. A. and Murzin, V. N. (1969). *Pisma Zh. Eksp. Teor. Fiz.* **10**, 304 [*Sov. Phys. JETP Lett.* **10**, 192].

Voos, M. and Benoît à la Guillaume, C. (1976). In *Optical Properties of Solids: New Developments* (ed. B. Seraphin), North-Holland, Amsterdam, p. 143.

Westervelt, R. M. (1976). *Phys. Status Solidi (b)* **76**, 31.

Westervelt, R. M. (1983). In (Jeffries and Keldysh, 1983), Chapter 3, p. 187.

Westervelt, R. M., Lo, T. K., Staehli, J. L. and Jeffries, C. D. (1974). *Phys. Rev. Lett.* **32**, 1051.

Wolfe, J. P. and Jeffries, C. D. (1983). In (Jeffries and Keldysh, 1983), Chapter 5, p. 317.

Wolfe, J. P., Markiewicz, R. S., Kittel, C. and Jeffries, C. D. (1975). *Phys. Rev. Lett.* **34**, 275.

Worlock, J. M., Damen, T. C., Shaklee, K. L. and Gordon, J. P. (1974). *Phys. Rev. Lett.* **33**, 771.

Zamkovets, N. V., Sibeldin, N. N., Tikhodeev, S. G. and Tsvetkov, V. A. (1985). *Zh. Eksp. Teor. Fiz.* **89**, 2206 [*Sov. Phys. JETP* **62**, 1274].

Zamkovets, N. V., Sibeldin, N. N. and Tsvetkov, V. A. (1994). *Zh. Eksp. Teor. Fiz.* **105**, 1066 [*Sov. Phys. JETP* **78**, 572].

17

COLLECTIVE STATE OF INTERWELL EXCITONS IN DOUBLE QUANTUM WELL HETEROSTRUCTURES

V. B. Timofeev
Institute of Solid State Physics, Russian Academy of Sciences, 142432
Chernogolovka, Russian Federation

Abstract
Bose-condensation of interwell excitons in coupled double quantum
wells of GaAs/AlGaAs Schottky-diode heterostructures is the focus
of discussions. It is demonstrated that luminescence of the Bose-
condensed collective state of interwell excitons in a circular lateral
trap is manifested in real space in a periodically patterned structure of
bright spots which are coherently bound. The collective exciton state
is characterized by the large coherence length and is destroyed on temperature increase due to fluctuations in the thermal order parameter
across the system of interacting interwell excitons.

17.1 Introduction

Hydrogen-like excitons in intrinsic semiconductors are known to be the electrically neutral and energetically lowest electronic excitations. For several
decades excitons have been used as a convenient and very efficient tool to
simulate the behavior of matter under the effect of temperature and density
variation and external impacts, such as pressure, magnetic and electric fields,
strain fields, etc.

Depending on the concentration of electron–hole (e-h) excitations and temperature, experiments with bulk intrinsic semiconductors may be used to realize a
situation of weakly interacting exciton gas, molecular exciton gas (gas of biexcitons and excitonic trions), metallic electron–hole liquid, and electron–hole
plasma.

An exciton is a sort of composite boson because it consists of two fermions,
an electron and a hole, and this accounts for the resultant exciton spin being
integer-valued. This led to the hypothesis formulated in theoretical works in
the early 1960s (Moskalenko, 1962; Blatt *et al.*, 1962; Casella, 1963) that
Bose–Einstein condensation (BEC) is possible to be observed, in principle,
in a weakly nonideal and sufficiently diluted excitonic gas in semiconductors
at rather low temperatures. In the limit of a diluted excitonic gas, one has
$na_{\mathrm{ex}}^d \ll 1$, where n is the exciton density, a_{ex} is the exciton Bohr radius and d
is the dimensionality of the system under consideration.

The Bose–Einstein condensation of excitons in three-dimensional systems suggests macroscopic occupation of the exciton ground state with zero momentum and the appearance of a spontaneous order parameter (coherence) in the condensate (Keldysh and Kopaev, 1964; Kozlov and Maksimov, 1965; Keldysh and Kozlov, 1968).

In the limit of high e-h density, $na_{\mathrm{ex}}^d \gg 1$, the exciton concept used to be regarded by Keldysh and his school (Keldysh and Kopaev, 1964; Kozlov and Maksimov, 1965; Keldysh and Kozlov, 1968) in direct analogy to Cooper pairs, and the condensed excitonic state (a sort of excitonic insulator state) was described in the mean-field approximation by analogy with the Bardeen–Cooper–Shrieffer superconducting state. The sole difference consisted in the fact that the pairing in an excitonic insulator was determined by the e-h interaction and the excitons themselves served as an analog of the Cooper pairs (Keldysh and Kopaev, 1964). Well-apparent Coulomb gaps in the excitonic insulator state can arise under nesting of the electron–hole Fermi-surfaces. Theoretical studies reported by Comte and Nozieres (1982) suggest a smooth transition between the low- and high-density limits.

BEC was discovered one decade ago in the diluted and deeply cooled gases of atoms whose resulting spin is integer-valued (Ketterle, 2002). This striking discovery was possible owing to the elegant realization of the laser and evaporative cooling techniques in application to diluted atomic gases and to the selective small-volume accumulation of atoms in magnetic traps. The transition temperatures T_C in the case of alkaline atomic gases turned out to be very low (around one μK and even lower), which is due to the large masses of atoms and their relatively low densities resulting from the inevitable losses during cooling and accumulation in traps.

Since BEC has been found in the diluted gases of bosonic atoms, this phenomenon has stimulated special interest and acquired significance with reference to excitons. Indeed, the translational effective masses of excitons in semiconductors are typically small, i.e., equal to or smaller than the mass of a free electron. Therefore, unlike BEC in diluted gases of atomic hydrogen, alkalis, and transition metals, BEC in an excitonic gas at experimentally achievable densities may occur at much higher temperatures (e.g., in liquid helium temperature range). However, for a photoexcited excitonic gas, which, in principle, is a nonequilibrium system, the problem of its cooling to temperatures of the crystal lattice serving as a thermostat is important. Under real experimental conditions, the temperature of a quasi-equilibrium excitonic gas is always somewhat higher than that of the crystal lattice in which excitons are "immersed" due to their finite lifetime. Such overheating of the excitonic system is especially pronounced at $T < 1$ K because of the low thermal capacity of the lattice and the presence of a "narrow" region (a sort of "bottle neck") at small transferred momenta insurmountable for one-phonon relaxation of excitons ($|\mathbf{K}| < ms/\hbar$, where $\mathbf{K}$ and m are the exciton momentum and effective translational mass, respectively, and s is the speed of sound). In this context, the most promising objects to search for BEC are those whose exciton annihilation rate is several orders of magnitude lower than the exciton relaxation rate along the energy axis. This criterion is fulfilled, for example, in indirect-gap semiconductors in which radiative recombination of excitons involves short-wavelength phonons and is therefore rather slow compared with the relaxation process that result in thermal equilibrium with the lattice. But because such semiconductors are characterized by strong orbital degeneracy of the electron (many valleys) and hole spectra and strong anisotropy of carrier effective masses, the lowest state in an interacting and sufficiently dense e-h system is the electron–hole liquid into which excitons, excitonic molecules and trions are condensed (Keldysh, 1968; Jeffries and Keldysh, 1983). The phenomenon of exciton condensation into droplets of electron–hole liquid was predicted by L. V. Keldysh as early as in 1968 (Keldysh, 1968) and most comprehensively studied in Ge and Si (Jeffries and Keldysh, 1983). Evidently, a gas of excitons and excitonic molecules in equilibrium with electron–hole liquid remains a classical, Boltzmann gas. In contrast, the exciton recombination rates in direct-gap semiconductors with dipole allowed interband optical transitions are very high. Moreover, other complications are intrinsic in such semiconductors related to the exciton-polariton dispersion and polariton dynamics near $\mathbf{K} = 0$ (Rashba and Sturge, 1982). In the case of two-dimensions the properties of excitonic polaritons in microcavities and the related nonlinear optics and polariton dynamics are considered in length, for example, in Gippius *et al.* (2005).

Some experiments have been designed to discover BEC in an excitonic gas in Cu_2O (Hulin *et al.*, 1980; Snoke *et al.*, 1990; Fortin *et al.*, 1993; Mysyrowicz *et al.*, 1994, 1996, 1997), where the optical decay from the ground state of para- and orthoexcitons is forbidden in the zero order in $\mathbf{K}$ and therefore the states are long-lived, and also in uniaxially compressed crystals of Ge (Kukushkin *et al.*, 1981; Timofeev *et al.*, 1983) under condition of broken symmetry with respect to the spin degrees of freedom. It was recently shown for Cu_2O that the Auger processes heat the excitonic system and restrict the exciton density from above (O'Hara *et al.*, 1999, 2000). For this reason, in optical pumping experiments conducted thus far, the exciton gas in these crystals was the classical Boltzmann gas (O'Hara *et al.*, 1999, 2000). Studies with Ge crystals examined excitons with the spin alignment, directly analogous to spin-aligned hydrogen atoms. The narrowing of the spontaneous exciton–photon emission line with increasing exciton concentration at given temperature has evidenced the well-apparent degenerate Bose-statistics of excitons at high densities. However, it is still impossible to realize the critical conditions for BEC of excitons in the above crystals, although studies of this effect are in progress.

In recent years, the BEC of excitons has been extensively investigated in two-dimensional (2D) systems based on semiconductor heterostructures. The studies are focused on 2D systems with spatially separated electron–hole layers. Researchers' attention to these objects was attracted by theoretical works done

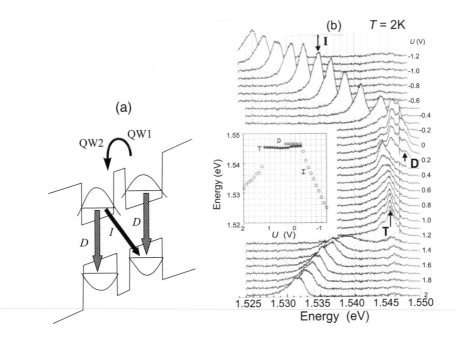

FIG. 17.1. (a) Schematic representation of the energy levels in coupled double
quantum wells under applied electric field. Arrows indicate intrawell (D) and
interwell (I) optical transitions. From (Gorbunov and Timofeev, 2006). (b)
Luminescence spectra of intrawell excitons (D), trions (T), and interwell excitons (I) in DQWs at different bias voltages U (U values in Volts are given near
corresponding curves). The inset shows the dependence of spectral positions
of maxima of corresponding lines on electric field. $T = 2$ K.

in the mid-1970s (Lozovik and Yudson, 1975; Shevchenko, 1976). In the context of the problem under consideration, double quantum wells (DQWs) and
superlattices turned out to be the most interesting among other 2D objects
based on semiconductor heterostructures, because they allow the photoexcited
electrons and holes to be spatially separated between adjacent QWs (Fukuzawa
et al., 1990; Golub *et al.*, 1990; Kash *et al.*, 1991; Butov, 1996; Timofeev *et
al.*, 1998a, 1998b; Butov *et al.*, 1999; Larionov *et al.*, 2000; Butov *et al.*, 2000;
Larionov *et al.*, 2002). In DQWs under applied bias in the growth direction,
tilting quantized subbands, it is possible to excite excitons whose electron and
hole are in different QWs separated by a transparent tunnel barrier. Such an
exciton is called a spatially indirect or interwell exciton (IE), as opposed to
a direct (D) or intrawell exciton in which electron and hole are placed within
the same QW (Fig. 17.1). A situation realized in GaAs/AlAs heterostructures
where excitons are not only spatially separated but also proved to be indirect
in momentum space was considered by Butov (1996). Interwell excitons have a

longer lifetime than intrawell ones because of limited overlap between electron and hole wavefunctions through the tunnel barrier. Characteristic times of the radiative recombination of indirect excitons are as large as dozens and even hundreds of nanoseconds. Therefore, it is easy for them to accumulate, and the gas of interwell excitons can be cooled to rather low temperatures close to the temperature of the crystal lattice. It is worthwhile to note that thermalization of two-dimensional excitons, and interwell excitons, particularly, involves bulk phonons. In this case, however, the law of momentum conservation along the quantization axis is relaxed, which accounts for the lack of a "bottle neck" for the recoil on small momenta and the feasibility of one-phonon relaxation with participation of acoustic phonons at $T < 1$K. As a result, the relaxation of IEs to the lattice temperature occurs a few orders of magnitude faster than their radiative decay (Ivanov *et al.*, 1997, 1999). Because of the broken inverse symmetry, interwell excitons have a large dipole moment even in the ground state. As a result, such excitons cannot be bound into molecules or other complicated complexes due to dipole–dipole repulsion.

It should be recalled, however, that in ideal and unconfined 2D system with constant one-particle density of states, BEC at finite temperature cannot proceed in principle because of divergence of the number of states in the case when the chemical potential $\mu \to 0$ (i.e., the states with moments $\mathbf{K} \geq 0$ can accumulate an unlimited number of Bose particles). Therefore, in this case, one can speak about Bose condensation only at $T = 0$ K, which has no physical sense. In this context, it is appropriate to recall the work by Hohenberg (1967) which provides a rigorous proof (based on N. N. Bogolubov's inequality) that unconfined and ideal 2D electron systems cannot have a nonzero order parameter, because it is destroyed by fluctuations. This proof is applicable to both superfluidity and superconductivity in ideal 2D systems. A similar theorem was proved for a 2D model of Heisenberg's ferromagnet by Mermin and Wagner (1966). The Kosterlitz–Thouless phase transition (Kosterlitz and Thouless, 1973) when the superfluid state in 2D systems develops as a result of vortex pairing, is not discussed here. Such a transition is topological and not at variance with the theorem of Hohenberg (1967).

This situation is changed in 2D spatially confined system where BEC is known to occur at finite temperatures. The critical temperature for the case of interwell excitons confined in a lateral trap with rectangular barriers is given by the expression

$$T_{\mathrm{C}} = \frac{2\pi\hbar^2}{g_{\mathrm{ex}} k_{\mathrm{B}} m_{\mathrm{ex}} \ln(N_{\mathrm{ex}} S / g_{\mathrm{ex}})} \quad , \tag{17.1}$$

i.e., it decreases logarithmically with increasing area S occupied by 2D Bose particles with density N_{ex}, translational effective mass in the plane m_{ex}, and degeneracy factor g_{ex} (k_{B} is the Bolzmann constant). For example, the critical temperature T_{C} of interwell exciton gas in GaAs/AlGaAs DQWs with density 10^{10} cm^{-2} and translational mass $0.2m_0$ (m_0 is the free electron mass) under micron-scale lateral trap is around 3 K.

The spatial confinement in the QWs plane may be due to large-scale random potential fluctuations related to variations of the QW width on heteroboundaries $w(R)$. Also related to these variations are changes in the effective lateral potential $U(r) = U(w(r))$ (Zhu et al., 1995). Under quasi-equilibrium conditions, the exciton density distribution is described by the equality $\mu(N(r)) + U(r) = \overline{\mu}$, where the chemical potential of interwell excitons $\overline{\mu}$ is related to their average density and $\mu(N)$ is the chemical potential of a homogeneous excitonic phase in the spatially confined region (in a trap). Evidently, $|\mu(r)| < |\overline{\mu}|$ because $\mu(N) = -|E_{\mathrm{ex}}| + |\delta U|$ (E_{ex} is the exciton binding energy). This means that excitons accumulate more easily in the lateral confinement region, where their density can be much in excess of the mean density in the QWs plane. Therefore, critical conditions corresponding to the Bose condensation of IEs are most readily realized in lateral domains that function as exciton traps.

A variety of scenarios of collective behavior in a sufficiently dense system of spatially separated electrons and holes were theoretically considered (Yoshioka and MacDonald, 1990; Chen and Quinn, 1991; Fernandez-Rossier and Tejedor, 1997; Lozovik and Berman, 1997; Lozovik and Ovchinnikov, 2001; Keeling et al., 2004). For example, it was claimed by Lozovik and Berman (1997) and Lozovik and Ovchinnikov (2001) that a liquid dielectric excitonic phase may constitute the metastable state in an e-h system despite the dipole–dipole repulsion of interwell excitons at certain critical parameters, such as the dipole moment, density and temperature. As shown by Zhu et al., 1995, the condensed dielectric excitonic phase can arise only in the presence of lateral confinement in the DQWs plane. Such confinement and the associated external compression facilitate the accumulation of interwell excitons to critical densities sufficient for the effects of collective exciton interaction to be manifested. In Fernandez-Rossier and Tejedor (1997) the role of spin degrees of freedom under Bose condensation is discussed.

In a recently published theoretical work (Keeling et al., 2004), it was shown that Bose condensation in a trap results in a dramatic change of the interwell exciton photoluminescence angular distribution. Due to long-range coherence of condensed phase a sharply focused peak of luminescence in the direction normal to DQWs plane should appear. The radiation peak should exhibit strong temperature dependence due to the thermal order parameter phase fluctuations across the system. The angular distribution of photoluminescence can also be used for imaging vortices in a trapped condensate, because vortex phase spatial variations lead to destructive interference of photoluminescence intensity in certain directions, creating nodes in intensity distribution that imprint the vortex configuration.

Real heterostructure systems with quantum wells always contain a fluctuating random potential due to the presence of residual impurities, both charged and neutral, as well as variety of structure defects. These imperfections create a random potential relief in the QWs plane; as a result, photoexcited electrons and holes spatially separated between adjacent QWs (as well as interwell excitons) may be strongly localized on these fluctuations at sufficiently low temperatures. Such an effect of strong localization in coupled quantum systems is manifested, for example, as thermally activated tunneling of localized particles (Timofeev *et al.*, 1998a, 1998b). In connection to this, investigations of the properties of delocalized (above "mobility edge") excitons are carried out in perfect structures with a minimum density of localized states (around or even less that 10^9 cm^{-2}).

17.2 Experimental

Now we consider a few recently conducted experiments using GaAs/AlGaAs heterostructures with DQWs that have large-scale random potential fluctuations, where interwell excitons exhibit collective behavior after the critical density and temperature are achieved (Larionov *et al.*, 2002; Dremin *et al.*, 2002; Gorbunov *et al.*, 2004; Gorbunov and Timofeev, 2004, 2006).

Firstly, we describe the architecture of the used n-i-n heterostructure with DQWs. 12 nm thick double quantum wells were separated by a narrow barrier of four AlAs monolayers. Similar AlAs epitaxial layers were grown on the borders between each QW and 150 nm thick AlGaAs insulating barriers. Build-in electrodes were 100 nm thick n$^+$-GaAs (Si-doped) layers. Large-scale random potential fluctuations were induced in these structures by the growth interruption technique on the boundaries of the AlAs and AlGaAs barriers.

The main information about interwell exciton properties has been obtained by analyzing the luminescence spectra as excitation power, temperature, electric field, and polarization of resonant optical pumping varied during steady state or pulsed optical excitation.

17.2.1 *Phase diagram*

Let us start with the illustration of the quality of the used structures with DQWs. The optical transitions of interest are schematically presented in Fig. 17.1(a). The luminescence spectra of IEs measured at different biases are depicted in Fig. 17.1(b). The intrawell luminescence region at zero electric bias exhibits two lines. One is $1s$ HH of free heavy-hole exciton (denoted as D) and the other is the line of bound, charged excitonic complexes (intrawell excitonic trion, T) (Timofeev *et al.*, 1999). At small bias, it is possible to change both the charge and the structure of the excitonic trion by varying the sign of gate voltage. The line of interwell excitons (I-line) emerges in the spectra after the electric field is switched on and causes the Stark shift of the dimensionally quantized bands in adjacent QWs to be $eF\Delta z \geq E_{\mathrm{D}} - E_{\mathrm{I}}$ (E_{D} and E_{I} are the intrawell and interwell exciton binding energies, respectively, F is the electric field, and Δz is the distance between electron and hole in the interwell exciton). The I-line shifts almost linearly upon a change in electric field at both positive and negative voltages between the electrodes (n$^+$-doped regions, see the inset in Fig. 17.1b). This line can be displaced along the energy scale over

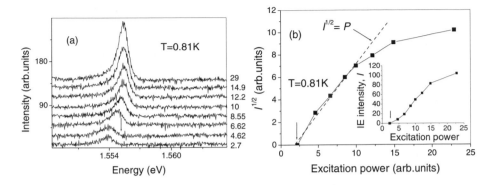

FIG. 17.2. (a) Luminescence spectra at growing excitation power, $T = 0.81$ K. Numbers show the excitation power in units of $P_0 = 60$ nW. (b) Intensity of the narrow line of interwell excitons plotted versus the excitation power P. The arrow indicates the threshold pumping value (about 200 nW) after which the narrow line becomes apparent in the spectrum.

the distance of almost a dozen times larger than the interwell exciton binding energy. At high electric voltage and steady-state excitation, the IE-line in the luminescence spectra predominates. Under the same conditions, luminescence of intrawell excitons (D) and charged excitonic complexes (T) is several orders of magnitude weaker. A large quantum yield of IE luminescence in the studied structures suggests their high quality. This ensues from the fact that a rise in the applied bias causes the IE lifetime to increase by factor of ten or even more, while the intensity of luminescence remains practically unaltered (see Fig. 17.1).

Now consider changes in the IE luminescence spectra at various pumping values under a steady-state excitation by He-Ne laser focused to a spot of about 20 μm at the sample surface. The IE-line at sufficiently low temperatures ($T = 0.81$ K) and small excitation powers around 100 nW is wide (≈ 2.5 meV) and asymmetric, with a large longwave "tail" and clear-cut violet boundary (Fig. 17.2a). Such properties of the IE photoluminescence line result from their strong localization in random potential fluctuations due to the presence of residual impurities. In this case, the linewidth reflects the statistical distribution of the chaotic potential amplitudes. As the pumping power increases, a narrow ($\approx$ meV wide) line starts to emerge in a threshold manner at the "blue" edge of the spectrum (see Fig. 17.2). The intensity of this line growths superlinearly near threshold in accordance with a near-quadratic law (Fig. 17.2b). It is only under high-power pumping that the superlinear growth of the intensity changes to linear one, and the line starts to broaden and extend to higher energies. The high energy shift of the line suggests screening of the applied electric field when IE density becomes sufficiently high. Then, the experimental measurements of this shift allow us to estimate the interwell exciton density

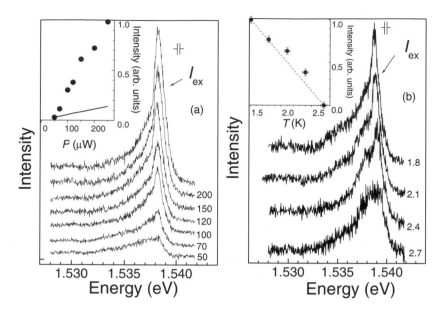

FIG. 17.3. Interwell exciton luminescence spectra in a sample with a metallic mask. Spectra were detected through the micron-scale windows: (a) At various excitation power given near the corresponding curves in microwatts ($T = 1.6$ K); (b) At various temperatures, shown to the right of the corresponding curves in kelvin. From (Larionov *et al.*, 2002). Adapted figure with permission from [A. V. Larionov, V. B. Timofeev, P. A. Ni, S. V. Dubonos, J. Hvam, and K. Soerensen, *Pisma Zh. Eksp. Teor. Fiz.* **75**, 689 (2002)], copyright (2002) by Pisma v Zhurnal Teoreticheskoi i Eksperimentalnoi Fiziki.

using the Gauss formula for spectral shift $\delta E = 4\pi e^2 n \Delta z / \varepsilon$ (n is the exciton density, and ε is the dielectric permittivity). The IE concentration appears to be $n = 3 \times 10^{10}$ cm^{-2} for the line shift less than 1.3 meV. A sufficiently narrow IE luminescence line can be seen at various negative bias voltages in the range from -0.5 to -1.2 V.

Direct evidence of the interwell exciton condensation was obtained in experiments with single trapping domains associated with large-scale fluctuations of the random potential. The experiments were conducted using n-i-n structures of the described architecture covered with a 100 μm thick metallic layer (aluminum film). The film was etched using electron beam lift-off photolithography to fabricate ring-shaped holes (windows) with a minimum diameter around one micron. These windows were used to excite and record luminescence signals. The aluminum film in such experiments was not contacted with n$^+$-doped contact layer of the heterostructure.

The results exhibited below were obtained by optical excitation and subsequent detection of luminescence directly through 1 μm diameter windows (Fig. 17.3).

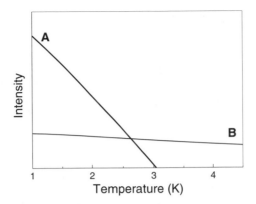

Fig. 17.4. Computed temperature dependences of the luminescence intensities of condensed (A) and supracondensed (B) interwell excitons. From (Gorbunov *et al.*, 2004). Adapted figure with permission from [A. V. Gorbunov, V. E. Bisti, V. B. Timofeev, *Zh. Eksp. Teor. Fiz.* **128**, 803 (2004)], copyright (2004) by Zhurnal Teoreticheskoi i Eksperimentalnoi Fiziki.

The experiments were carried out under resonant excitation of intrawell excitons with heavy holes (1*s* HH-intrawell excitons) by tunable Ti:Sap-phire laser in order to minimally overheat the excitonic system relative to the lattice temperature. At small excitation densities (below 40 μW), the luminescence spectra exhibit a relatively broad (some 2 meV) asymmetric IE band (Fig. 17.3a). This inhomogeneously broadened line is due to strongly localized excitons. As the pumping increases ($\geq$ 50 μW), a narrow line emerges in a threshold manner at the "blue" edge of the broad band. The intensity of this line grows superlinearly with increasing pumping (see inset to Fig. 17.3a). The minimal measured total width of this line in the experiments was about 250 μeV. It was close to the lattice temperature. A further rise in pumping power (to over 0.5 mW) led to a monotonic broadening of the narrow IE-line and its shift towards higher energies (the external electric field screening effect).

In these experiments, the narrow IE-line disappeared from the spectrum at $T \geq 3.6$ K. Figure 17.3(b) illustrates the typical behavior of the IE-line upon temperature variations and fixed pumping. One can see that at $T = 1.8$ K and the excitation power of 250 μW, this high-intensity line evidently rises above the structureless background of localized excitons. Its intensity drops with increasing T without a substantial change in the width; at $T = 3.6$ K it becomes practically indiscernible against the background of structureless spectrum.

It is noteworthy that a decrease in the intensity of the IE-line with increasing temperature is not of an activating origin. Measurements of the temperature dependence of the IE-line intensity in different pumping regimes yielded the following relation for its temperature behavior:

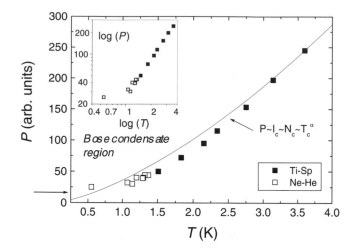

FIG. 17.5. Phase diagram of the Bose condensation of interwell excitons in the sample with large-scale random potential fluctuations. Open squares and circles correspond to optical excitation by Ti:sapphire laser, black ones to that by a He-Ne laser. From (Dremin *et al.*, 2002). Adapted figure with permission from [A. A. Dremin, V. B. Timofeev, A. V. Larionov, J. Hvam, and K. Soerensen, *Pisma Zh. Eksp. Teor. Fiz.* **76**, 526 (2002)], copyright (2002) by Pisma v Zhurnal Teoreticheskoi i Eksperimentalnoi Fiziki.

$$I_{\mathrm{T}} \propto 1 - \frac{T}{T_{\mathrm{C}}} \;, \tag{17.2}$$

where I_{T} is the line intensity at given T and T_{C} is the critical temperature corresponding to the disappearance of this line from the spectrum at a given excitation power.

We believe that the above experimental findings are evidence of the Bose condensation of interwell excitons in a lateral trap about one micron size that originates from large-scale fluctuations in random potential. At small excitation powers and sufficiently low temperatures, photoexcited IEs turn out to be strongly localized due to the presence of imperfections (for instance, residual impurities). Corresponding to this situation is a wide, inhomogeneously broadened luminescence band at small excitation powers. But strong dipole–dipole repulsion forbids localization of more than one exciton on defect. These account for saturation of this luminescence channel in considered high-quality samples at concentrations around or even below 3×10^9 cm^{-2}. A further above-threshold increase in the pumping intensity in the trap results in the delocalization of interwell excitons (excitation of IEs above "mobility edge"). Then, after the critical density values are reached, excitons undergo condensation (macroscopic occupation) to the lowest delocalized state in the trap. In our experiments, this transformation is apparent as the appearance, in the threshold manner, of a narrow luminescence line and its superlinear growth. The most convincing argument in favor of the exciton condensation is the critical temperature dependence of their properties. It is possible to calculate the change in the luminescence intensity of the condensed and supracondensed fractions of the excitons in a micron-scale trapping domain as the temperature increases. The results of such calculations are presented in Fig. 17.4 (Gorbunov *et al.*, 2004). It can be seen that in framework of the model used by Gorbunov *et al.* (2004), the linear behavior of the intensity of the narrow luminescence line at varying temperature, up to its disappearance from the spectrum, is realized for condensed excitons. At the same time, luminescence of supracondensate excitons displays very low sensitivity to temperature variations in the range under study.

We evaluated the threshold for the appearance of the narrow IEs luminescence line corresponding to the onset of macro-filling by excitons of the lowest state in the lateral trap. The results were used to construct a phase diagram out-lining the region of the exciton Bose condensate (Fig. 17.5). For this aim, we examined pumping dependences of the luminescence spectra in the temperature range 0.5–4 K. For each temperature value in this range, we determined the threshold powers P_C at which the narrow spectral line corresponding to excitonic condensate either appeared for the first time or began to disappear. In other words, the phase diagram was built in the coordinates $P_C - T$. The density of interwell excitons was estimated from the "violet" shift of the line associated with screening of the applied electric voltage at large excitation powers. The threshold exciton density found was $n_C = 3 \times 10^{-9}$ cm^{-2} at $T = 0.55$ K (indicated by the arrow in Fig. 17.5). The line intensity and exciton concentrations on the pump scale of the phase diagram were linearly related. In the temperature range 1–4 K, the critical density and temperature values at which condensation occurs are related by the power law

$$N_C \propto T^\alpha, \qquad (17.3)$$

where $\alpha \geq 1$. At $T < 1$K, the phase boundary cannot be described by a simple power law.

17.2.2 *Luminescence kinetics and spin relaxation*

Condensed excitons must be spatially coherent. The spatial coherence must be manifested at least on the scale of the de Broglie interwell exciton wavelength $\lambda_{ex} = h/\sqrt{\pi m_{ex} k_B T} = 1.5 \times 10^3$ Å, at $T = 2$ K, which is one order of magnitude larger than the exciton Bohr radius. The exciton density in the above experiments corresponded to the dimensionless parameter $r_\lambda = n\lambda_{ex}^2 = 4$. At such special scales, coherent excitons must be phased, i.e., described by the same wavefunction. The development of a collective excitonic state may be accompanied by a rise in the radiative decay rate of condensed excitons compared with that of supracondensed ones, as well as by a longer spin relaxation time of condensed excitons than single-particle excitonic spin relaxation time.

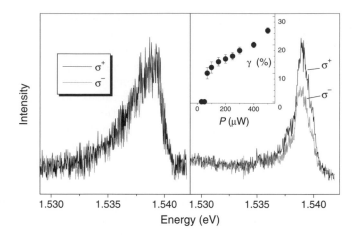

FIG. 17.6. Luminescence of interwell excitons resonantly excited by circularly polarized light, excitation power 60 and 150 μW (left and right panels, respectively). Black and grey curves correspond to different circular polarizations. The luminescence line in the left panel (excitation power 60 μW) is not polarized: line contours measured for clockwise and counterclockwise polarization coincide within the measured noise range. The inset shows the excitation power dependence of the degree of circular polarization γ of the interwell exciton line. From (Larionov *et al.*, 2002). Adapted figure with permission from [A. V. Larionov, V. B. Timofeev, P. A. Ni, S. V. Dubonos, J. Hvam, and K. Soerensen, *Pisma Zh. Eksp. Teor. Fiz.* **75**, 689 (2002)], copyright (2002) by Pisma v Zhurnal Teoreticheskoi i Eksperimentalnoi Fiziki.

These hypotheses were verified in experiments on the resonant excitation of excitons by polarized light. We recall that the ground state of interwell excitons is not a simple Kramers doublet but a four-fold degenerate state with respect to angular momentum projection ($M = \pm 1, \pm 2$). "Bright" and "dark" excitonic states are characterized by the angular momentum projections $M = \pm 1$ and $M = \pm 2$, respectively. In the case of resonant steady-state excitation of intrawell 1s HH exciton by circularly polarized light, the degree of circular polarization of the IEs line corresponding to condensed phase was found to increase in a threshold manner (Fig. 17.6). Enhanced pumping caused a rise in the degree of circular polarization up to 40%. This effect indirectly evidenced a rise in the radiative recombination rate in the condensate and, as well as, a possible increase of the spin relaxation time. Both were confirmed by direct measurements of the evolution and kinetics of the luminescence spectra under pulsed resonant excitation by circularly polarized light (Fig. 17.7). It follows from Fig. 17.7(a) that the narrow line of the condensed excitonic phase emerged in the spectra with time delay of about 4 nsec relative to the excitation pulse. Its decay time was 20 nsec, whereas the background contin-

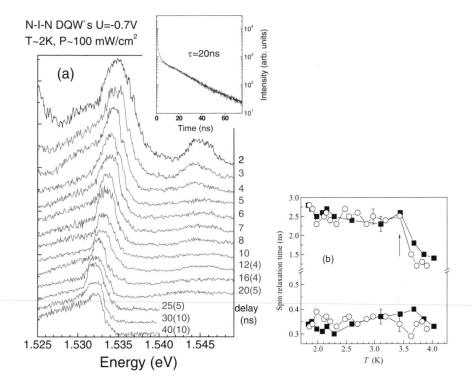

FIG. 17.7. (a) Time evolution of the interwell exciton luminescence spectra under pulsed excitation. Numbers to the right of the curves are delay times
and signal integration times (in parentheses) in nanoseconds. $T = 2$ K. From
(Larionov *et al.*, 2000). (b) The temperature dependence of spin-relaxation
time for intrawell excitons (the lower pair of curves) and interwell excitons
(the upper pair of curves). The applied voltage $U = 0.6$ V (circles) and 0.55 V
(squares). The arrow indicates the region of abrupt change of spin-relaxation
time. From (Larionov *et al.*, 2005). Adapted figures with permission from [L. V. Larionov, M. Bayer, J. Hvam, and K. Soerensen, *Pisma Zh. Eksp. Teor. Fiz.* **81**, 139 (2005)],
copyright (2005) by Pisma v Zhurnal Teoreticheskoi i Eksperimentalnoi Fiziki.

uum under the narrow line "lived" over 100 nsec. Direct measurements of the
degree of circular polarization demonstrated that the spin relaxation time of
condensed excitons was almost twice as long as the single-particle excitonic
spin relaxation measured at $T > T_C$ (Fig. 17.7b from Larionov *et al.*, 2005).
These results may be regarded as indirect evidence of the enlarged coherent
volume of excitonic condensate (Kagan *et al.*, 2000).

In considering the enhanced coherence of the condensed excitonic phase it is
worth recalling the interesting experiments by Butov *et al.* (1999, 2000) in
which the authors observed an unusual behavior of IEs photoluminescence

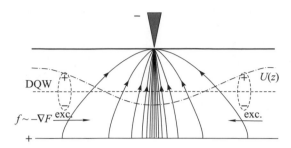

FIG. 17.8. Schematic presentation of a nonuniform electric field in DQW heterostructure; the current flow through the point contact into the bulk of
heterostructure. The arrows are electric field lines close to the needle; the
dashed-dotted line shows the potential well for excitons; the dashed line is
the DQW cross-section. From (Gorbunov and Timofeev, 2004). Adapted figure
with permission from [A. V. Gorbunov and V. B. Timofeev, *Pisma Zh. Eksp. Teor. Fiz.*
80, 210 (2004)], copyright (2004) by Pisma v Zhurnal Teoreticheskoi i Eksperimentalnoi
Fiziki.

kinetics under pulsed laser excitation. At sufficiently high excitation powers
and low enough temperature, the kinetics of the IEs radiative decay cannot be
described by a simple exponential law. Instead, the intensity of photoluminescence increases in a jump manner immediately after the pulse excitation and
thereafter decreases nonexponentially rapidly.

Such unusual behavior cannot be observed at a low excitation power, at high
temperatures, in the presence of strong disorder associated with a chaotic potential, or under applied magnetic field perpendiculars to heterolayers. Under
such conditions, the photoluminescence kinetics is monoexponential and characterized by a large times. We should emphasize that only those delocalized
excitons whose translational motion momenta are of the order of the light
momentum undergo radiative annihilation, i.e., $|\mathbf{K}| \leq E_g/\hbar c$ (c is the light
velocity in the heterostructure). The rise in the interwell exciton radiative
annihilation observed by Butov *et al.* (1999, 2000) is regarded as a result of
two effects. One is related to the enlargement of the IE coherence area under conditions of exciton condensation to states with smaller momenta than
the light ones. The other is assumed to be due to superlinear filling of the
optically active excitonic states induced by the stimulated exciton scattering,
when the filling numbers in the final exciton states $n \gg 1$ (i.e., as a result of
the degenerate Bose statistics of interwell excitons).

17.2.3   *Condensation of interwell excitons in a nonuniform electric field*

In the case of lateral traps due to chaotic potential, the particular potential
trap shape, the actual depth and the lateral size of traps remain to be determined; this leads to the actual problem of artificially creating lateral traps

for IEs by means of controlled external impacts, with properties amenable to reliable management and monitoring. Nonuniform electric fields appear to be most suitable for this purpose (Gorbunov and Timofeev, 2004), in comparison with the method based on nonuniform deformation suggested by Snoke *et al.* (2004). The action of such a field on interwell excitons is schematically illustrated in Fig. 17.8. As the current flows through the point contact between the needle of tunneling microscope and the surface of heterostructure containing DQWs, with electric bias applied between the conducting needle and built-in gate (n^+-layer), it generates a strongly nonuniform electric field. Naturally, the maximum field strength can be expected to be found directly beneath the needle tip. It can be monitored experimentally from the spectral shift of the IE luminescence line with respect to intrawell exciton line.

Furthermore, interwell excitons are dipoles. When placed in a nonuniform electric field, they are subject to the action of an electrostatic force, $f \approx -e\nabla F$, which makes them move to the center of the potential well functioning as a IEs trap. The distance over which the interwell excitons can drift depends on the field gradient, exciton lifetime, and mobility. The qualitative aspect of the potential well (shown in Fig. 17.8) is depicted by the dashed-dotted line.

The nonuniform electric field inside the heterostructure was created using the needle of a tunneling microscope (a commercial silicon cantilever was used for this purpose). The entire device was placed in liquid helium inside an optical cryostat (Gorbunov and Timofeev, 2004).

Firstly, we consider the shape that the potential well had when the current flowed through the above structure. For this, the exciton luminescence spectra were measured with regard to the position of the finely focused excitation laser spot. The luminescence was detected directly inside the exciting spot. Figures 17.9(a) and (b) show the luminescence spectra of interwell and intrawell excitons depending on the distance from the point contact. The figures also illustrate the behavior of the spectral position of the luminescence maxima of interwell excitons as a function of the same distance. These figures essentially depict the shape of the lateral potential well for IEs: $U(x) \simeq eF(x)\Delta z + E_{\mathrm{I}}(x)$. It can be seen that the largest spectral shift in IE luminescence relative to the line of intrawell excitons occurs in the immediate proximity of the point contact, where the well has the largest depth (see Fig. 17.9) and the spectral shift amounts to 13.5 meV. This shift is a good indicator of the maximum electric field strength, which equals 1.1×10^4 V cm^{-1} for the case under consideration. It decreases with distance from the point contact, and the intrawell exciton spectral line begins to appear at distances of the order of 500 μm.

The above electric field distribution around the point contact is close to an axially symmetric one, but the radial distribution of the resulting potential is not a monotonic function of the distance from the contact. First, the potential well shows a deep and narrow fall in the immediate proximity to the contact (see Fig. 17.10a and b). The most surprising and unexpected finding is that such narrow and rather deep falls in the potential curve are located far from

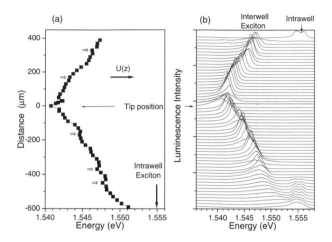

FIG. 17.9. The profile of the lateral potential well created by a nonuniform electric field. Panel (b) shows the luminescence spectra of interwell and intrawell excitons measured at different distanced between the exciting laser spot and the point contact (the focused spot size is around 20 μm, the luminescence is detected only from excitation region). Panel (a) shows the position of the IE luminescence maximum depending on the same distance. $T = 2$ K From (Gorbunov and Timofeev, 2004). Adapted figure with permission from [A. V. Gorbunov and V. B. Timofeev, *Pisma Zh. Eksp. Teor. Fiz.* **80**, 210 (2004)], copyright (2004) by Pisma v Zhurnal Teoreticheskoi i Eksperimentalnoi Fiziki.

the center of the well (as shown by the wide arrows in the left part of Fig. 17.9). These narrow falls always occur when the current exceeds 1 μA and depart from the well center as the current grows further. The origins of such axially positioned falls on the potential curve still await explanation. They may result from the nonlinear screening of the electric field by photoexcited carriers inside the structure. However, narrow and rather deep falls in the potential curve can be themselves functioning as a lateral trap for IEs.

Of interest is the behavior of the IEs luminescence excited directly in the traps created by a nonuniform electric field. A wide interwell exciton luminescence band (about 2–3 meV) is produced in the central part located just beneath the needle due to strong overheating of the near-contact region with maximum current density (see Fig. 17.10). Also interesting in this context are the narrow falls on the potential curve located far away from the contact site, in which overheating can be neglected. These regions are also potential traps for IEs. Therefore, it is worth examining exciton luminescence excited in one of the traps remote from the center where the gas of interwell excitons is assumed to be rather "cold". Weak pumping (below 300 nW) induces a wide asymmetric band of interwell exciton luminescence with long-wavelength "tail", whose intensity gradually decreases (see Fig. 17.11). Under low-excitation pumping

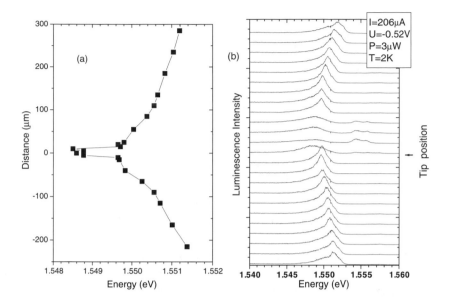

FIG. 17.10. Lateral potential well created by a nonuniform electric field close to the cantilever needle: (a) position of IE luminescence maximum measured at different distances between the exciting laser spot and the point contact; (b) luminescence spectra of interwell excitons depending on the same distances. From (Gorbunov and Timofeev, 2004). Adapted figures with permission from [A. V. Gorbunov and V. B. Timofeev, *Pisma Zh. Eksp. Teor. Fiz.* **80**, 210 (2004)], copyright (2004) by Pisma v Zhurnal Teoreticheskoi i Eksperimentalnoi Fiziki.

this inhomogeneously broadened band is due to radiative decay of strongly localized excitons. On excitation power increase, a rather narrow line of delocalized excitons arises in a threshold manner at the blue end of this band. Its intensity near the threshold behaves superlinearly as the pumping grows. Only at large pumping powers (more than $3~\mu W$) IEs luminescence line begins to broaden and shift towards larger energies due to electric field screening. As the excitation power varies, the behavior of interwell exciton luminescence line in a lateral trap "prepared" by the electric field is qualitatively similar to the behavior of IEs during their condensation to the lowest state in the lateral trapping domains associated with large-scale random potential fluctuations.

17.2.4 *Spatially resolved collective state in a Bose gas of interacting interwell excitons*

In this section we consider a new class of experiments related to highly spatially resolved luminescence of a Bose-gas of interacting interwell excitons. These experiments have been performed with the use of Schottky-photodiode heterostructures with DQWs of the same architecture, as described above. The surface of the samples was coated with metallic Al-mask (thickness around 100

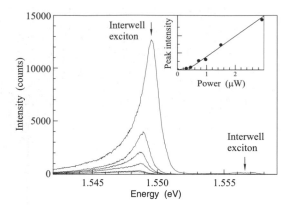

FIG. 17.11. The interwell luminescence spectra upon variations of excitation
power and measured directly out of the center of the potential well (spaced
80 μm from the needle). The inset shows the power dependence of maximum
intensity. $T = 2$ K. From (Gorbunov and Timofeev, 2004). Adapted figure with
permission from [A. V. Gorbunov and V. B. Timofeev, *Pisma Zh. Eksp. Teor. Fiz.* **80**, 210
(2004)], copyright (2004) by Pisma v Zhurnal Teoreticheskoi i Eksperimentalnoi Fiziki.

nm) that had circle windows (diameters of windows were 2, 5, 10 and 20 μm)
prepared by electron beam "lift-off" lithography. Through such windows the
photoexcitation and simultaneous microscopic observation of photoluminescence with spatial resolution of around one micron were performed (Gorbunov
and Timofeev, 2006). In the used Schottky-diode heterostructures the build-in
electrode was n^+-GaAs layer (Si-doped) and a metallic mask was used as a
Schottky gate-electrode. An electrical bias was applied between the metallic
mask and the n^+-layer.

Let us start with experiments, when a window of circular shape on the top of
the heterostructures was projected on the entrance split of the spectrometer.
This slit cuts only central part of the projected window along its diameter.
First of all, it was found that under applied bias and photoexcitation the
radial distribution of the electric field inside the heterostructure is strongly
inhomogeneous. Scattered field is minimal at the center of window and increases radially to the window edges. But more important is that close to the
circle edge of the window electric field in radial direction behaves nonmono-
tonically, namely, a ring-like potential trap for interwell excitons appears along
the window edge perimeter. The existence of such a ring-like potential trap
for interwell excitons along the circle edge of the window in the metallic mask
was justified by analyzing of the spectral shifts of the interwell exciton line under scanning of strongly focused excitation spot in the radial direction in the
vicinity of the window edge. The reason for such a behavior of the electric field
in the close vicinity of the window in metallic mask, as well as the behavior

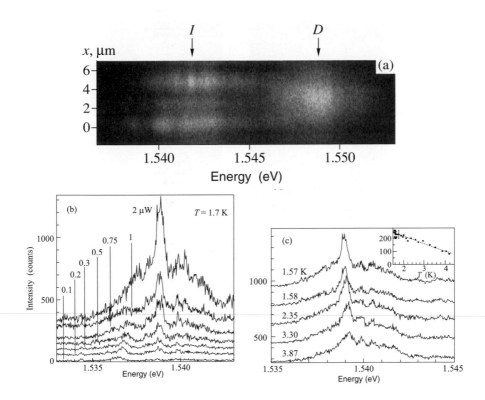

FIG. 17.12. Luminescence spectra measured directly from the 5 μm window in the metallic mask on top of the heterostructure with DQWs. (a) Space distribution (vertical direction) of photoluminescence within the window; entrance slit of the spectrometer cut only central part of the window. (b) Behavior of narrow luminescence line corresponding to interwell excitons on excitation powers at $T = 1.7$ K; spectra from bottom to top correspond to excitation powers 0.1, 0.2, 0.3, 0.5, 0.75, 1 and 2 μW. (c) Behavior of the narrow luminescence line under temperature variation in the range (1.6–4) K and excitation power 5 μW. Spectra from top to bottom are measured at temperatures: 1.57, 1.58, 2.35, 3.30 and 3.87 K respectively. From (Gorbunov and Timofeev, 2006). Adapted figures with permission from [A. V. Gorbunov and V. B. Timofeev, *Pisma Zh. Eksp. Teor. Fiz.* **83**, 178 (2006)], copyright (2006) by Pisma v Zhurnal Teoreticheskoi i Eksperimentalnoi Fiziki.

of the shape and the depth of the circle potential trap on applied bias and on conditions of photoexcitation will be published elsewhere (see also Gorbunov and Timofeev, 2006).

The luminescence spectrum from a window of 5 μm in size, which is projected on the entrance slit of the spectrometer, is illustrated in Fig. 17.12. The im-

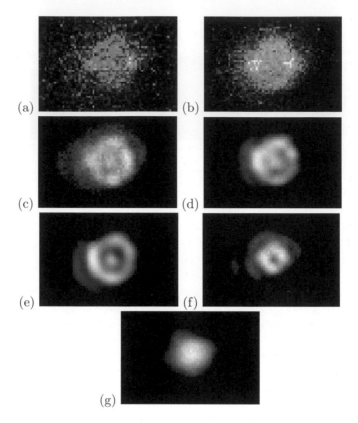

(a) (b) (c) (d) (e) (f) (g)

FIG. 17.13. Spatial structure of interwell exciton luminescence measured from
5 μm window at $T = 1.7$ K. Images (a)–(e) (from the top to the bottom in the
left column) correspond to excitation powers 1, 1.5, 5, 70 and 300 μW, the
size of excitation spot is equal to 50 μm. (f) Spatial structure of luminescence
in the 2 μm window and image (g) corresponds to luminescence of intrawell
excitons. From (Gorbunov and Timofeev, 2006). Adapted figures with permission
from [A. V. Gorbunov and V. B. Timofeev, *Pisma Zh. Eksp. Teor. Fiz.* **83**, 178 (2006)],
copyright (2006) by Pisma v Zhurnal Teoreticheskoi i Eksperimentalnoi Fiziki.

age of interwell exciton luminescence from the window, observed from the exit
slit, is presented in Fig. 17.12(a). One can see bright spots of interwell exciton luminescence (spot sizes around 1.5 μm) located at the upper and lower
edges of the window. Furthermore, structureless luminescence of the intrawell
excitons (D) is seen in the center of the window where the scattered electric
field is low. Figure 17.12(b) demonstrates how the narrow interwell exciton
line, corresponding to a bright spot, starts to appear and is growing on excitation power increase. This line is placed above the luminescence continuum
background connected with localized excitons. Figure 17.12(c) illustrates the

behavior of this narrow line at given excitation power on temperature increase. One can see almost linear decreasing of the intensity of narrow interwell exciton line on temperature increase in the range 1.7–4.2 K. The data presented in Fig. 17.12 are completely equivalent to the observations discussed in previous sections.

Now let us turn to experiments on spatially resolved luminescence when a window of corresponding size was directly projected on the CCD-camera avoiding spectrometer. Under these experiments the luminescence of interwell and intrawell excitons was spectrally and separately selected with the use of interference filters. Figure 17.13 presents a series of images of the spatially resolved lateral structures of luminescence connected with interwell excitons measured through 5 μm window at different excitation powers. One can see that at minimal excitation powers, corresponding to average exciton concentration around or less than 10^9 cm^{-2}, the luminescence spot is structureless and its intensity is almost homogeneous within the window under investigation (Fig. 17.13a). With excitation power increase, a discrete luminescence structure starts to appear in a threshold manner along the perimeter of the window. Firstly, two luminescence spots start to appear (Fig. 17.13b), then four spots (Fig. 17.13c), then six spots (Fig. 17.13d), consequently with excitation power increase. Spot sizes are around 1.5 μm. At excitation powers around 150 μW the number of spots is equal to eight. Finally, at excitation powers more that 200 μW only structureless luminescence with a ring shape remains (Fig. 17.13e).

In the case of a 2 μm window the well resolved lateral structure with four equidistantly placed spots is observed (Fig. 17.13f). In 10 μm window the lateral structure of luminescence is more complicated: besides of the radially symmetric structure of spots, radially symmetric structures of different rings with spots fragmentation start to be visible. In the case of 20 μm windows we did not observe (perhaps, did not resolve) any structures. We should emphasize that luminescence intensity distributions for intrawell excitons under the same experimental conditions remain homogeneous and do not manifest any spatial structure (Fig. 17.13g).

For the 5 μm window, taken as an example, the behavior of a discrete configuration consisting of six equidistant spots was studied under temperature variation at a given excitation power (Fig. 17.14). It was established that on temperature increase the discrete luminescence structure starts to wash off (pairs of spots are merging) at $T > 4$ K (see Fig. 17.14d). Finally, the whole discrete structure of equidistant spots is merged into a continuous luminescent ring at $T \geq 15$ K.

It is useful to recall that spatial luminescence distributions of interwell excitons manifested in the form of lateral rings with some fragmentized structure along the ring were observed by Butov *et al.* (2002a, 2002b) for the case of DQWs and by Snoke *et al.* (2002) and Rapaport *et al.* (2004) for wide single QW. Such ring-like luminescence structures arose under the effect of rather powerful laser excitation. Their origin is related to the depletion of electrons and electric field

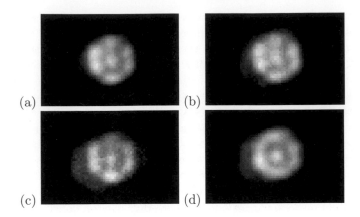

FIG. 17.14. Spatial structure of interwell exciton luminescence in the 5 μm
window measured at different temperatures: (a) 1.74 K, (b) 3.9 K, (c) 4.25 K,
(d) 4.35 K. Excitation power is 50 μW. From (Gorbunov and Timofeev, 2006).
Adapted figures with permission from [A. V. Gorbunov and V. B. Timofeev, *Pisma Zh.
Eksp. Teor. Fiz.* **83**, 178 (2006)], copyright (2006) by Pisma v Zhurnal Teoreticheskoi i
Eksperimentalnoi Fiziki.

screening just in the optical pumping region, as well as to the oncoming drift
of electrons and holes that occurs under these conditions. Therefore there is no
direct connection between the experiments presented and discussed here and
published by Butov *et al.* (2002a, 2002b), Snoke *et al.* (2002) and Rapaport
et al. (2004).

The above described experiments were performed on a dozen circular windows
with sizes 2, 5, 10 and 20 μm. But lateral spatial configurations of equidistantly
placed luminescence spots in windows of given size and at similar experimental conditions (temperature, excitation power) were always reproduced. This
means that the origin of spatially periodical structures of spots seen in the
luminescence of laterally confined in a ring trap interwell excitons is not determined by the random potential fluctuations which definitely exist in the
investigated samples. In connection to this statement it is worth emphasizing
that the spatially resolved luminescence of intrawell excitons, measured under
the same experimental conditions, was always structureless. Because the observed phenomenon is sensitive to the temperature and excitation powers it
cannot be mediated by such effects as "phonon wind" (Keldysh and Tikhodeev,
1975; Keldysh, 1976), exciton density waves (Chen and Quinn, 1991) or polariton effects connected with surface plasmons in the vicinity of the metallic
mask edges.

We assume that the observed phenomenon is a manifestation of collective coherent properties of interacting 2D interwell excitons. It is an effect of the
interwell exciton Bose-condensation in a lateral trap of the ring shape. The
collective exciton state is characterized by a large coherence length (in the

considered case around and even more than one micrometer) and is destroyed on temperature increase due to the thermal order parameter (coherence) fluctuations across the system of interacting interwell excitons (in the considered case along the ring structure of luminescence spots). We assume that the bright luminescence spots of the observed periodically patterned structure are coherently bound and as a consequence, a temporal quantum beating under pulsed excitation is expected. In the recently published paper by Keeling *et al.* (2004) it was shown that a vortex character of the Bose-condensate of the interwell excitons in a lateral trap is manifested in a peculiar angle distribution of luminescence intensity due to destructive interference. It is interesting to emphasize that the expected vortex configurations in real space according to Keeling *et al.* (2004) coincide qualitatively with the above presented experimental observations (see Figs. 17.13 and 17.14).

17.3 Conclusion

The phenomenon of Bose condensation discovered in the studied structures emerges in a limited range of the exciton concentration scale: $N_{\text{loc}} < N_{\text{ex}} < N_{\text{I-M}}$. In the low-density region ($N_{\text{loc}} \leq 3 \times 10^9$ cm^{-2}), the limitation is imposed by the effects of strong localization in imperfections and in high-density region ($N_{\text{I-M}}$), by the disappearance of bound exciton state caused by the metal–insulator transition. Indeed, a rise in the excitation power above 0.5 mW leads to the broadening of the IE luminescence line in our structures, which continues to be extended and shifts in the spectrum to the region of higher energy. The broadening of the IE-line results from the overlap between the exciton wavefunction in the QW planes and the associated Fermi repulsion between electrons in one QW and holes in another. At an estimated density of $N_{\text{I-M}} \simeq 8 \times 10^{10}$ cm^{-2}, interwell excitons lose their individual characteristics and the e-h plasma is formed, with electrons and holes spatially separated between adjacent QWs. Corresponding to this density is the dimensionless parameter $r_s = 1/(\pi N_{\text{I-M}})^{1/2} a_{\text{B}} = 1.8$ (the interwell exciton Bohr radius $a_{\text{B}} \approx$ 150 Å was found from the diamagnetic shifts of IEs). The e-h plasma begins to screen the externally applied electric field, and the interwell luminescence band shifts towards higher energies. It is this shift that was used to find the e-h density.

We believe that the discovered collective state of a Bose gas of interacting interwell excitons under condensation in the circular lateral traps will stimulate further experiments in this interesting area. For instance, preliminary measurements of the optically Fourier-transformed images of the regular periodic lateral structure of collective state (from real to K-space at a given frequency), which arises in a threshold manner and is temperature-dependent, show complicated angular distributions of luminescence intensity due to destructive interference. These findings are further evidence of Bose condensate coherence and an indication that luminescence of the condensed phase is spatially directed. We assume that due to space coherence of the observed periodic structures a temporal beating of luminescence intensity under pulsed excitation can be found (an analog of Josephson oscillations). Besides, experiments on Bose-condensation in lateral traps under pulse resonant photoexcitation with the use of circularly polarized light are expected to be very efficient and informative.

Acknowledgements

I am grateful to A. V. Gorbunov and A. V. Larionov for significant contributions in the course of these investigations.

References

Blatt, J. M., Boer, K. W. and Brandt, W. (1962). *Phys. Rev.* **126**, 1691.

Butov, L. V. (1996). *Proceedings of the 23th International Conference on the Physics of Semiconductors*,Vol.3. (ed. Scheffer and Zimmermann R.), World Scientific, Singapure, p. 1927.

Butov, L. V., Imamoglu, A., Mintsev, A. V., Campman, K. L. and Gossard, A. C. (1999). *Phys. Rev.* B **59**, 1625.

Butov, L. V., Mintsev, A. V., Lozovik, Yu. E., Campman, K. L. and Gossard, A. C. (2000). *Phys. Rev.* B **62**, 1548.

Butov, L. V., Lai, C. W., Ivanov, A. L., Gossard, A. C. and Chemla, D. S. (2002). *Nature (London)* **417**, 47.

Butov, L. V., Gossard, A. C. and Chemla, D. S. (2002). *Nature (London)* **418**, 751.

Casella, R. C. (1963). *J. Appl. Phys.* **34**, 1703.

Chen, X. M. and Quinn, J. J. (1991). *Phys. Rev.* B **67**, 895.

Comte, C. and Nozieres, P. (1982). *J. Phys. (Paris)* **43**, 1069; 1083.

Dremin, A. A., Timofeev, V. B., Larionov, A. V., Hvam, J. and Soerensen, K. (2002). *Pisma Zh. Eksp. Teor. Fiz.* **76**, 526 [*JETP Lett.* **76**, 450].

Fernandez-Rossier, J. and Tejedor, C. (1997). *Phys. Rev. Lett.* **78**, 4809.

Fortin, E., Fafard, S. and Mysyrowicz, A., (1993). *Phys. Rev. Lett.* **70**, 3951.

Fukuzawa, T., Mendez, E. E. and Hong, J. M. (1990). *Phys. Rev. Lett.* **64**, 3066.

Gippius, N. A., Tikhodeev, S. G., Keldysh, L. V. and Kulakovskii, V. D. (2005). *Usp. Fiz. Nauk* **175**, 327 [*Phys. Usp.* **48**, 306].

Golub, J. E., Kash, K., Harbison, J. P. and Florez, L. T. (1990). *Phys. Rev.* B **41**, 8564.

Gorbunov, A. V. and Timofeev, V. B. (2004). *Pisma Zh. Eksp. Teor. Fiz.* **80**, 210 [*JETP Lett.* **80**, 185].

Gorbunov, A. V., Bisti, V. E. and Timofeev, V. B. (2004). *Zh. Eksp. Teor. Fiz.* **128**, 803 [*JETP* **101**, 693].

Gorbunov, A. V. and Timofeev, V. B. (2006). *Pisma Zh. Eksp. Teor. Fiz.* **83**, 178 [*JETP Lett.* **83**, 146].

Hulin, D., Mysyrowicz, A., Benoit a la Guillaume, C. (1980). *Phys. Rev. Lett.* **45**, 1970.

Hohenberg, P. C. (1967). *Phys. Rev.* **158**, 383.

Ivanov, A. L., Ell, C. and Haug, H. (1997). *Phys. Rev.* E **55**, 6363.

Ivanov, A. L., Littlewood, P. B. and Haug, H. (1999). *Phys. Rev.* B **59**, 5032.

Jeffries, C. D. and Keldysh, L. V., (1983) *Electron-Hole Droplets in Semiconductors*, North-Holland, Amsterdam.

Kagan, Yu., Kashurnikov, V. A., Krasavin, A. V., Prokofev, N. V. and Svistunov, B. V. (2000). *Phys. Rev.* A **61**, 043608.

Kash, J., Zachau, A. M., Mendez, E. E., Hong, J. and Fukuzawa, M. T. (1991). *Phys. Rev. Lett.* **66**, 2247.

Keeling, J., Levitov, L. S. and Littlewood, P. B. (2004). *Phys. Rev. Lett.* **92**, 176402.

Keldysh, L. V. and Kopaev, Yu. V. (1964). *Fiz. Tverd.Tela* **6**, 279 [(1965). *Sov. Phys. Solid State* **6**, 2219].

Keldysh, L. V. (1968). In *Proceedings of the Ninth International Conference on the Physics of Semiconductors.* (ed. by S. M. Ryvkin and V. V. Shmastev), Nauka Leningrad, p. 1303.

Keldysh, L. V. and Kozlov, A. N. (1968). *Zh. Eksp. Teor. Fiz.* **54**, 978 [*Sov. Phys. JETP*, **27**, 521].

Keldysh L. V. and Tikhodeev, S. G. (1975). *Pisma Zh. Eksp. Teor. Fiz.* **21**, 582 [*Sov. Phys. JETP Lett.* **21**, 273].

Keldysh, L. V. (1976). *Pisma Zh. Eksp. Teor. Fiz.* **23**, 100 [*Sov. Phys. JETP Lett.* **23**, 86].

Ketterle, W. (2002). *Rev. Mod. Phys.* **74**, 1131.

Kosterlitz, J. M. and Thouless, D. J. (1973). *J. Phys. C. Solid State Phys.* **6**, 1181.

Kozlov, A. N. and Maksimov, L. A. (1965). *Zh. Eksp. Teor. Fiz.* **48**, 1184 [*Sov. Phys. JETP* **21**, 790].

Kukushkin, I. V., Kulakovskii, V. D. and Timofeev, V. B. (1981). *Pisma Zh. Eksp. Teor. Fiz.* **34**, 36 [*Sov. Phys. JETP Lett.* **34**, 34].

Larionov, A. V., Timofeev, V. B., Hvam, J. and Soerensen, C. (2000). *Pisma Zh. Eksp. Teor. Fiz.* **71**, 174 [*JETP Lett.* **71**, 117].

Larionov, L. V., Timofeev, V. B., Ni, P. A., Dubonos, S. V., Hvam, J. and Soerensen, C. (2002). *Pisma Zh. Eksp. Teor. Fiz.* **75**, 689 [*JETP Lett.* **75**, 570].

Larionov, A. V., Bayer, M., Hvam, J. and Soerensen, C. (2005). *Pisma Zh. Eksp. Teor. Fiz.* **81**, 139 [*JETP Lett.* **81**, 108].

Lozovik, Yu. E. and Yudson, V. I. (1975). *Pisma Zh. Eksp. Teor. Fiz.* **22**, 556 [*JETP Lett.* **22**, 274].

Lozovik, Yu. E. and Berman, O. L. (1997). *Zh. Eksp. Teor. Fiz.* **111**, 1879 [*JETP* **84**, 1027].

Lozovik, Yu. E. and Ovchinnikov, I. V. (2001). *Pisma Zh. Eksp. Teor. Fiz.* **74**, 318 [*JETP Lett.* **74**, 288].

Mermin, N. D. and Wagner, H. (1966). *Phys. Rev. Lett.* **17**, 1133.

Moskalenko, S. A. (1962). *Fiz. Tv. Tela* **4**, 276 [*Sov. Phys. Solid State* **4**, 199].

Mysyrowicz, A., Fortin, E., Benson, E., Fafard, S. and Hanamura, E. (1994). *Solid State Commun.* **92**, 957.

Mysyrowicz, A., Benson, E. and Fortin, E. (1996). *Phys. Rev. Lett.* **77**, 896.

Mysyrowicz, A., Benson, E. and Fortin, E. (1997). *Phys. Rev. Lett.* **78**, 3226.

O'Hara, K. E., O'Suilleabhain, L. and Wolfe, J. P. (1999). *Phys. Rev. B* **60**, 10565.

O'Hara, K. E. and Wolfe, J. P. (2000). *Phys. Rev. B* **62**, 12909.

Rapaport, R., Gang, C., Snoke D., Simon, S. H., Pfeiffer, L., West, K., Liu, Y. and Denev, S. (2004). *Phys. Rev. Lett.* **92**, 117405.

Rashba, E. I. and Sturge, M. D., eds. (1982). *Excitons*, North-Holland, Amsterdam.

Shevchenko, S. I. (1976). *Fiz. Nizk. Temp.* **2**, 505 [*Sov. J. Low Temp. Phys.* **44**, 389].

Snoke, D. W., Wolfe, J. P. and Mysyrowicz, A. (1990). *Phys. Rev. B* **41**, 11171.

Snoke, D., Denev, S., Lui, Y., Pfeiffer, L. and West, K. (2002). *Nature (London)* **418**, 754.

Snoke, D. W., Lui, Y. and Voros, Z. (2004). *Cond-mat/0410298*.

Timofeev, V. B., Kulakovskii, V. D. and Kukushkin, I. V. (1983). *Physica* B+C **117-118**, 327.

Timofeev, V. B., Filin, A. I., Larionov, A. V., Zeman, J., Martinez, G., Hvam, J. M., Birkedal, D. and Sorensen, C. B. (1998). *Europhys. Lett.* **41**, 535.

Timofeev, V. B., Larionov, A. V., Ioselevich, A. S., Zeman, J., Martinez, G., Hvam, J. and Soerensen, K. (1998). *Pisma Zh. Eksp. Teor. Fiz.* **67**, 580 [*JETP Lett.* **67**, 613].

Timofeev, V. B., Larionov, A. V., Grassi Alessi, M., Capizzi, M., Frova, A. and Hvam, J. M. (1999). *Phys. Rev. B* **60**, 8897.

Yoshioka, D. and MacDonald, A. H. (1990). *J. Phys. Soc. Japan* **59**, 4211.

Zhu, Z., Littlewood, P. B., Hibersen, M. S. and Rice, T. M. (1995). *Phys. Rev. Lett.* **74**, 1633.

18

BOSE–EINSTEIN CONDENSATION OF EXCITONS: PROMISE AND DISAPPOINTMENT

R. Zimmermann
Institut für Physik der Humboldt-Universität zu Berlin, Newtonstrasse 15, 12489 Berlin, Germany

Abstract

The Bose–Einstein condensation of excitons has a long history with seminal contributions from Leonid V. Keldysh. Despite numerous efforts, however, compelling experimental evidence is still missing. A brief survey of attempts to realize exciton condensation in different semiconductor systems is given. Specific problems compared with atomic Bose condensation are highlighted. More details are given on coupled quantum wells as a possible candidate in the search for exciton condensation. While here extremely long radiative lifetimes of indirect excitons can be achieved, their strong dipole–dipole repulsion leads to a genuine nonideal behavior. Theoretical results from a dynamical T-matrix theory are presented which allow us to explain the blue-shift and line broadening under strong excitation which have been seen in recent high-density photoluminescence experiments using a lateral trap.

18.1 A short history of exciton condensation

My first encounter with exciton physics was not reading one of the famous textbooks like "Knox" (Knox, 1963), but trying to understand the "Excitonic Insulator" from a review written by Halperin and Rice (1968). Soon we realized that this branch of solid state theory was shaped and brought forward by (at that time) Soviet physicists, with cornerstone papers by Keldysh and coworkers (Keldysh and Kopaev, 1964; Keldysh and Kozlov, 1968). The basic idea was to look at a small-gap semiconductor whose energy gap could be tuned to values below the exciton binding energy. Then, the ground state must become unstable against formation of excitonic correlations. The theoretical description is very close to the famous BCS (Bardeen–Cooper–Schrieffer) theory of superconductivity, only that here the attraction between electron and hole is the standard Coulomb force, while in the BCS the dynamical screening by acoustic phonons provides the necessary attraction. More important, however, are two other facts. In the exciton case, anomalous propagators have a simple physical meaning – they are the optically driven interband polarization. Therefore, what needs sophisticated experiments using Josephson junctions in superconductors, can be easily done in the excitonic case via interband optics.

Secondly, outside the excitonic ground state (or condensate), we find excitons as normal bound states, being thermally excited.

There is a long-standing claim that the intermediate valent compound TmSeTe fits to the original idea of the excitonic insulator (Bucher *et al.*, 1991; Wachter, 2005). When driving the bandgap through zero by applying hydrostatic pressure, the material exhibits a phase transition with a critical temperature as high as 250 K, which is deduced from measuring thermal properties. Further specific experiments would be needed to clarify if excitons are the main players in this game – ruling out competitive mechanisms like charge density wave or lattice instability.

A new turn came in when the excitonic insulator was identified as a possible phase transition in wide-gap semiconductors, too. If optical recombination is slow enough, excited electrons and holes may reach quasi-equilibrium (with common temperature, but different chemical potentials). Although more speculative than the small-gap case with strict equilibrium, a lot of interesting physics was to be expected. However, the pressing question "Does it work, even in principle?" was always lurking behind this scenario. This feeling of uncertainty was often seen already in the paper titles, as, e.g., "Possibility of the excitonic phase in insulators" (which was my first scientific publication, Zimmermann, 1970).

The story with the excitonic insulator was coming to rest for some time when a new phase transition – from dilute exciton gas to the electron–hole liquid – was shown to dominate the physics in highly excited semiconductors. Again, Leonid V. Keldysh was among the pioneers of this new "electron–hole droplet physics" (Keldysh, 1968), which is mainly electron-gas theory extended to two species (electrons and holes). In fact, the ground-state energy minimum for multivalley semiconductors (Si or Ge) is rather deep and lies at relatively high densities. Therefore, the excitonic character of the dilute gas around the droplets was of not much importance, and a random-phase-like approximation for the two-component electron–hole plasma did rather well. Indeed, at these high densities, strong screening prevents the formation of excitons as bound states (Mott transition). Things were expected to be different for single-valley semiconductors (GaAs, CdS). Theoretical work revealed that in this case even the high-density side of the first-order phase diagram bears excitonic signatures (Zimmermann, 1976). However, these materials have a direct bandgap with dipole-allowed optical transitions, and the much shorter radiative lifetime prevents the build-up of quasi-equilibrium under high excitation. Still it was general belief that here a quite interesting sequence could be expected with rising density: Formation of excitons, their Bose–Einstein condensation (BEC), strong nonideal effects due to the underlying fermionic structure (excitonic insulator), and a high-density plasma which looses the excitonic correlations gradually (Zimmermann, 1988). When including polar optical phonons into the electron–hole plasma theory, Leonid V. Keldysh was once more paving the way (Keldysh and Silin, 1975), which gave me the first chance to come into personal contact with him, and to learn quite a lot (which continued to be the case at all further meetings).

How to overcome the exciton lifetime problem? One way was to use wide-gap semiconductors with dipole-forbidden optical transitions. The paradigm material is here cuprous oxide which has indeed a long history on the search for exciton BEC. Following earlier claims on biexciton condensation in CuCl (Nagasawa *et al.*, 1975), pioneering work on Cu_2O was done by Wolfe (Lin and Wolfe, 1993) and Mysyrowicz (Mysyrowicz *et al.*, 1996). However, several findings were not really conclusive, for instance to read off directly the bosonic distribution function from the (phonon-assisted) photoluminescence. Other BEC claims found a different and much less spectacular explanation, as, e.g., exciton super transport being driven by phonon wind (Bulatov and Tikhodeev, 1992; Tikhodeev, 1997). Quite recently, new findings on Cu_2O give new hopes, as the exploration of 1s-2p transitions using infrared femtosecond spectroscopy (Kuwata-Gonokami, 2005).

Another way to achieve extremely long radiative lifetimes of excitons is using coupled quantum wells. Application of a static electric field in the growth direction allows to tilt the confinement potentials such that the lowest state has electrons and holes residing in different quantum wells. A spatially indirect exciton is formed which acquires microsecond lifetime since the overlap between electron and hole is exponentially small. Already in 1976, Lozovik made a first theoretical prediction for the excitonic insulator in such a system (Lozovik and Yudson, 1976). The first experimental hint on a possible BEC was reported quite early, too (Fuzukawa *et al.*, 1990). But here again, spectacular findings as the nearly millimeter-sized ring emission has been considered first as BEC evidence (Butov, 2002; Snoke *et al.*, 2002), but are now discussed in terms of a dynamic p-n junction. Still, the quite regular fragmentation of this ring emission is waiting for a conclusive explanation. Another interesting issue is to take the fluctuations (noise) of the photoluminescence as indicator of BEC (Butov, 2003). In Sections 18.3 and 18.4 we give more details and own theoretical results for this very promising system, focussing on the emission lineshape.

It is quite common knowledge that at a given density, the BEC critical temperature is larger if the Bose particles have a lighter mass (a specific example is given in equation (18.3) below). Usually, the exciton mass is dictated by the underlying semiconductor material. However, if exciton-polaritons are formed within a microcavity, the dispersion is dominated by the cavity mode, and the relevant polariton mass can be orders of magnitude smaller than the exciton mass. However, there is always a price to be paid! Here, the system has to be pumped hard in order to get reasonable polariton densities. Furthermore, the cooling is slowed down since only the exciton part in the polariton is able to transfer energy to the lattice (Doan *et al.*, 2005). Both effects hinder to establish quasi-equilibrium, which we consider to be one condition for classification as BEC. At least bosonic stimulation has been seen (Deng *et al.*, 2003), and

TABLE 18.1. Bose–Einstein condensation of excitons is rather improbable – if not impossible – since:

Excitons are:	Due to:	While for atoms,
instable	radiative decay	stability is given
hard to equilibrate within lifetime	slow cooling	sophisticated cooling in use
composite bosons	electron–hole pairs, Mott transition!	electron–ion plasma is far away
strongly nonideal	dipolar repulsion (hinders condensation)	repulsion is weak

interesting features like parametric scattering under resonant pumping have been reported, too (Baumberg *et al.*, 2000). For the latter, a description as "driven condensate" would be probably right. In a genuine BEC, we hope to see a condensate whose phase coherence builds up spontaneously, and is not triggered from outside.

A more exotic version of exciton condensation is discussed for the quantum Hall effect in electron bilayer systems. At half filling, a description in terms of electrons and holes within the Landau level can be applied. Using the exciton terminology provides a new look on this system (Eisenstein and MacDonald, 2004), but it may be questioned if a classification as exciton BEC really works. Quite a good overview on experimental attempts to find exciton BEC and the ambiguities involved in their interpretation has been given by David Snoke at the NOEKS-7 conference in 2003 (Snoke, 2003). The paper title "When should we say we have observed Bose condensation of excitons?" speaks on its own. A broader survey of relevant work is compiled in a special issue of Solid State Communications (Snoke, 2005), being an offspring of a workshop held in 2004 near Pittsburgh (USA).

The actual strong interest in exciton BEC – often revitalizing old concepts – is surely triggered by the tremendous success to reach experimentally Bose–Einstein condensation of atomic systems (Cornell and Wiemann, 2002; Ketterle, 2002). In contrast, clear-cut proofs of excitonic BEC in solid state systems are still missing. The small exciton mass is an advantage, but other facts are less promising. Several of them are listed in Table 18.1 and compared with their counterparts in atomic systems. In spite of the many cons and only few pros, I am rather confident that we will see clear evidence of excitonic BEC in semiconductor systems at some time. However, before being too optimistic on a too short time-scale, we should try to learn more from the hard work which was needed to reach BEC in atomic systems.

To formulate tasks for theory, we should concentrate more on the way towards exciton BEC, while the study of the condensate itself might be deferred to a later stage. A particular problem is to understand how the excitons can manage to cool down. Nonequilibrium Green's functions and the "Keldysh technique" (Keldysh, 1964) provide the proper machinery, which has been used in the present context by Haug and coworkers (Schmitt *et al.*, 2001). Even when assuming that quasi-equilibrium has been established (as I will do in the following sections), we face another complication: Since excitons are composite quantum particles with a sizable interaction, the formulation and solution of the many-exciton problem is more challenging than treating the nearly ideal atomic Bose gas. However, in doing so we will learn a lot which might be of use for other problems in interacting solid state systems.

18.2 The ideal Bose gas in a trap

Let us start with compiling results being valid for ideal (i.e., noninteracting) bosons. The general expression for the boson number in dependence on temperature T and chemical potential μ is

$$N(\mu, T) = N_0 + \int_0^\infty dE \, \rho(E) \, n_{\mathrm{B}}(E - \mu), \quad n_{\mathrm{B}}(E) = \frac{1}{\exp(\beta E) - 1}, \quad (18.1)$$

where the Bose distribution function $n_{\mathrm{B}}(E)$ has been introduced as usual ($\beta = 1/k_{\mathrm{B}}T$). With rising boson number N, the chemical potential moves toward the lowest available states (onset of the Boson density of states $\rho(E)$, here taken at $E = 0$). If $\mu = 0$ leads to a finite value of the integral, all remaining bosons have to occupy the lowest state, forming the condensate part N_0. Therefore, the shape of $\rho(E)$ at small energies is crucial. A behavior $\rho(E) \propto E^p$ with $p > 0$ gives a finite integral. Therefore, a two-dimensional system with $\rho(E) \propto \Theta(E)$ ($p = 0$) does not allow for condensation – any boson number can be accommodated in the normal phase at $E > 0$. This would apply for excitons in the lowest sublevel state of a quantum well. However, introducing a trap changes the situation completely. For instance, a parabolic trap profile

$$V(\mathbf{R}) = \frac{\alpha}{2} R^2 \qquad (18.2)$$

leads to a density of states which rises linearly with energy, $\rho(E) \propto E$. Therefore, BEC is possible in principle. If at a given temperature the total exciton number N exceeds the critical number

$$N_c = (k_{\mathrm{B}}T)^2 \, \frac{\pi^2}{6} \, \frac{g_{\mathrm{S}} M}{\hbar^2 \alpha}, \qquad (18.3)$$

a condensate forms, $N_0 = N - N_c$ (g_{S} is the spin degeneracy).

To get numbers, we adopt for the (kinetic) exciton mass M of a typical GaAs QW a value of $M = 0.3 \, m_0$ (Siarkos *et al.*, 2000), and take for the trap strength $\alpha = 0.31 \, \mu\mathrm{eV}/\mu\mathrm{m}^2$, a value which was extracted from the experimental data by Snoke (Vörös *et al.*, 2005). At a temperature of $T = 2\,\mathrm{K}$, equation (18.3) gives $N_c = 2.5 \times 10^6$, which seems to be not out of reach.

In a trap, only a finite number of Bose particles can be placed. What about the general theorem that a true phase transition can be realized only in the

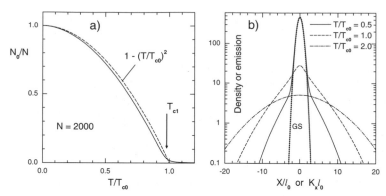

FIG. 18.1. Ideal Bose gas in a 2D trap. (a) Condensate fraction N_0/N versus temperature for $N = 2000$ bosons (full curve), and its thermodynamic limit (dashed). (b) Density distribution in the trap at different temperatures. The ground-state wavefunction squared is marked by "GS".

thermodynamic limit (both the system volume and particle number going to infinity)? Indeed, using the density of states as a continuous function of energy, we have implicitly made such a limiting process already in equation (18.1). However, in the present rather simple case we can do the finite-volume problem from scratch: In the parabolic trap potential of equation (18.2), the eigenenergies are $\epsilon_j = j\hbar\omega_0$ $(j = 1, 2, \ldots)$ with $\omega_0 = \sqrt{\alpha/M}$. Accounting properly for the degeneracy of levels, we obtain for the total exciton number

$$N(\mu, T) = g_S \sum_{j=1}^{\infty} \frac{j}{e^{\beta(\hbar\omega_0 j - \mu)} - 1}. \qquad (18.4)$$

The numerator resembles the density of states (linear increase with energy), but now completely discretized. The occupation of the lowest state $(j = 1)$ gives the condensate number N_0. Obviously, for any finite N, the chemical potential will never reach the lowest state, which a purist would call "absence of phase transition". However, even for the unrealistically small number of $N = 2000$ bosons, the dependence of the condensate number on temperature (Fig. 18.1a) resembles closely the "thermodynamic" relation from equation (18.3), here written as $N_0 = N(1 - (T/T_{c0})^2)$, and $k_B T_{c0} = \hbar\omega_0 \sqrt{6N/\pi^2 g_S}$. The point of maximum curvature may serve as a practical definition of the critical temperature. This can be defined a bit more accurately: Leave out in the sum over states equation (18.4) the lowest state $(j \neq 1)$, and put $\mu = \hbar\omega_0$. The resulting critical exciton number (or better the corresponding temperature at $N = 2000$) is marked by a vertical arrow in Fig. 18.1(a).

With the the harmonic oscillator eigenfunctions $\phi_j(X)$ at hand, we can calculate the spatial distribution of bosons in the trap quite easily,

$$n(\mathbf{R}) = g_S \sum_{j,k=0}^{\infty} \phi_j^2(X)\, \phi_k^2(Y)\, \frac{1}{e^{\beta(\hbar\omega_0(j+k+1)-\mu)} - 1}. \qquad (18.5)$$

Integration over X, Y gives the total Boson number of equation (18.4), as it should be. In Fig. 18.1(b) we display the density profile for a few different temperatures. After reaching the condensation threshold, a strong central peak evolves which is due to the macroscopic number of bosons sitting in the ground state (marked by GS). Thus, for ideal bosons, the evolution of a sharp central peak in the trap density signals condensation. However, we can read the same figure in reciprocal space as well: Then, the plot gives the emission intensity in dependence on the (in-plane) photon momentum. For this to work, we employ that the Fourier transform of the harmonic oscillator functions have the same functional form as their originals. A sharply peaked angular emission normal to the quantum well plane is to be expected, which might be used as another indication of exciton BEC. The numbers involved here are $\hbar\omega_0 = 0.3\,\mu\text{eV}$ (really tiny), and a length-scale of $l_0 = \sqrt{\hbar\omega_0/\alpha} = 1\,\mu\text{m}$.

However, results obtained for ideal bosons are of rather limited value for the spatially indirect excitons in coupled quantum wells, which feel a strong dipole–dipole repulsion. The trap potential is going to be flattened out, which will obscure the sharp central peak. On the other hand, the directional emission will persist. More details will be presented in the following sections.

18.3   Excitons in coupled quantum wells

In a recent experiment by Snoke and coworkers (Vörös *et al.*, 2005), a sample with two coupled GaAs quantum wells of $10\,\text{nm}$ width each has been used, being separated by a $4\,\text{nm}$ wide barrier. A static electric field is applied in the growth (z-) direction which allows to tune the indirect exciton state below the direct exciton. The band edge diagram is shown in Fig. 18.2. Pressing with a needle on the sample allowed to form a lateral trap which is guiding the indirect excitons towards the trap center. With increasing excitation, a blue-shift of up to $5\,\text{meV}$ and a broadening of the exciton emission line has been seen.

In order to model the system we have solved the Schrödinger equation for the confinement functions $u_a(z)$ ($a = e_{\text{dir}}, e_{\text{ind}}, h$), as displayed in Fig. 18.2. For calculating the single-exciton states, we have split off the confinement functions and introduced for the in-plane motion relative ($\mathbf{r}$) and center-of-mass ($\mathbf{R}$) coordinates, writing

$$\Psi(\mathbf{r}_e, \mathbf{r}_h) = u_e(z_e)\, u_h(z_h)\, \phi(\mathbf{r})\, \psi(\mathbf{R}). \qquad (18.6)$$

The Schrödinger equation for the 1s exciton wavefunction $\phi(\mathbf{r})$ is solved numerically (Zimmermann, 2005) with the attractive electron–hole potential $-v_{eh}(\mathbf{r})$. The in-plane Coulomb potentials between different carriers are given by

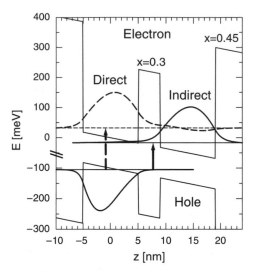

FIG. 18.2. Band edge diagram for the coupled quantum well system used in (Vörös *et al.*, 2005). Two GaAs quantum wells of 10 nm width are separated by a $\mathrm{Al}_x\mathrm{Ga}_{1-x}\mathrm{As}$ barrier of 4 nm. A static electric field of $F = 36\,\mathrm{kV/cm}$ is tilting the band edges. Each confinement wavefunction is plotted with a vertical offset being its energy. From (Zimmermann, 2006). Reprinted figure with permission from: [R. Zimmermann, *phys. stat. sol.* (b) **243**, 2358 (2006)], copyright (2006) Wiley-VCH Verlag GmbH & Co. KGaA.

$$v_{\mathrm{ab}}(\mathbf{r}) = \int dz\,dz' \frac{e^2}{4\pi\varepsilon_0\varepsilon_s\sqrt{r^2 + (z - z')^2}}\, u_{\mathrm{a}}^2(z)\,u_{\mathrm{b}}^2(z') \,. \qquad (18.7)$$

The indirect exciton has a binding energy of $E_{\mathrm{B}} = 3.5\,\mathrm{meV}$, not much different from the GaAs bulk value. Due to the extremely small overlap of the confinement functions, the radiative lifetime of the indirect exciton at zero momentum is quite long, $\tau_{\mathrm{ind}} = 0.45\,\mu\mathrm{s}$. Consequently, an equilibration of the indirect exciton gas at low bath temperatures can be expected. For the sake of comparison, we give the values for the direct exciton as well: $E_{\mathrm{B}} = 19.1\,\mathrm{meV}$, $\tau_{\mathrm{dir}} = 31\,\mathrm{ps}$.

Let us now investigate the forces acting between two indirect excitons which are at a distance $\mathbf{R}$ apart from each other. Reducing for the moment the excitons to point charges, we can easily add those four Coulomb potentials which are not already "eaten up" by forming the excitons itself, with the result

$$U_{\mathrm{d0}}(\mathbf{R}) = \frac{e^2}{4\pi\varepsilon_0\varepsilon_s}\left[\frac{2}{R} - \frac{2}{\sqrt{R^2 + d^2}}\right] \,. \qquad (18.8)$$

Here, d is the effective charge separation between electron and hole, which comes out a bit larger ($d = 15.5\,\mathrm{nm}$) compared to the distance between the quantum wells (14 nm) – an effect induced by the static electric field. A strong

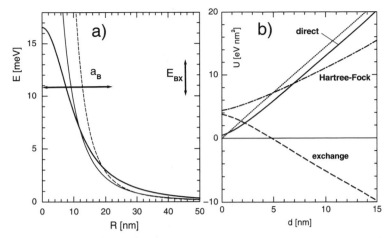

Fig. 18.3. (a) Direct repulsion between two indirect excitons: Full calculation (full curve) compared with the point-charge exciton model (thin curve) and the dipole–dipole limit (dashed curve). The arrows denote the size of the exciton and its binding energy. (b) Strengths of the contact interaction: Direct, exchange and Hartree–Fock combination $U_{\mathrm{HF}} = (5/4)U_{\mathrm{d}} + U_{\mathrm{x}}$. The dotted straight line is the point-charge result $U_{\mathrm{d}0}$.

repulsive potential is obtained (thin full curve in Fig. 18.3a). For large exciton separation, the famous dipole–dipole law

$$\lim_{R \to \infty} U_{\mathrm{d}0}(\mathbf{R}) = \frac{e^2}{4\pi\varepsilon_0\varepsilon_{\mathrm{s}}} \frac{d^2}{R^3} \qquad (18.9)$$

appears (dashed curve). However, the real charge distribution of the exciton leads to a somewhat different shape (thick full curve in Fig. 18.3a). It can be best formulated in reciprocal space as

$$U_{\mathrm{d}}(\mathbf{q}) = v_{ee}(\mathbf{q})\,\chi^2(\beta\mathbf{q}) + v_{hh}(\mathbf{q})\,\chi^2(\alpha\mathbf{q}) - 2v_{eh}(\mathbf{q})\,\chi(\beta\mathbf{q})\,\chi(\alpha\mathbf{q}), \quad (18.10)$$

where the mass ratios $\alpha = m_e/M, \beta = m_h/M$ $(M = m_e + m_h)$ appear, and $\chi(\mathbf{q})$ is the Fourier transform of $\phi^2(\mathbf{r})$.

Now, the composite nature of excitons consisting of two fermions comes into play (Hanamura and Haug, 1977). This leads to a nonlocal potential due to fermionic exchange between the constituents. Forcing this into a local potential $U_{\mathrm{x}}(\mathbf{R})$, one can construct the following Hamiltonian for many center-of-mass excitons (we closely follow Laikhtman and coworkers (Ben-Tabou de-Leon and Laikhtman, 2001)):

$$H = \int d\mathbf{R} \sum_s \Psi_s^\dagger(\mathbf{R}) \left[-\frac{\hbar^2\nabla^2}{2M} + V(\mathbf{R}) \right] \Psi_s(\mathbf{R}) + \qquad (18.11)$$

$$+ \frac{1}{2} \int d\mathbf{R}\, d\mathbf{R}' \left[U_{\mathrm{d}}(\mathbf{R} - \mathbf{R}') + U_{\mathrm{x}}(\mathbf{R} - \mathbf{R}') \right] \sum_{ss'} \Psi_s^\dagger(\mathbf{R}) \Psi_{s'}^\dagger(\mathbf{R}') \Psi_{s'}(\mathbf{R}') \Psi_s(\mathbf{R})$$

$$+ \int d\mathbf{R}\, d\mathbf{R}' U_{\mathrm{x}}(\mathbf{R} - \mathbf{R}') \sum_{ss'} \left(\frac{1}{4} - \delta_{ss'} \right) \Psi_s^\dagger(\mathbf{R}) \Psi_{-s}^\dagger(\mathbf{R}') \Psi_{-s'}(\mathbf{R}') \Psi_{s'}(\mathbf{R}) .$$

Here, it was essential to take into account the spin structure of excitons which are composed from spin $1/2$ conduction electrons and spin $3/2$ heavy hole states. The exciton spin label s denotes the four exciton states, which are $s = \pm 1$ (bright) and $s = \pm 2$ (dark). The one-exciton potential $V(\mathbf{R})$ can model a lateral trap confinement, including disorder if present.

There is an intense discussion in the literature on to what degree it is allowed to treat excitons as bosons, as done in equation (18.11). Note that in equation (18.11), both effects are implemented, namely the Coulomb interaction between two excitons as well as their departure from being true bosons. However, this is done in a perturbative fashion only (linear in the Coulomb potential). An improvement has been formulated by Okumura and Ogawa (2001), taking into account the composite-particle effect properly. Restricting to the extreme mass ratio ($m_e/m_h \rightarrow 0$), their result agrees nicely with a Heitler–London type treatment. Another route is followed by Combescot and coworkers (Combescot and Betbeder-Maribet, 2002), who try to avoid Bose commutation rules completely. Coming up with an overcomplete set of exciton operators, however, leads to a series of contradicting results, which makes their approach questionable, if not useless.

To simplify matters we reduce in what follows the exciton–exciton (XX) potential to contact form, $U_{\mathrm{d}}(\mathbf{R}) \Rightarrow U_{\mathrm{d}}\, \delta(\mathbf{R} - \mathbf{R}')$, where the strength U_{d} is given by integrating the full potential over $\mathbf{R}$, or taking its zero-momentum value. Explicitly, we arrive at

$$U_{\mathrm{d}} = \int d\mathbf{R}\, [v_{ee}(\mathbf{R}) + v_{hh}(\mathbf{R}) - 2v_{eh}(\mathbf{R})] \qquad (18.12)$$

$$= \lim_{q \rightarrow 0} [v_{ee}(\mathbf{q}) + v_{hh}(\mathbf{q}) - 2v_{eh}(\mathbf{q})] .$$

The point-charge result is simply

$$U_{\mathrm{d}0} = \frac{e^2}{\varepsilon_0 \varepsilon_{\mathrm{s}}}\, d . \qquad (18.13)$$

The exchange strength is

$$U_{\mathrm{x}} = \sum_{\mathbf{k},\mathbf{k}'} \left[2v_{eh}(\mathbf{k} - \mathbf{k}')\, \phi_{\mathbf{k}}^3\, \phi_{\mathbf{k}'} - (v_{ee}(\mathbf{k} - \mathbf{k}') + v_{hh}(\mathbf{k} - \mathbf{k}'))\, \phi_{\mathbf{k}}^2\, \phi_{\mathbf{k}'}^2 \right] .$$

$$(18.14)$$

A calculation for the CQW of Fig. 18.2 gave the values of $U_{\mathrm{d}} = 18.8\,\mathrm{eV\,nm}^2$ and $U_{\mathrm{x}} = -8.9\,\mathrm{eV\,nm}^2$.

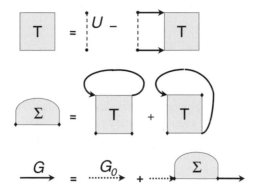

FIG. 18.4. Diagrammatic representation of the dynamical T-matrix scheme. The full line with arrow denotes the single-exciton Green's function $G_{\mathbf{k}}(z)$, while the dotted arrow represents the noninteracting Green's function. The dashed line is the XX interaction $U^{\pm}$ (for simplicity, we have omitted the spin degree of freedom). The two terms in the self-energy are the direct and (boson) exchange contributions.

In Fig. 18.3(b), the dependence on charge separation d is plotted. In order to reach smoothly the direct exciton, we have taken here single potential wells for electron and hole, displaced by d. It is interesting to see how the exchange interaction (dashed) which is dominant and repulsive at $d = 0$, reduces and changes sign at larger well separation. On the other hand, the direct repulsion (full) is taking over, so that the total effect is always repulsive.

We have added to Fig. 18.3(b) as dash-dotted curve the combination $U_{\mathrm{HF}} = (5/4)U_{\mathrm{d}} + U_{\mathrm{x}}$ which is the central quantity appearing in a Hartree–Fock decoupling (and assuming spin equilibration). The (local) shift of the bare dispersion $\epsilon_{\mathbf{k}} = \hbar^2 k^2 / 2M$ is given by

$$\Delta(\mathbf{R}) = U_{\mathrm{HF}}\, n(\mathbf{R})\,. \tag{18.15}$$

In a smooth trap potential, the exciton density follows from

$$n(\mathbf{R}) = g_{\mathrm{s}} \sum_{\mathbf{k}} n_{\mathrm{B}}(\epsilon_{\mathbf{k}} + V(\mathbf{R}) + \Delta(\mathbf{R}) - \mu)\,, \tag{18.16}$$

which needs to be solved self-consistently.

18.4 Dynamical T-matrix theory

The Hartree–Fock decoupling of the many-exciton problem as formulated in equation (18.15) and equation (18.16) can easily explain the blue-shift. However, while the Hartree–Fock approach (including a c-number term) is at the heart of the Bogolubov theory of BEC (Griffin, 1996), it is not able in principle to describe line broadenings. Going one step further, we outline here a dynamical T-matrix theory which we have formulated recently (Zimmermann, 2006). Note that this theory cannot be applied to the BEC state itself, but gives a reasonable insight into the route towards condensation.

In diagrammatic language, multiple XX scattering events are summed up and form the dynamical T-matrix $T_{\mathbf{q}}(z)$ (Fig. 18.4, first line). Plugging this into the one-exciton self-energy $\Sigma_{\mathbf{k}}(z)$ (second line) gives via the Dyson's equation (third line) an improved one-exciton propagator $G_{\mathbf{k}}(z)$. Self-consistency has to be achieved by repeated iteration.

The specific spin structure of equation (18.11) allows to split the T-matrix into a bonding/antibonding part $T^{\pm}$, which obey

$$T_{\mathbf{q}}^{\pm}(\Omega) = \frac{U^{\pm}}{1 + U^{\pm} G_{\mathbf{q}}(\Omega)}, \qquad U^{\pm} = U_{\mathrm{d}} \pm U_{\mathrm{x}}, \qquad (18.17)$$

where the exciton pair propagator (two parallel arrows in Fig. 18.4) is given by

$$G_{\mathbf{q}}(\Omega) = \sum_{\mathbf{k}} \int \frac{d\omega \, d\omega'}{\pi \; \pi} A_{\mathbf{k}}(\omega) \, A_{\mathbf{k}-\mathbf{q}}(\omega') \frac{1 + 2n_{\mathrm{B}}(\hbar\omega - \mu)}{\omega + \omega' - \Omega}. \qquad (18.18)$$

The real part of equation (18.18) has a logarithmic divergency coming from the integration over $\mathbf{k}$. This is a well-known shortcoming of using in two dimensions a contact potential. We have chosen to cut off the integration at $\epsilon_{\mathbf{k}} = 100\,\mathrm{meV}$, and postpone a proper treatment of the full $\mathbf{k}$-dependence to further work.

For the contact interaction used, both diagrams of the self-energy can be combined into one term, with $T \equiv (9/2)T^{+} + (1/2)T^{-}$, resulting in

$$\Sigma_{\mathbf{k}}(z) = \sum_{\mathbf{q}} \int \frac{d\omega}{\pi} A_{\mathbf{k}-\mathbf{q}}(\omega) \times \qquad (18.19)$$

$$\times \left[T_{\mathbf{q}}(z + \omega)\, n_{\mathrm{B}}(\hbar\omega - \mu) + \int \frac{d\omega'}{\pi} \frac{\mathrm{Im}T_{\mathbf{q}}(\omega' - i0)\, n_{\mathrm{B}}(\hbar\omega' - 2\mu)}{\omega' - z - \omega} \right].$$

This enters the exciton Green's function resp. its spectral function

$$A_{\mathbf{k}}(\omega) = \mathrm{Im}G_{\mathbf{k}}(\omega - i0) = \frac{\mathrm{Im}\Sigma_{\mathbf{k}}(\omega - i0)}{[\omega - \epsilon_{\mathbf{k}} - \mathrm{Re}\Sigma_{\mathbf{k}}(\omega - i0)]^2 + [\mathrm{Im}\Sigma_{\mathbf{k}}(\omega - i0)]^2}. \qquad (18.20)$$

For the exciton system, the spectrally and directionally resolved spontaneous optical emission is given by

$$I(\mathbf{k}, \omega) = \mu_{\mathrm{cv}}^2 \int d\mathbf{R}\, d\mathbf{R}' \, e^{i\mathbf{k}(\mathbf{R}-\mathbf{R}')} \int dt\, e^{i\omega t} \left\langle \Psi_s^{\dagger}(\mathbf{R}, t)\Psi_s(\mathbf{R}', 0) \right\rangle, \quad (s = \pm 1) \qquad (18.21)$$

which shows clearly the importance of contributions $\mathbf{R} \neq \mathbf{R}'$ (off-diagonal long-range order) for the directional characteristic (dependence on $\mathbf{k}$). Further,

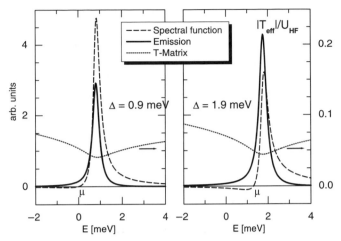

FIG. 18.5. Spectral function at $\mathbf{k} = 0$ and emission line shape of indirect excitons for two different excitation levels at $T = 10\,\text{K}$. The effective reduction of the T-matrix coupling compared to the Hartree–Fock value is shown, too (right scale).

equation (18.21) can be simply expressed via the single-exciton propagator resp. the spectral function,

$$I(\mathbf{k}, \omega) \propto i\, G_{\mathbf{k}}^{<}(\omega) \equiv n_{\text{B}}(\hbar\omega - \mu)\, A_{\mathbf{k}}(\omega)\,. \tag{18.22}$$

For understanding the spectral shape, it is important to note that the spectral function changes sign at the chemical potential μ where the Bose distribution function $n_{\text{B}}(\hbar\omega - \mu)$ has a pole, resulting in a strictly positive emission. Portions below μ are due to photon emission accompanied by XX scattering. Results of the full dynamical T-matrix calculation (at zero momentum) are shown in Fig. 18.5. Nothing special happens if the chemical potential is located within the luminescence line. Obviously, the classical argument for BEC onset which we have formulated in Section 18.2 has to be refined here: Instead of "μ is touching the energy of the lowest state", a phase transition can only happen if the chemical potential hits the quasiparticle dispersion, $\text{Re}\Sigma_0(\hbar\omega = \mu) = \mu$. The sharpening of the spectrum (Fig. 18.6a) is related to the undamping of XX scattering close to the phase transition. Technically, the imaginary part of the self-energy must be zero at the chemical potential. As shown in Fig. 18.6(b), the initial increase in linewidth due to XX scattering turns into a shrinkage, a finding which can be used to locate in a photoluminescence experiment the approach towards condensation. For the given density range (blue-shift up to $2\,\text{meV}$) and an exciton temperature of $T = 5\,\text{K}$, however, condensation is not achieved. The directional characteristics (not shown here) evolves into a sharp angular peak, too. Similar conclusions on the condensation signature in the

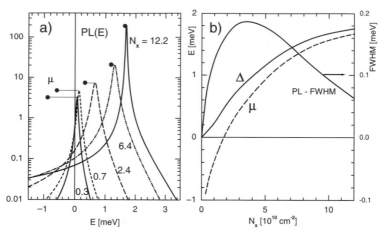

FIG. 18.6.   (a) Emission line shape of indirect excitons for different excitation
    levels (logarithmic plot). The lines are getting sharper as the chemical poten-
    tial μ (dots) approaches the emission maximum (quasiparticle position Δ).
    The temperature is held fixed at $T = 5\,\mathrm{K}$, and exciton densities N_X are given
    in units of $10^{10}\,\mathrm{cm}^{-2}$. (b) The linewidth (right scale) shows a nonmonotonic
    behavior on exciton density.

angular emission have been drawn by Littlewood and coworkers (Keeling *et
al.*, 2004).

The exciton densities given in Fig. 18.6 are calculated with the standard
expression using the momentum- and frequency-dependent spectral function
equation (18.20) and the Bose distribution function. It is worth noting that
the T-matrix comes out appreciably smaller than the bare interaction $(U^{\pm})$,
as shown by the dotted lines in Fig. 18.5. Consequently, at a given density, the
blue-shift of the emission is much less than the simple Hartree–Fock argument
would predict. Obviously, the strong dipole–dipole repulsion and its dynamical
character hinder an easy build-up of coherence. Refined calculations are needed
before reliable predictions for exciton condensation in coupled quantum wells
can be made.

References

Baumberg, J. J., Savvidis, P. G., Stevenson, R. M., Tartakovskii, A. I., Skol-
    nick, M. S., Whittaker, D. M. and Roberts, J. S. (2000). *Phys. Rev. B* **62**,
    R16247.
Ben-Tabou de-Leon, S. and Laikhtman, B. (2001). *Phys. Rev. B* **63**, 125306.
Bucher, B., Steiner, P. and Wachter, P. (1991). *Phys. Rev. Lett.* **67**, 2717.
Bulatov, A. E. and Tikhodeev, S. G. (1992). *Phys. Rev. B* **46**, 15058.
Butov, L. V., Lai, C. W., Ivanov A. L., Gossard A. C. and Chemla D. S.
    (2002). *Nature* **417**, 47.

Butov, L. V. (2003). *Solid State Commun.* **127**, 89.

Combescot, M. and Betbeder-Matibet, O. (2002). *Europhys. Lett.* **58**, 87.

Cornell, E. A. and Wiemann, C. E. (2002). *Rev. Mod. Phys.* **74**, 875.

Deng, H., Weihs, G., Snoke, D., Bloch, J. and Yamamoto, Y. (2003). *Proceedings of the National Academy of Sciences USA* **100**, 15318.

Doan, T. D., Cao, H. T., Tran, D. B. and Haug, H. (2005). *Phys. Rev.* B **72**, 085301.

Eisenstein, J. P. and MacDonald, A. H. (2004). *Nature* **432**, 691.

Fukuzawa, T., Mendez, E. E. and Hong, J. M. (1990). *Phys. Rev. Lett.* **64**, 3066.

Griffin, A. (1996). *Phys. Rev.* B **53**, 9341.

Halperin, B. I. and Rice, T. M. (1968). *Rev. Mod. Phys.* **40**, 755.

Hanamura, E. and Haug, H. (1977). *Phys. Rep.* **33**, 209.

Keldysh, L. V. (1964). *Zh. Eksp. Teor. Fiz.* **47**, 1515 [(1965). *Sov. Phys. JETP* **20**, 1018].

Keldysh, L. V. and Kopaev, Yu. V. (1964). *Fiz. Tv. Tela* **6**, 279 [*Sov. Phys. Solid State* **6**, 2219 (1965)].

Keldysh, L. V. (1968). *Proceedings of the Ninth International Conference on the Physics of Semiconductors* (ed. by S.M. Ryvkin and V. V. Shmastev), Nauka Leningrad, p. 1303.

Keldysh, L. V. and Kozlov, A. N. (1968). *Zh. Eksp. Teor. Fiz.* **54**, 978 [*Sov. Phys. JETP* **27**, 521].

Keldysh, L. V. and Silin, A. P. (1975). *Zh. Eksp. Teor. Fiz.* **69**, 1053 [*Sov. Phys. JETP* **42**, 535].

Keeling, J., Levitov, L. S. and Littlewood, P. B. (2004). *Phys. Rev. Lett.* **92**, 176402.

Ketterle, W. (2002). *Rev. Mod. Phys.* **74**, 1131.

Knox, R. S. (1963). *Theory of excitons*, Academic Press, New York.

Kuwata-Gonokami, M. (2005). *Solid State Commun.* **134**, 127.

Lin, J. L. and Wolfe, J. P. (1993). *Phys. Rev. Lett.* **71**, 1222.

Lozovik, Yu. E. and Yudson, V. I. (1976). *Zh. Eksp. Teor. Fiz.* **71**, 738 [*Sov. Phys. JETP* **44**, 389].

Mysyrowicz, A., Benson, E. and Fortin, E. (1996). *Phys. Rev. Lett.* **77**, 896.

Nagasawa, N., Doi, Y. and Ueta, M. (1975). *J. Phys. Soc. Jpn.* **38**, 593.

Okumura, S. and Ogawa, T. (2001). *Phys. Rev.* B **65**, 035105.

Schmitt, O. M., Tran Thoai, D. B., Bànyai, L., Gartner, P. and Haug, H. (2001). *Phys. Rev. Lett.* **86**, 3839.

Siarkos, A., Runge, E. and Zimmermann, R. (2000). *Phys. Rev.* B **61**, 10854
.

Snoke, D., Denev, S., Liu, Y., Pfeiffer, L. and West, K. (2002). *Nature* **418**, 754.

Snoke, D. W. (2003). *Phys. stat. sol.* (b) **238**, 389.

Snoke, D. W. (2005). Special issue on *Spontaneous Coherence in Excitonic Systems*, *Solid State Commun.* **134**, 1–164.

Tikhodeev, S. G. (1997). *Phys. Rev. Lett.* **78**, 3225.

Vörös, Z., Berman, O., Snoke, D. W., Pfeiffer, L. and West, K. (2005). *Solid State Commun.* **134**, 37.

Wachter, P. (2005). *Solid State Commun.* **134**, 73.

Zimmermann, R. (1970). *Phys. stat. sol.* (b) **39**, 95.

Zimmermann, R. (1976). *Phys. stat. sol.* (b) **76**, 191.

Zimmermann, R. (1988). *Many-particle theory of highly excited semiconductors*, Teubner-Texte zur Physik Bd. 18, BGB, Teubner-Verlag, Leipzig.

Zimmermann, R. (2005). *Solid State Commun.* **134**, 43.

Zimmermann, R. (2006). *Phys. stat. sol.* (b), **243**, 2358.

19

ACOUSTICALLY INDUCED SUPERLATTICES: FROM PHOTONS AND ELECTRONS TO EXCITONS, TO-PHONONS AND POLARITONS

A. L. Ivanov

School of Physics and Astronomy, Cardiff University, Queens Buildings,
The Parade, Cardiff CF24 3AA, United Kingdom

Abstract

We review the recently proposed concept of resonant acousto-optics, when the interaction between the light and acoustic fields is mediated by the dipole-active polarization, associated with either electronic (e.g., exciton) or lattice (e.g., TO-phonon) transitions which give rise to a solitary absorption line in the transparency band of the host medium. For exciton polaritons, the resonant acousto-optic response is mediated and strongly enhanced by the exciton states resonant with the acoustic and optic fields in the intraband and interband transitions, respectively. In this case, in sharp contrast with conventional acousto-optics, the resonantly enhanced Bragg spectra reveal the multiple orders of diffracted light, i.e., a Brillouin band structure of parametrically driven polaritons can be visualized. In particular, for GaAs-based microcavities driven by a surface acoustic wave of $\nu_{\rm saw} \sim 1\,{\rm GHz}$ and $\langle I_{\rm saw} \rangle \sim 100\,{\rm W/cm^2}$, the main acoustically-induced bandgap in the polariton spectrum can be as large as $\Delta_{\rm ac}^{\rm MC} \sim 1\,{\rm meV}$. The resonant acousto-optic effect can also be effectively realized for TO-phonon polaritons. In this case the acoustically changed restrahlen band and the acoustically induced spectral gaps in the polariton spectrum can be used for frequency resolved detection of THz radiation.

19.1 Introduction

The seminal paper by L. Brillouin (1922) gave rise to acousto-optics, nowadays a well-established field of physics (Born and Wolf, 1970; Wilson and Hawkes, 1983; Korpel, 1997). A textbook definition of the acousto-optic effect is "the change in the refractive index of a medium caused by the mechanical strains accompanying the passage of an acoustic wave through the medium" (Wilson and Hawkes, 1983). The acousto-optic effect can be visualized as optical diffraction of the incident light field by the acoustically induced grating of the wavelength period $\lambda_{\rm ac}$. The acoustically induced diffraction grating is due to the photoelastic effect, which gives rise to a small, nonresonant change of the dielectric constant, $|\delta\varepsilon_{\rm b}| \ll \varepsilon_{\rm b}$. Here $\delta\varepsilon_{\rm b}$ is given by $\delta\varepsilon_{\rm b} = 4\pi\chi_{\gamma-{\rm ac}}^{(2)}I_{\rm ac}^{1/2}$, where $I_{\rm ac}$ is the acoustic intensity and $\chi^{(2)}_{\gamma-\text{ac}}$ is a second-order nonresonant acousto-optical susceptibility. The nonresonant acousto-optic nonlinearities are small, so that usually $\delta\varepsilon_{\text{b}} \sim 10^{-4}$–$10^{-3}$ (for GaAs, e.g., $\delta\varepsilon_{\text{b}} \simeq 1.6 \times 10^{-3}$ for $I_{\text{ac}} = 100\,\text{W/cm}^2$). In conventional acousto-optics, the strong coupling limit of photon–phonon nonresonant interaction refers to nonperturbative solutions of the Maxwell wave equation for the dielectric constant ε harmonically modulated by an acoustic wave $\{\mathbf{k}, \Omega^{\text{ac}}_{\mathbf{k}}\}$ (Berry, 1966; Korpel, 1997):

$$\left[\varepsilon(\mathbf{r}, t)\frac{1}{c^2}\frac{\partial^2}{\partial t^2} - \Delta\right]\mathbf{E}(\mathbf{r}, t) = 0\,, \tag{19.1}$$

where $\varepsilon = \varepsilon(\mathbf{r}, t) = \varepsilon_{\text{b}} + \delta\varepsilon_{\text{b}}\cos(\Omega^{\text{ac}}_{\mathbf{k}}t - \mathbf{kr})$ and $\mathbf{E}$ is the electric component of the light field. In 1933, Léon Brillouin described for the first time nonperturbative solutions of equation (19.1) (Brillouin, 1933). The eigenstates of this equation are given in terms of the Mathieu functions and yield an infinite sequence of the alternating allowed and forbidden energy bands. The energy structure is akin to the electron energy bands, which appear in the presence of a periodic (atomic) potential. However, a particular feature of the harmonic potential is that the forbidden energy bands, i.e., the acoustically induced energy gaps $\Delta^{(n)}_{\text{ac}}$ ($n = 1, 2, \ldots$) in the photon spectrum, extremely effectively close up with increasing energy band number n: $\Delta^{(n)}_{\text{ac}} \propto (\delta\varepsilon_{\text{b}})^n$. This is an inherent property of the Mathieu wavefunctions (Blanch, 1965) which can also be realized in terms of the n-phonon-assisted transitions. As a result, the conventional acousto-optics, both fundamental and applied, deals with the one-phonon transition only, i.e., with the main acoustically induced gap $n = 1$ (Born and Wolf, 1970; Chu and Tamir, 1976; Chu and Kong, 1980; Korpel, 1997).

In 1962, Leonid Keldysh proposed to change the spectrum of free electrons in a bulk crystal by applying an ultrasonic wave (Keldysh, 1962). In this case the corresponding Schrödinger equation is approximated by

$$i\hbar\frac{\partial}{\partial t}\psi = -\frac{\hbar^2}{2m_{\text{e}}}\Delta\psi + \left[V_0\cos(\Omega^{\text{ac}}_{\mathbf{k}}t - \mathbf{kr})\right]\psi\,. \tag{19.2}$$

Here, ψ is the electron envelope wavefunction and m_{e} is the electron effective mass. The potential V_0 is given by $V_0 = M^{\text{PE(DP)}}_{\text{e−ac}}\sqrt{N^{\text{ph}}_0}$, where $M^{\text{PE(DP)}}_{\text{e−ac}}$ is the matrix element of piezoelectric (deformation potential) electron – TA-phonon (LA-phonon) interaction and the concentration $N^{\text{ph}}_0 \propto I_{\text{ac}}$ of coherent phonons, associated with the acoustic pump wave, is determined by

$$N^{\text{ph}}_0 = \frac{I_{\text{ac}}}{\hbar v_{\text{s}}\Omega^{\text{ac}}_{\mathbf{k}}}\,. \tag{19.3}$$

Here, v_{s} is the sound velocity. Similarly to equation (19.1), which deals with acoustically induced scattering of the light field, the eigen-wavefunctions of

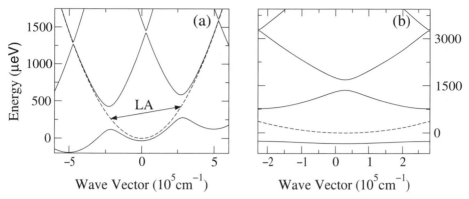

Fig. 19.1. The quasi-energy spectrum (solid lines) of conduction-band electrons in undoped bulk GaAs parametrically driven by a longitudinal acoustic wave, $\nu_{ac} = \Omega_{\mathbf{k}}^{ac}/(2\pi) = 38.2\,\text{GHz}$ and $k_{ac} = 5 \times 10^5\,\text{cm}^{-1}$ $(h\nu_{ac} = \hbar\Omega_{\mathbf{k}}^{ac} = 158\,\mu eV)$. The arrow indicates the electron states resonantly coupled by bulk LA phonons (one-phonon transition). The dashed lines show the initial parabolic dispersion of the conduction band, $E_c = (\hbar^2 p^2)/(2m_e)$. The quasi-energy spectrum for (a) the extended Brillouin zone, $I_{ac} = 42.2\,\text{W}/\text{cm}^2$ and (b) the reduced Brillouin zone, $I_{ac} = 500\,\text{W}/\text{cm}^2$.

equation (19.2), which describes diffraction of electrons by the ultrasonic wave, are also given by the Mathieu functions. However, in the latter case the efficiency of the diffraction process is much higher than in the former one. While, to the best knowledge of the author, the acoustically induced quasi-energy spectrum of free electrons has not yet been observed, the very concept of acoustically induced superlattices preceded the MBE-grown nanostructures. The acoustically induced spectrum, calculated with equation (19.2) for conduction-band electrons in bulk GaAs driven by a longitudinal acoustic wave, is plotted in Fig. 19.1.

The work by L. V. Keldysh (1962) has also initiated research on acoustically induced transport of electrons in solids. In particular, a single-electron acousto-electric current induced by a surface acoustic wave (SAW) of $\nu_{saw} = 1\text{–}3\,\text{GHz}$ in a quasi-one-dimensional GaAs channel has recently been observed (Shilton *et al.*, 1996; Ebbecke *et al.*, 2000; Cunningham *et al.*, 2005). The surface acoustic waves are equally relevant to the effective realization of the acousto-optic effect. Nowadays, the SAW technique, based on interdigital transducer (IDT) lithography, routinely operates with the acoustic frequency $\nu_{saw} \leq 1\,\text{GHz}$ (Campbell, 1989). The IDT-induced surface acoustic wave is a Rayleigh wave, which is evanescent in the z-direction (direction normal to the surface of a bulk semiconductor or semiconductor nanostructure), with a characteristic decay length ℓ_{saw}^z of about the SAW wavelength, $\lambda_{ac} = \lambda_{saw} = v_{saw}/\nu_{saw}$ (Farnell, 1978; Hess, 2002). The SAW velocity v_{saw} is less than the bulk v_s by about

30 %. The IDT generation of ultrahigh-frequency surface acoustic waves in GaAs-based ($\nu_{\text{saw}} \simeq 24\,\text{GHz}$) and AlN-based ($\nu_{\text{saw}} \simeq 32\,\text{GHz}$) structures has been reported (Kukushkin *et al.*, 2004; Takagaki *et al.*, 2004).

In this paper we review *resonant acousto-optics*, a concept which was recently proposed and developed (Ivanov and Littlewood, 2001; Ivanov and Littlewood, 2003; Cho *et al.*, 2005; Ivanov, 2005). The resonant acousto-optic effect deals with quantum diffraction of optically-active excitons (or TO-phonons, see the last section) by an ultrasonic coherent wave. In this case the interaction of two "matter" waves, the excitonic polarization and the acoustic field, is much more effective than the nonresonant photoelastic coupling of photons with acoustic phonons. At the same time the photon component of optically-dressed excitons, i.e., of polaritons, can be large enough to ensure an efficient resonant conversion "photon $\leftrightarrow$ exciton". This qualitatively explains the origin of the giant acousto-optic nonlinearity, $|\delta\varepsilon_{\text{x-ac}}| \gg |\delta\varepsilon_{\text{b}}|$, which is mediated and strongly enhanced by the relevant exciton states resonant with the acoustic and optic fields in the intraband and interband transitions, respectively. For GaAs-based structures at low temperatures, the resonant acousto-optic effect gives rise to $\delta\varepsilon_{\text{x-ac}} \sim 0.1$ for $I_{\text{ac}} = 100\,\text{W/cm}^2$. For the resonant acousto-optics of polaritons, both resonant interactions, the exciton–photon coupling and the interaction of excitons with the acoustic pump wave, should be treated nonperturbatively (strong coupling regime) and on an equal basis. In this case, the resonant acousto-optical response of bulk and microcavity polaritons cannot generally be given in terms of the Mathieu functions.

19.2   The acoustically induced quasi-energy spectrum of bulk polaritons

The Hamiltonian of an exciton–photon system driven by an intense coherent acoustic wave is approximated by

$$
H = \sum_{\mathbf{p}} \left[\hbar\omega_{\mathbf{p}}^{\text{x}} B_{\mathbf{p}}^{\dagger} B_{\mathbf{p}} \; + \; \hbar\omega_{\mathbf{p}}^{\gamma} \alpha_{\mathbf{p}}^{\dagger} \alpha_{\mathbf{p}} \; + \; i\hbar\frac{\Omega_{\text{c}}}{2}\left(\alpha_{\mathbf{p}}^{\dagger} B_{\mathbf{p}} - B_{\mathbf{p}}^{\dagger} \alpha_{\mathbf{p}}\right) \right.
$$
$$
\left. + im_{\mathbf{k}}^{\text{x}}\left(B_{\mathbf{p}}^{\dagger} B_{\mathbf{p}-\mathbf{k}} e^{-i\Omega_{\mathbf{k}}^{\text{ac}} t} - B_{\mathbf{p}-\mathbf{k}}^{\dagger} B_{\mathbf{p}} e^{i\Omega_{\mathbf{k}}^{\text{ac}} t}\right) \right], \tag{19.4}
$$

where $B_{\mathbf{p}}$ are $\alpha_{\mathbf{p}}$ are the exciton and photon operators, respectively, $\hbar\omega_{\mathbf{p}}^{\text{x}} = \hbar\omega_{\text{T}} + \hbar^2 p^2/2M_{\text{x}}$ and $\hbar\omega_{\mathbf{p}}^{\gamma} = \hbar c p/\sqrt{\varepsilon_{\text{b}}}$ are the corresponding energies, M_{x} is the exciton translational mass, and Ω_{c} is the polariton Rabi frequency. The acoustically induced coupling of the polariton states occurs via their exciton component, so that in equation (19.4) the matrix element $m_{\mathbf{k}}^{\text{x}}$ is given by $m_{\mathbf{k}}^{\text{x}} = M_{\text{x-ph}}^{\text{DP(PE)}}(\mathbf{k})\sqrt{N_0^{\text{ph}}}$. For a purely longitudinal acoustic wave, the deformation potential interaction with $M_{\text{x-ph}}^{\text{DP}} = D_{\text{x}}[(\hbar k)/(2\rho v_{\text{s}})]^{1/2}$ (D_{x} is the deformation potential of exciton – LA-phonon interaction and ρ is the mass density) is dominant, while for a purely transverse acoustic pump wave the piezoelectric matrix element $M^{\text{PE}}_{\text{x-ac}}$ is $\propto k^{3/2}\left[(m_{\text{e}} - m_{\text{h}})/M_{\text{x}}\right]e^{\text{em}}$ (e^{em} is the relevant component of an electromechanical tensor of the electron – TA-phonon piezoelectric interaction).

Because $\hbar\omega_{\text{T}} \gg \hbar\Omega^{\text{ac}}_{\mathbf{k}}$, an acoustically driven undoped semiconductor remains in its ground electronic state. As a result, no many-body effects due to electron (hole) – electron (hole) Coulomb scattering, screen the resonant acousto-optic effect and weaken the exciton–photon interaction, i.e., the polariton effect.

In order to solve Hamiltonian (19.4), at first the explicit time dependence of H is removed by the canonical transformation:

$$S = \exp\left[it\sum_{\mathbf{p}}(\mathbf{v}_{\text{s}} \cdot \mathbf{p})\left(B^{\dagger}_{\mathbf{p}}B_{\mathbf{p}} + \alpha^{\dagger}_{\mathbf{p}}\alpha_{\mathbf{p}}\right)\right],$$

$$B_{\mathbf{p}} \rightarrow SB_{\mathbf{p}}S^{\dagger} = B_{\mathbf{p}}e^{-i(\mathbf{v}_{\text{s}}\cdot\mathbf{p})t},$$

$$\alpha_{\mathbf{p}} \rightarrow S\alpha_{\mathbf{p}}S^{\dagger} = \alpha_{\mathbf{p}}e^{-i(\mathbf{v}_{\text{s}}\cdot\mathbf{p})t}, \tag{19.5}$$

where $\mathbf{v}_{\text{s}} = v_{\text{s}}(\mathbf{k}/k)$ and $(\mathbf{v}_{\text{s}} \cdot \mathbf{p}) = \Omega^{\text{ac}}_{\mathbf{k}}[(\mathbf{k} \cdot \mathbf{p})/k^2]$. The transformation (19.5) means the use of the coordinate system which moves with the acoustic wave. In this case the quadratic Hamiltonian $\tilde{H} = SHS^{\dagger} - iS(\partial S^{\dagger}/\partial t)$ is time-independent. The diagonalization of $\tilde{H}$ yields the dispersion equation for polaritons driven by the coherent acoustic wave (Ivanov and Littlewood, 2001):

$$\omega^2 - (\tilde{\omega}^{\text{x}}_{\mathbf{p}})^2 - \frac{(\omega\Omega_{\text{c}})^2}{\omega^2 - (\tilde{\omega}^{\gamma}_{\mathbf{p}})^2}$$

$$- \frac{4\omega^2_{\text{T}}|m^{\text{x}}_{\mathbf{k}}|^2}{\omega^2 - (\tilde{\omega}^{\text{x}}_{\mathbf{p+k}})^2 - \frac{(\omega\Omega_{\text{c}})^2}{\omega^2 - (\tilde{\omega}^{\gamma}_{\mathbf{p+k}})^2} - M_{\mathbf{p+2k}}}$$

$$- \frac{4\omega^2_{\text{T}}|m^{\text{x}}_{\mathbf{k}}|^2}{\omega^2 - (\tilde{\omega}^{\text{x}}_{\mathbf{p-k}})^2 - \frac{(\omega\Omega_{\text{c}})^2}{\omega^2 - (\tilde{\omega}^{\gamma}_{\mathbf{p-k}})^2} - M_{\mathbf{p-2k}}} = 0, \tag{19.6}$$

where $M_{\mathbf{p}\pm 2\mathbf{k}}$ is given through $M_{\mathbf{p}\pm 3\mathbf{k}}$, $M_{\mathbf{p}\pm 3\mathbf{k}}$ through $M_{\mathbf{p}\pm 4\mathbf{k}}$, etc. by the recurrent formula valid for a positive integer $n \geq 1$:

$$M_{\mathbf{p}_{\parallel}\pm n\mathbf{k}} = \frac{4\omega^2_{\text{T}}|m^{\text{x}}_{\mathbf{k}}|^2}{\omega^2 - (\tilde{\omega}^{\text{x}}_{\mathbf{p}\pm n\mathbf{k}})^2 - \frac{(\omega\Omega_{\text{c}})^2}{\omega^2 - (\tilde{\omega}^{\gamma}_{\mathbf{p}\pm n\mathbf{k}})^2} - M_{\mathbf{p}\pm(n+1)\mathbf{k}}}, \tag{19.7}$$

and $\hbar\tilde{\omega}^{\text{x},\gamma}_{\mathbf{p}\pm n\mathbf{k}} = \hbar\omega^{\text{x},\gamma}_{\mathbf{p}\pm n\mathbf{k}} \mp n\hbar\Omega^{\text{ac}}_{\mathbf{k}} = \hbar\omega^{\text{x},\gamma}_{\mathbf{p}\pm n\mathbf{k}} \mp n\hbar(\mathbf{v}_{\text{s}} \cdot \mathbf{k})$ are the exciton and photon quasi-energies, respectively.

The quasi-energy spectrum (19.6) can be interpreted in terms of the Brillouin energy bands: the initial polariton dispersion is modulated by the periodic potential associated with the coherent acoustic wave. The latter potential is periodic not only in space, with "lattice" constant λ_{ac}, but also in time, with period $1/\nu_{\text{ac}}$. This gives rise to the quasi-energies $\hbar\tilde{\omega}^{\text{x},\gamma}_{\mathbf{p}_{\parallel}\pm n\mathbf{k}}$ in equations (19.6)–(19.7).

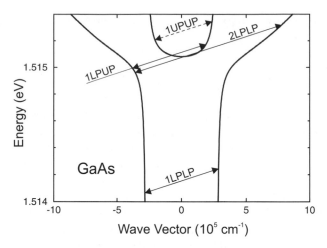

FIG. 19.2. The energy diagram of the polariton states in acoustically-pumped GaAs. The solid arrows 1LPLP, 1LPUP, and 2LPLP show the polariton states resonantly coupled by the bulk LA wave of frequency $\nu_{ac} = 43.4$ GHz (1LPLP and 2LPLP refer to the one- and two-phonon resonant transitions between the states of the lower polariton dispersion branch, respectively, 1LPUP indicates the one-phonon transition between the lower and upper polariton branches). The dashed arrow 1UPUP shows the one-phonon transition within the upper polariton branch, $\nu_{ac} = 33.6$ GHz. From (Ivanov and Littlewood, 2001). Reprinted figure with permission from [A. L. Ivanov and P. B. Littlewood, *Phys. Rev. Lett.* **87**, 136403 (2001)]. Copyright (2001) by the American Physical Society.

The concept of quasi-energies was introduced by Ya. B. Zeldovich (1966). Thus not only the (quasi-) momenta $\hbar(\mathbf{p} \pm n\mathbf{k})$ are equivalent for the acoustically driven polariton states, but, in a similar way, the energies $\hbar(\omega_{\mathbf{p}_\parallel \pm n\mathbf{k}}^{x,\gamma} \mp n\Omega_{\mathbf{k}}^{ac})$ are also indistinguishable for $n = 1, 2, \ldots$. This is valid only in the absence of incoherent scattering of excitons, when $\gamma_x = 0$ (see the next section).

In the dispersion equation (19.6), the interaction between excitons and photons (polariton effect) is treated in the nonresonant approximation. With decreasing acoustic intensity, $I_{ac} \to 0$, the acoustically induced coupling disappears, so that $|m_{\mathbf{k}}^x|^2 \propto I_{ac} \to 0$ and equation (19.6) reduces to the polariton dispersion relationship:

$$\frac{c^2 p^2}{\varepsilon_b} = \omega^2 + \frac{\omega^2 \Omega_c^2}{\omega_T^2 + \hbar\omega_T(p^2/M_x) - \omega^2} \ . \tag{19.8}$$

Equations (19.6)–(19.7) are akin to the famous Raman–Nath relationship (Korpel, 1997), which is frequently used in conventional acousto-optics.

For polaritons parametrically driven by a bulk LA wave, the acoustically induced coupling matrix element $m_{\mathbf{k}}^x$ is due to the exciton – LA-phonon deformation potential interaction, so that in equations (19.4)–(19.7) one has

$M_{\text{x-ph}} \equiv M^{\text{DP}}_{\text{x-ph}}$. For SAW-driven polaritons ($\mathbf{k} = \mathbf{k}_{\text{saw}}$, $\Omega^{\text{ac}}_{\mathbf{k}} = \Omega_{\text{saw}}$ and $v_s = v_{\text{saw}}$), $m^{\text{x}}_{\mathbf{k}}$ is given by

$$m^{\text{x}}_{\mathbf{k}} = I^{1/2}_{\text{ac}} \frac{|u_x|}{(|u_x|^2 + |u_z|^2)^{1/2}} \, |M^{\text{DP}}_{\text{x-ac}} + i\gamma_{\text{saw}} M^{\text{PE}}_{\text{x-ac}}| \, , \qquad (19.9)$$

where $\gamma_{\text{saw}} = |u_z|/|u_x| \exp(i\delta_{\text{saw}})$. Equation (19.9) refers to the one-dimensional scattering geometry with SAW wavevector $\mathbf{k}\|\mathbf{p}\|x$-axis and to the exciton and light fields linearly polarized along the y-axis (the x-y plane guides the SAW). In this case the surface acoustic wave, which has both transverse and longitudinal displacement components, u_x and u_z, is elliptically-polarized in the x-z plane (the SAW sagittal plane) (Campbell, 1989). The phase shift δ_{saw} between u_x and u_z is nearly $\pi/2$, so that both interaction channels interfere constructively. For the Rayleigh surface acoustic wave propagating along a (100) surface in the [011] direction of bulk GaAs one has $|u_z|/|u_x| \simeq 1.4$ (Simon, 1996). In this case, because $M^{\text{DP}}_{\text{x-ac}} \gg M^{\text{PE}}_{\text{x-ac}}$, the deformation potential mechanism is absolutely dominant over the piezoelectric one, as follows from equation (19.9) (Ivanov, 2005).

According to equation (19.6), with increasing I_{ac} the optical bandgaps, $\Delta^{(n)}_{\text{ac}}$, open up and develop in the polariton spectrum. The gaps $\Delta^{(n)}_{\text{ac}} \propto I^{n/2}_{\text{ac}}$ arise for the polariton states resonantly coupled through $n = 1, 2, \ldots$ phonon-assisted transitions induced by the acoustic pump wave (see Fig. 19.2 for bulk polaritons and Fig. 19.5 for microcavity polaritons). A typical acoustically induced quasi-energy spectrum $\omega = \omega_{\mathbf{p}}(I_{\text{ac}})$, calculated with equations (19.6)–(19.7) as the quasiparticle solution ($\mathbf{p}$ is real), is plotted in Figs. 19.3(a) and 19.3(b) for the upper and lower polariton dispersion branches in bulk GaAs, respectively. For lower-branch bulk polaritons, the one-dimensional interaction geometry can only be realized for high-frequencies of the acoustic pump wave, $\nu_{\text{ac}} = \nu_{\text{saw}} > 20\,\text{GHz}$. However, for the upper dispersion branch of bulk polaritons, a low-frequency acoustic pump wave can be used to resonantly couple two polariton states, $\mathbf{p}$ and $\mathbf{p} - \mathbf{k}$, even if $\mathbf{p}\|\mathbf{k}$. Furthermore, because for small wavevectors $\mathbf{p}$ the upper-branch polariton dispersion is quadratic (see Fig. 19.3a), in this case the quasi-energy spectrum can also be approximately described in terms of the Mathieu equation.

The first (main) bandgap, $\Delta_{\text{ac}} = \Delta^{(n=1)}_{\text{ac}}$, arises from the polariton states $\{\mathbf{p}, \omega_{\text{pol}}(\mathbf{p})\}$ and $\{\mathbf{p} - \mathbf{k}, \omega_{\text{pol}}(\mathbf{p} - \mathbf{k})\}$ resonantly coupled by the one-phonon transition induced by the acoustic pump wave $\{\mathbf{k}, \Omega^{\text{ac}}_{\mathbf{k}}\}$. The main bandgap $\Delta_{\text{ac}} \propto \sqrt{I_{\text{ac}}}$ is approximated by

$$\Delta_{\text{ac}} = 2|m^{\text{x}}_{\mathbf{k}}| \, [\varphi^{\text{x}}(\mathbf{p})\varphi^{\text{x}}(\mathbf{p} - \mathbf{k})]^{1/2} = \Delta^{\text{x}}_{\text{ac}} [\varphi^{\text{x}}(\mathbf{p})\varphi^{\text{x}}(\mathbf{p} - \mathbf{k})]^{1/2} \, , \qquad (19.10)$$

where $\varphi^{\text{x}} = \varphi^{\text{x}}(\mathbf{p})$ is the exciton component of the polariton state $\mathbf{p}$, and $\Delta^{\text{x}}_{\text{ac}} = 2|m^{\text{x}}_{\mathbf{k}}|$ is the main bandgap relevant to optically-undressed excitons. For the exciton–phonon deformation potential interaction one estimates:

$$\hbar\Delta_{\mathrm{ac}}^{\mathrm{x}} = D_{\mathrm{x}}\left(\frac{2I_{\mathrm{ac}}}{\rho v_{\mathrm{s}}^3}\right)^{1/2} = 2\sqrt{2}\pi D_{\mathrm{x}}\frac{\delta a_{\mathbf{k}}}{\lambda_{\mathrm{ac}}} , \qquad (19.11)$$

where $\delta a_{\mathbf{k}}$ and λ_{ac} are the amplitude and wavelength of the acoustic pump wave, respectively. According to equation (19.11), $\Delta_{\mathrm{ac}}^{\mathrm{x}}$ is independent of the LA-phonon wavevector k. The acoustically induced gap $\Delta_{\mathrm{ac}}^{\mathrm{x}}$ in the polariton spectrum is due to coherent phonon-mediated coupling of virtual, optically-dressed excitons $\mathbf{p}$ and $\mathbf{p} - \mathbf{k}$ via one-phonon transition.

19.3   Macroscopic equations of resonant acousto-optics

Optics of the acoustically driven bulk polaritons is given by the macroscopic equations for the electric field $\mathbf{E}(\mathbf{r}, t)$ and excitonic polarization $\mathbf{P}(\mathbf{r}, t)$ (Ivanov and Littlewood, 2001):

$$\left[\frac{\varepsilon_{\mathrm{b}}}{c^2}\frac{\partial^2}{\partial t^2} - \Delta\right]\mathbf{E}(\mathbf{r}, t) = -\frac{4\pi}{c^2}\frac{\partial^2}{\partial t^2}\mathbf{P}(\mathbf{r}, t) ,$$

$$\left[\frac{\partial^2}{\partial t^2} + 2\gamma_{\mathrm{x}}\frac{\partial}{\partial t} + \omega_{\mathrm{T}}^2 - \frac{\hbar\omega_{\mathrm{t}}}{M_{\mathrm{x}}}\Delta - 4\frac{m_{\mathbf{k}}^{\mathrm{x}}}{\hbar}\omega_{\mathrm{T}}\cos(\Omega_{\mathbf{k}}^{\mathrm{ac}}t - \mathbf{k}\mathbf{r})\right]\mathbf{P}(\mathbf{r}, t)$$

$$= \frac{\varepsilon_{\mathrm{b}}\Omega_{\mathrm{c}}^2}{4\pi}\mathbf{E}(\mathbf{r}, t) , \qquad (19.12)$$

where $2\gamma_{\mathrm{x}} = 1/T_1 = 2/T_2$ is the rate of incoherent scattering of excitons. The macroscopic equations (19.12) refer to an optically isotropic bulk crystal, and $m_{\mathbf{k}}^{\mathrm{x}} \propto I_{\mathrm{ac}}^{1/2}$ can be time-dependent through $I_{\mathrm{ac}}(t)$. If $I_{\mathrm{ac}} \to 0$ and, therefore, $m_{\mathbf{k}}^{\mathrm{x}} \to 0$, equations (19.12) reduce to the polariton macroscopic equations (Agranovich and Ginzburg, 1984). For the plane CW acoustic pump wave, equations (19.12) yield the same quasi-energy spectrum that is given by equations (19.6)–(19.7) where $(\tilde{\omega}_{\mathbf{p}-n\mathbf{k}}^{\mathrm{x}})^2$ is now replaced by $(\tilde{\omega}_{\mathbf{p}-n\mathbf{k}}^{\mathrm{x}})^2 - 2i\gamma_{\mathrm{x}}\omega$.

The exciton-mediated, giant second-order acousto-optical susceptibility $\chi_{\gamma-\mathrm{ac}}^{(2)}$ $= \chi_{\gamma-\mathrm{x-ac}}^{(2)}$ of polaritons parametrically driven by a surface acoustic wave or bulk LA-wave gives rise to a large modulation of ε_{b} in a frequency band of the exciton resonance (Ivanov, 2005):

$$\delta\varepsilon_{\mathrm{x-ac}} = 4\pi\chi_{\gamma-\mathrm{x-ac}}^{(2)}(\omega)I_{\mathrm{ac}}^{1/2} = \varepsilon_{\mathrm{b}}\left(\frac{\Omega_{\mathrm{c}}}{\omega_{\mathrm{T}} - \omega}\right)^2\frac{M_{\mathrm{x-ac}}^{\mathrm{DP}}I_{\mathrm{ac}}^{1/2}}{\omega_{\mathrm{T}}} , \qquad (19.13)$$

where $|\omega - \omega_{\mathrm{T}}| \gg \gamma_{\mathrm{x}}$ and $|\omega - \omega_{\mathrm{T}}| \gg (\hbar^2p^2)/(2M_{\mathrm{x}})$ are assumed. The large values of $\delta\varepsilon_{\mathrm{x-ac}}$ stem from the resonant denominator on the right-hand side of equation (19.13) and from the large deformation potential interaction in GaAs. For bulk GaAs or GaAs-based nanostructures driven by an acoustic pump wave of $\nu_{\mathrm{ac}} \sim 1\,\mathrm{GHz}$ and $I_{\mathrm{ac}} \sim 100\,\mathrm{W/cm^2}$, one estimates $\delta\varepsilon_{\mathrm{x-ac}} \sim 0.1 - 1$. Note that while equation (19.13) refers to $\delta\varepsilon_{\mathrm{x-ac}}$ associated with the second-order resonant acousto-optic susceptibility $\chi_{\gamma-\mathrm{x-ac}}^{(2)}$, equations (19.6)–(19.7) and equations (19.12) are valid beyond low-order perturbation theory.

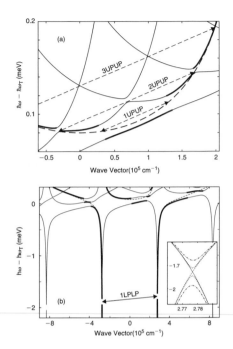

FIG. 19.3. The acoustically induced quasi-energy spectrum (solid lines) of bulk polaritons in GaAs. The initial polariton dispersion is shown by the dashed lines. The thick solid lines visualize the acoustically induced gaps in the polariton spectrum. (a) The quasi-energy spectrum of the upper polariton branch induced by a bulk LA wave of ν_{ac}=7.6 GHz and I_{ac}=0.1 W/cm^2. The dashed arrows 1UPUP, 2UPUP, and 3UPUP indicate one-, two-, and three-LA-phonon-assisted resonant transitions. (b) The quasi-energy spectrum of bulk GaAs driven by a SAW of $\nu_{saw} = 25.3$ GHz and $\langle I_{saw} \rangle = 2.14$ W/cm^2. Inset: The main optical bandgap (1LPLP transition) $\Delta_{ac}^{(n=1)} = \Delta_{ac}(p)$ for $\langle I_{saw} \rangle = 2.14$ W/cm^2 (dotted line) and 21.4 W/cm^2 (dot-dashed line), respectively. From (Ivanov and Littlewood, 2001). Reprinted figure with permission from [A. L. Ivanov and P. B. Littlewood, *Phys. Rev. Lett.* **87**, 136403 (2001)]. Copyright (2001) by the American Physical Society.

The acoustically induced gaps $\Delta_{ac}^{(n)} \propto I_{ac}^{n/2}$ in the polariton spectrum drastically change the spectral shape of the exciton line, in particular, the excitonic reflection coefficient $R = |r(\omega, I_{ac})|^2$. Because these spectral changes dynamically follow the acoustic intensity I_{ac}, the quasi-energy spectrum can also be interpreted as an acoustically induced Stark effect for optically-active excitons. In order to model numerically the influence of the acoustic Stark effect on the reflectivity, the exciton damping rate $\hbar\gamma_x \simeq 10 - 30\,\mu$eV is used in evaluations. These values refer to bulk GaAs at helium temperatures,

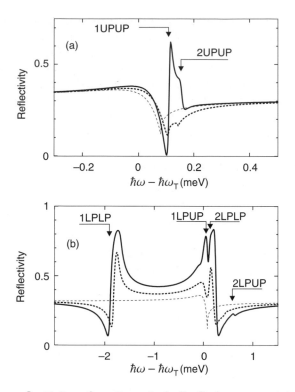

FIG. 19.4. The reflectivity of excitons in bulk GaAs parametrically driven by a coherent acoustic pump wave, i.e., the acoustically induced Stark effect in the excitonic reflectivity. The initial excitonic reflectivity, $R = |r(\omega)|^2$, is shown by the thin dashed lines. (a) $I_{ac} = 0.1\,\text{W/cm}^2$ (dashed line) and $0.5\,\text{W/cm}^2$ (solid line), and $\hbar\gamma_x = 10\,\mu\text{eV}$. The other parameters are the same as for Fig. 19.3(a). (b) $\langle I_{saw} \rangle = 2.14\,\text{W/cm}^2$ (dashed line) and $10.7\,\text{W/cm}^2$ (solid line), and $\hbar\gamma_x = 30\,\mu\text{eV}$. The other parameters are the same as for Fig. 19.3(b). From (Ivanov and Littlewood, 2001). Reprinted figure with permission from [A. L. Ivanov and P. B. Littlewood, *Phys. Rev. Lett.* **87**, 136403 (2001)]. Copyright (2001) by the American Physical Society.

$T \simeq 2 - 4\,\text{K}$ (Nägel *et al.*, 2001). Furthermore, a finite value of γ_x allows us to select from the complicated quasi-energy spectrum two harmonic-forced quasi-eigenwaves $\mathbf{p} = \mathbf{p}_{i=1,2}(\omega, I_{ac})$ which are excited by the probe light of frequency ω: the optically induced waves should attenuate in the propagation direction ($\text{Im}\{\mathbf{p}_i\} > 0$).

The reflection coefficient $R = |r(\omega, I_{ac})|^2$, changed by the acoustic wave, is plotted in Fig. 19.4 for normal incidence of the probe light (the one-dimensional geometry with $\mathbf{p} \| \mathbf{k}$) and calculated by using the Pekar's additional boundary condition (Agranovich and Ginzburg, 1984). The sharp spikes and dips, which are associated with the optical bandgaps, arise and develop with increasing I_{ac} in the reflection spectrum of acoustically-driven excitons. In particular, for $I_{ac} \sim 1\,\mathrm{W/cm^2}$ the spectral width of the main S-shaped spectral feature (coupled dip and spike in the R-profile), which corresponds to the one-phonon resonant coupling between two polariton states, $\mathbf{p}$ and $\mathbf{p} - \mathbf{k}$, is already comparable with that of the acoustically unperturbed exciton line (see Fig. 19.4b). The spectral position of the acoustically induced bandgaps can easily be tuned by sub-GHz changes in frequency $\Omega_{\mathbf{k}}^{ac}$ of the acoustic pump wave.

19.4 Microcavity polaritons parametrically driven by a SAW

The resonant acousto-optic effect can effectively be realized for microcavity (MC) polaritons in quantum well (QW) λ-cavities: semiconductor microcavities are compatible with SAW technique, and the one-dimensional geometry for resonant, phonon-mediated interaction of two counterpropagating MC polaritons can be used by applying a surface acoustic wave of relatively low frequency, $0.1\,\mathrm{GHz} \leq \nu_{saw} \leq 3\,\mathrm{GHz}$ (Ivanov and Littlewood, 2003; Cho *et al.*, 2005).

The quasi-energy spectrum of SAW-driven MC polaritons can be calculated by using equations (19.6)–(19.7) with the following substitutions: $\mathbf{p} \to \mathbf{p}_{\parallel}$, $\Omega_c \to \Omega_x^{MC}$, $(\tilde{\omega}_{\mathbf{p}-n\mathbf{k}}^x)^2 \to (\tilde{\omega}_{\mathbf{p}_{\parallel}-n\mathbf{k}}^x)^2 - 2i\gamma_x\omega$, and $(\tilde{\omega}_{\mathbf{p}-n\mathbf{k}}^\gamma)^2 \to (\tilde{\omega}_{\mathbf{p}_{\parallel}-n\mathbf{k}}^{\gamma,MC})^2 - 2i\gamma_R\omega$. Here, $\mathbf{p}_{\parallel}$ is the in-plane wavevector of MC polaritons, $\omega_{\mathbf{p}_{\parallel}}^{\gamma,MC} = (c^2 p_{\parallel}^2/\varepsilon_b + \omega_0^2)^{1/2}$ with $\omega_0 = (2\pi c)/(L_z\sqrt{\varepsilon_b})$ (L_z is the microcavity thickness) is the frequency of the 1λ MC photon eigenmode, Ω_x^{MC} is the microcavity Rabi frequency, and $2\gamma_R$ is the inverse radiative lifetime of microcavity photons, due to their leakage into external bulk photon modes.

A typical SAW-induced quasi-energy spectrum, which is calculated for a zero-detuning λ-microcavity ($\omega_0 = \omega_T$) with 100 % optical confinement ($\gamma_R = 0$) and no damping of the exciton states ($\gamma_x = 0$), i.e., for completely coherent interaction of QW excitons with the light and SAW fields, is plotted in Fig. 19.5. In these calculations we assume the one-dimensional scattering geometry with SAW wavevector $\mathbf{k}\|\mathbf{p}_{\parallel}\|x$-axis and that the exciton and light fields are linearly polarized along the in-plane y-axis. Similarly to the case of polaritons in a bulk semiconductor, the quasi-energy spectrum shows the acoustically induced bandgaps $\Delta_{ac}^{MC(n)} \propto I_{ac}^{n/2}$, due to the $n = 1, 2, \ldots$ – phonon-assisted transitions. In this case, however, the energy of a SAW phonon is rather small, $h\nu_{saw} \simeq 4.1\,\mu\mathrm{eV}$ for $\nu_{saw} = 1\,\mathrm{GHz}$, so that the acoustically induced spectral gaps refer to the counterpropagating MC polaritons $\{nk/2, \omega_{LB}^{MC}(nk/2)\}$ and $\{-nk/2, \omega_{LB}^{MC}(nk/2)\}$ (see the nearly horizontal arrows in Fig. 19.5), where $\omega_{LB}^{MC}(p_{\parallel})$ with $p_{\parallel} = \pm nk/2$ is the lower-branch dispersion of MC polaritons. In calculations of the quasi-energy spectrum shown in Fig. 19.5, $|u_z|/|u_x| \simeq 1.4$ and $D_x = 12\,\mathrm{eV}$ are used.

According to equation (19.10), the main acoustically induced bandgap in the MC polariton spectrum is given by $\Delta_{ac}^{MC} = \Delta_{ac}^{MC(n=1)} = 2|m_{\mathbf{k}}^x|\varphi_{MC}^x(k/2)$,

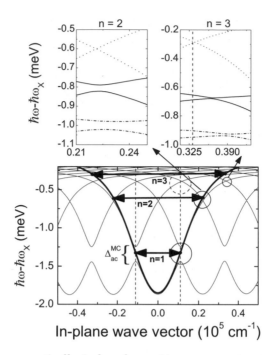

FIG. 19.5. The acoustically induced quasi-energy spectrum of polaritons in a
   GaAs λ-microcavity driven by the surface acoustic wave of $\nu_{\text{saw}} = 1\,\text{GHz}$ and
   $\langle I_{\text{saw}} \rangle = 10\,\text{W/cm}^2$ (thin solid lines). The initial, acoustically-unperturbed
   lower dispersion branch of MC polaritons is shown by the bold solid line.
   The MC Rabi energy is $\hbar\Omega_{\text{x}}^{\text{MC}} = 3.7\,\text{meV}$. The magnified energy bands relevant to the resonant $n = 2$ and 3 transitions are plotted in the left-hand
   side and right-hand side insets, respectively, for $\langle I_{\text{saw}} \rangle = 0$ (dotted lines),
   $\langle I_{\text{saw}} \rangle = 50\,\text{W/cm}^2$ (solid lines), and $\langle I_{\text{saw}} \rangle = 100\,\text{W/cm}^2$ (dash-dotted lines).
   The vertical dashed lines show the boundaries of the acoustically induced
   first Brillouin zone. From (Cho *et al.*, 2005). Reprinted figure with permission
   from [K. Cho, K. Okumoto, N. I. Nikolaev, and A. L. Ivanov, *Phys. Rev. Lett.* **94**, 226406
   (2005)]. Copyright (2005) by the American Physical Society.

where $\varphi_{\text{MC}}^{\text{x}}(k/2) = (\Omega_{\text{x}}^{\text{MC}})^2/\{(\Omega_{\text{x}}^{\text{MC}})^2 + 4[\omega_{\text{T}} - \omega_{\text{LB}}^{\text{MC}}(k/2)]^2\}$ is the exciton com-
ponent of the MC polariton states $p_{\|} = \pm k/2$ resonantly coupled via one-SAW-
phonon-induced transition. In the time domain, the bandgaps $\Delta_{\text{ac}}^{\text{MC}(n)}$ can be
visualized as acoustically induced coherent oscillations of MC polaritons, back
and forth between the states $nk/2$ and $-nk/2$, with angular frequency $\Delta_{\text{ac}}^{\text{MC}(n)}$.
Even for modest SAW intensities the main SAW-induced gap is large, so that
$\Delta_{\text{ac}}^{\text{MC}} \gg 2\pi\nu_{\text{saw}}$. For example, $\langle I_{\text{saw}} \rangle = 10\,\text{W/cm}^2$ yields $\Delta_{\text{ac}}^{\text{MC}}/(2\pi) \simeq 42\,\text{GHz}$.
Furthermore, the acoustic bandgaps associated with the two- and three-phonon
transitions in the MC polariton spectrum can also be well developed: e.g., for

$\langle I_{ac} \rangle = 100 \, \text{W/cm}^2$ one estimates $\Delta_{ac}^{\text{MC}(n=1)} : \Delta_{ac}^{\text{MC}(n=2)} : \Delta_{ac}^{\text{MC}(n=3)} = 1 : 0.08 : 0.03$ and $\Delta_{ac}^{\text{MC}(n=2)}/(2\pi) \simeq 10.2 \, \text{GHz}$. In contrast, the conventional acousto-optics deals with the acoustically induced bandgap $\Delta_{\gamma-ac}$ which is less than the SAW frequency $\Omega_k^{ac} = 2\pi \nu_{\text{saw}}$ and, due to small values of the nonresonant acousto-optic nonlinearities, uses the one-phonon-assisted transition only (Sapriel, 1979; Campbell, 1989; Korpel, 1997).

19.5 SAW-induced Bragg scattering of microcavity polaritons

Due to the periodicity of the acoustic wave, the quasi-energy spectrum can also be interpreted in terms of an acoustically induced extended Brillouin zone for MC polaritons, with the band boundaries at $p_\parallel \simeq \pm nk/2$. The first SAW-induced Brillouin zone is shown in Fig. 19.5 by the vertical dashed lines. The boundaries of the acoustically induced energy bands, where the spectral gaps arise and develop with increasing $I_{ac} = I_{\text{saw}}$, can be probed in Bragg scattering, by changing an incidence angle α of the external optical wave (Cho *et al.*, 2005). In this case the acousto-optical macroscopic equations for the Bragg scattering of MC polaritons are given by

$$\left[\frac{\partial^2}{\partial x^2} + \frac{\partial^2}{\partial z^2} - \frac{\varepsilon_b(z)}{c^2} \frac{\partial^2}{\partial t^2} \right] E = \frac{4\pi}{c^2} \frac{\partial^2}{\partial t^2} P + J_{\text{ext}},$$

$$\left[\frac{\partial^2}{\partial t^2} + 2\gamma_x \frac{\partial}{\partial t} + \omega_T^2 - \frac{\hbar \omega_T}{M_x} \frac{\partial^2}{\partial x^2} - 4m_k^x \omega_T \cos(\Omega_k^{ac} t - kx) \right] P$$
$$= \Omega_{x-\gamma}^2 E, \qquad (19.14)$$

where $\Omega_{x-\gamma}$ is the matrix element of the QW exciton – MC photon interaction, J_{ext} is a source of the external optical wave necessary for the Bragg scattering process, and the z-dependent background dielectric function $\varepsilon_b(z)$ refers to a stack of layers, "DBR – λ-cavity with an embedded QW – DBR", which form the MC structure (the z-axis is normal to the planar microcavity, so that $z \perp \mathbf{k} \| \mathbf{p}_\parallel$). The Bragg reflectors give rise to the transverse optical confinement and, at the same time, ensure an optical coupling of the external light wave with the in-plane polariton quasi-eigenstates. If the transverse optical confinement by the distributed Bragg reflectors (DBRs) is perfect, the microcavity Rabi frequency is given by $\Omega_x^{\text{MC}} = 2(\pi/\varepsilon_b)^{1/2}\Omega_{x-\gamma}$.

In order to calculate the SAW-induced Bragg diffraction spectrum of optically excited MC polaritons, the Green function technique developed in (Cho, 2003) is applied (Cho *et al.*, 2005). Because the in-plane wavevector $p_\parallel$ is conserved in the propagation of the light field through an optically transparent planar structure, one defines the MC photon Green function $g(z, z'; \omega)$ by the equation:

$$\frac{d^2}{dz^2} g(z, z'; \omega) + \kappa^2 g(z, z'; \omega) = -4\pi \delta(z - z'), \qquad (19.15)$$

where $\kappa^2 = (\omega/c)^2 \varepsilon_b(z) - p_\parallel^2$. The Green function $g(z, z'; \omega)$, which satisfies the Maxwellian boundary conditions, can be calculated numerically with the use of $\varepsilon_b(z)$ relevant to an analyzed MC structure. After substitution in equations (19.14) a Fourier expansion of the E and P fields in terms of the in-plane quasiwavevector $p_\parallel + \ell k$ and quasifrequency $\omega + 2\pi\ell\nu_{\text{saw}}$, one evaluates the SAW-mediated acousto-optic susceptibility matrix $\chi_{\ell,\ell'}$: $P_\ell = \sum_{\ell'} \chi_{\ell,\ell'} E_{\ell'}$ ($\ell = 0, \pm 1, \ldots$, i.e., $\ell = \pm n$ corresponds to the n-phonon-assisted transition). The acoustically induced polarization harmonics P_ℓ give rise to the Bragg signal:

$$E_\ell \;=\; E_\ell^{(0)} \;+\; q_\ell^2 \int \mathrm{d}z' g(z,z',\omega_\ell) \sum_{\ell'} \chi_{\ell,\ell'} E_{\ell'}(z') \;, \qquad (19.16)$$

where $q_\ell^2 = (\omega + 2\pi\ell\nu_{\text{saw}})^2/c^2$, $E_\ell = E_\ell(z)$ is the signal light field, and $E_\ell^{(0)} = E_\ell^{(0)}(z)$ is the incoming light field induced by the incident field, E_{inc}. The field $E_\ell^{(0)}(z)$ is calculated by using the photon Green function defined by equation (19.15). The integration on the right-hand side of equation (19.16) is over the QW thickness, so that one can approximate $E_{\ell'}(z')$ by $E_{\ell'}(z'=0)$. The latter amplitude is evaluated from equation (19.16) by putting $z = 0$. Finally, the outgoing, diffracted field E_ℓ is calculated from the completely defined right-hand side of equation (19.16).

The SAW-induced Bragg spectra calculated for a GaAs λ-cavity with an In-GaAs quantum well in the middle and symmetrically sandwiched between two identical AlGaAs/AlAs DBRs (34 alternating AlAs and $\text{Al}_{0.13}\text{Ga}_{0.87}\text{As}$ $\lambda/4$-layers, so that the effective radiative width for $p_\parallel = 0$ is given by $2\hbar\gamma_{\text{R}} \simeq 45\,\mu\text{eV}$), are plotted in Fig. 19.6. The Bragg angle Θ_{B} corresponds to the effective SAW-induced diffraction of MC polaritons with $p_\parallel = \pm k/2$ and is given by $\sin\Theta_{\text{B}} \simeq k/(2k_{\text{opt}})$, where $k_{\text{opt}} = \omega/c$ is the wavevector of the incident optical wave. The reflection coefficient $R_{\text{MC}} = |r(\omega, I_{\text{saw}})|^2$, calculated for $\alpha = \Theta_{\text{B}}$ and $\nu_{\text{saw}} = 1\,\text{GHz}$, is shown in Fig. 19.6(a). In this case the boundary of the acoustically induced first Brillouin zone at $p_\parallel = +k/2$ is probed (see Fig. 19.5). In the vicinity of $\omega = \omega_{\text{LB}}^{\text{MC}}(k/2)$ (the area marked in Fig. 19.5 by the large solid-line circle), with increasing I_{saw} the reflectivity changes its single-line shape for $I_{\text{saw}} = 0$ (the initial reflection spectrum is shown in Fig. 19.6a by the thick solid line) to a high-contrast double-line shape with the separation $\Delta_{\text{ac}}^{\text{MC}}$ between two dips. Furthermore, as shown in the inset of Fig. 19.6(a), a similar double-line structure, which refers to the three-phonon-assisted transition, appears and develops with increasing I_{saw} for the reflectivity at $\omega \simeq \omega_{\text{LB}}^{\text{MC}}(3k/2)$ (see the area marked in Fig. 19.5 by the large dashed-line circle). Because the wavevector band $-k/2 \le p_\parallel \le k/2$ can also be interpreted in terms of the SAW-induced reduced Brillouin zone, the incident optical wave probes the odd-order SAW-induced energy gaps. The corresponding frequency-down-shifted -1 and $-1-1+1$ Bragg replicas give rise to the outgoing optical signal in the specular reflection direction (see Fig. 19.6b). The camel-back shape of the replicas follows the bandgaps

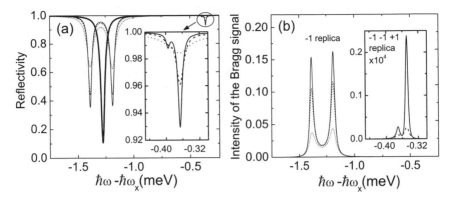

FIG. 19.6. The Bragg spectra of polaritons in GaAs-based microcavity parametrically driven by a surface acoustic wave of $\nu_{saw} = 1\,\text{GHz}$. The incident light
field excites the polariton state $p_\parallel = k/2$, i.e., the right-hand side boundary
of the acoustically induced first Brillouin zone (see Fig. 19.5). The Bragg
angle $\Theta_B \simeq 8.2°$. (a) The reflection coefficient $R_{MC} = |r(\omega, \langle I_{saw}\rangle)|^2$ of the
optical wave against the frequency ω. The thick solid line shows the initial polariton-mediated MC reflectivity for $\langle I_{saw}\rangle = 0$ and $\hbar\gamma_x = 10\,\mu\text{eV}$.
The thin lines refer to $|r(\langle I_{saw}\rangle = 10\,\text{W/cm}^2)|^2$ for spectral vicinity of
$n = 1$ SAW-phonon transition. Inset: $R_{MC} = |r|^2$ for spectral vicinity of
$n = -1-1+1$ SAW-phonon transition. (b) The one-SAW-phonon ($\hbar\omega - h\nu_{saw}$)
and three-SAW-phonon ($\hbar\omega - h\nu_{saw} - h\nu_{saw} + h\nu_{saw}$) Bragg diffraction replicas, normalized by the intensity of the incident light wave, in the specular
reflection direction. The solid, dashed and dotted lines refer to the excitonic
damping $\hbar\gamma_x = \hbar/T_2 = 5, 10$, and $30\,\mu\text{eV}$, respectively. From (Cho *et al.*,
2005). Adapted figure with permission from [K. Cho, K. Okumoto, N. I. Nikolaev, and
A. L. Ivanov, *Phys. Rev. Lett.* **94**, 226406 (2005)]. Copyright (2005) by the American
Physical Society.

Δ_{ac}^{MC} and $\Delta_{ac}^{MC(n=3)}$, respectively, and correlates with the acoustically induced
double-dip structure in the reflectivity.

For the analyzed scattering of microcavity polaritons by the surface acoustic
wave, the Klein–Cook parameter (Klein and Cook, 1967; Korpel, 1997) $Q = Q(\omega)$, which distinguishes the Raman–Nath (transmission-type) regime from
the Bragg (reflection-type) mode, is given by

$$Q = \frac{v_{pol}^{MC}}{\gamma_{pol}^{MC}} \frac{k^2}{k_{opt}}, \qquad (19.17)$$

where $v_{pol}^{MC}(\omega) = \partial\omega_{pol}^{MC}/\partial p_\parallel$ and $\gamma_{pol}^{MC} = \varphi_{MC}^x\gamma_x + (1 - \varphi_{MC}^x)\gamma_R$ are the MC polariton velocity and damping, respectively. For Figs. 19.5 and 19.6 we estimate
$Q \simeq 6.5$, i.e., the Bragg diffraction regime is realized.

The Bragg diffraction of microcavity photons in SAW-driven GaAs-based λ-cavities has been observed for the one-SAW-phonon-assisted transition (de Lima *et al.*, 2005; de Lima and Santos, 2005). Because in these experiments the MC detuning $\hbar(\omega_T - \omega_0) \simeq 170\,\text{meV}$ is rather large, the Bragg diffraction effect is not directly mediated by the excitonic resonance. As a result, the high acoustic intensities $I_{ac} \sim 1-10\,\text{kW/cm}^2$ were used. However, very recently the same group has observed an effective scattering of polaritons in GaAs-based microcavities driven by a surface acoustic wave of $\lambda_{saw} = 5.6\,\mu\text{m}$ at cryostat temperature $T = 20\,\text{K}$ (de Lima *et al.*, 2006). The latter work clearly confirms the concept of resonant acousto-optics.

The large, resonantly-enhanced acousto-optic nonlinearities result in a short "SAW – microcavity polariton" interaction length $\ell_{\gamma-x-ac}$ needed for the Bragg diffraction. By using equation (19.6) one estimates $\ell_{\gamma-x-ac}$ associated with the main acoustically induced gap (one-phonon transition):

$$\ell_{\gamma-x-ac} \simeq \frac{4\pi v_{pol}^{MC}(k/2)}{\Delta_{ac}^{MC}(I_{ac}, k/2)} \propto \frac{1}{I_{ac}^{1/2}} . \tag{19.18}$$

For example, for the case analyzed in Figs. 19.5–19.6 ($\hbar\Omega_x^{MC} = 3.7\,\text{meV}$, $\nu_{saw} = 1\,\text{GHz}$, and $\langle I_{saw} \rangle = 10\,\text{W/cm}^2$), equation (19.18) yields $\ell_{\gamma-x-ac} \simeq 56\,\mu\text{m}$. The interaction length $\ell_{\gamma-x-ac} \sim 10 - 100\,\mu\text{m}$ nicely fits the length-scale needed for the effective work of an interdigital transducer, which generates the SAW (Ivanov, 2006). The dynamical response of the acousto-optic effect is mainly determined by a time $\tau_{on/off}$ required to build up/reduce the coherent acoustic field. Thus a design, where the acousto-optic interaction occurs within the IDT-covered area or very close to it, is extremely favorable. In this case, the piezoelectric coupling between excitons, the SAW, and an interdigital transducer reduces $\tau_{on/off}$ to a nanosecond time-scale (the time needed to build up/reduce the Rayleigh evanescent acoustic field in the z-direction). This is in sharp contrast with conventional acousto-optic schemes, which require a long-distance SAW propagation along an acoustically-passive track of the length $\ell_{\gamma-ac} \sim 1 - 10\,\text{cm}$ and, therefore, operate with $\tau_{on/off}$ in a microsecond time-scale (Campbell, 1989; Östling and Engan, 1995).

19.6   Acoustically-driven TO-phonons for THz spectroscopy

Instead of the exciton states, dipole-active transverse optical (TO) phonons can also be used for effective realization of resonant acousto-optics. The latter case refers to 5–60 meV spectral band, i.e., to far infrared, terahertz spectroscopy. Note that the polariton effect associated with TO-phonons, i.e., the coherent coupling between the light field and transverse optical phonons, can persist even at room temperature, e.g., in bulk GaP (Juhasz and Bron, 1989; Stepanov *et al.*, 2001). The anharmonicity of the lattice forces gives rise to the resonant acousto-optic effect associated with the TO-phonon resonance. For TO-phonon polaritons parametrically driven by a coherent acoustic wave, the phonon nonlinearity refers to the "TO-phonon + acoustic phonon $\longleftrightarrow$ TO-phonon" scattering channel. The TO-phonon polariton states resonantly coupled via the one-acoustic-phonon transition are schematically shown in Fig. 19.7(a).

For a given TO-phonon dispersion branch of a compound semiconductor with two atoms in the unit cell, the third-order anharmonic effect, i.e., the third-order term of a Taylor expansion of the atomic lattice energy U (Gurevich, 1980), yields the following interaction Hamiltonian:

$$H_{\text{int}}^{\text{anh}} = \frac{\hbar}{4\mu_{\text{at}}\Omega_{\text{TO}}} K^{(3)}(0,0)\, u_{\text{ac}}(t) \sum_{\mathbf{p}} \left(b_{\mathbf{p}} + b_{-\mathbf{p}}^{\dagger}\right)\left(b_{-\mathbf{p}-\mathbf{k}} + b_{\mathbf{p}+\mathbf{k}}^{\dagger}\right), \quad (19.19)$$

where μ_{ac} is the reduced atomic mass, Ω_{TO} is the TO-phonon frequency, $b_{\mathbf{p}}$ is the TO-phonon operator, and $u_{\text{ac}} = \delta a_{\mathbf{k}} \cos(\Omega_{\mathbf{k}}^{\text{ac}} t - \mathbf{k}\mathbf{r})$ is the coherent displacement of the atomic lattice associated with the acoustic pump wave $\{\mathbf{k}, \Omega_{\mathbf{k}}^{\text{ac}}\}$. The anharmonicity parameter $K^{(3)}$ is given by $K^{(3)}(\mathbf{q}_1, \mathbf{q}_2) = \sum_{\mathbf{m},\mathbf{n}}(\partial^3 U/\partial u^3)e^{i\mathbf{q}_1\mathbf{m}} \times e^{i\mathbf{q}_2\mathbf{n}}$, where the real space vectors $\mathbf{m}$ and $\mathbf{n}$ run over the atomic site positions. The quadratic $H_{\text{int}}^{\text{anh}}$ includes only an acoustic-pump-wave induced part of the total third-order anharmonic lattice energy. The rest of the energy gives rise, in particular, to the damping rate of TO-phonons, $2\gamma_{\text{TO}} = 1/T_1 = 2/T_2$. The total Hamiltonian H of TO-phonon polaritons parametrically driven by the acoustic pump wave is given by $H = H_0 + H_{\text{int}}^{\text{anh}}$, where $H_0 = \sum_{\mathbf{p}} \left[\hbar\Omega_{\text{TO}}b_{\mathbf{p}}^{\dagger}b_{\mathbf{p}} + \hbar\omega_{\mathbf{p}}^{\gamma}\alpha_{\mathbf{p}}^{\dagger}\alpha_{\mathbf{p}} + (i\hbar\Omega_{\text{c}}^{\text{TO}}/2)(\alpha_{\mathbf{p}}^{\dagger}b_{\mathbf{p}} - b_{\mathbf{p}}^{\dagger}\alpha_{\mathbf{p}})\right]$ is the TO-phonon polariton Hamiltonian and $\Omega_{\text{c}}^{\text{TO}}$ is the corresponding Rabi frequency. Similarly to equations (19.4)–(19.7), the quadratic Hamiltonian H can be straightforwardly diagonalized. This gives the acoustically induced quasi-energy spectrum of TO-phonon polaritons.

The macroscopic equations for acoustically driven TO-phonon polaritons, which are rather similar to equations (19.12), are given by

$$\left[\frac{\varepsilon_{\text{b}}}{c^2}\frac{\partial^2}{\partial t^2} - \Delta\right]\mathbf{E}(\mathbf{r},t) = -\frac{4\pi}{c^2}\frac{\partial^2}{\partial t^2}\mathbf{P}(\mathbf{r},t),$$

$$\left[\frac{\partial^2}{\partial t^2} + 2\gamma_{\text{TO}}\frac{\partial}{\partial t} + \Omega_{\text{TO}}^2 - \kappa_{\text{anh}}^{(3)}u_{\text{ac}}(t)\right]\mathbf{P}(\mathbf{r},t) = \frac{\varepsilon_{\text{b}}(\Omega_{\text{c}}^{\text{TO}})^2}{4\pi}\mathbf{E}(\mathbf{r},t), \quad (19.20)$$

where $\mathbf{P}$ is the polarization associated with TO-phonons, $\kappa_{\text{anh}}^{(3)} = K^{(3)}(0,0)/\mu_{\text{at}}$, and generally the coherent displacement $u_{\text{ac}} = u_{\text{ac}}(t)$ associated with the acoustic pump wave $\{\mathbf{k}, \Omega_{\mathbf{k}}^{\text{ac}}\}$ has a time dependent amplitude, $\delta a_{\mathbf{k}} = \delta a_{\mathbf{k}}(t)$. For a CW pump wave, equations (19.20) yield the quasi-energy spectrum.

The third-order anharmonicity function $K^{(3)} = K^{(3)}(\mathbf{q}_1, \mathbf{q}_2)$ and, therefore, the coupling parameter $\kappa_{\text{anh}}^{(3)}$ in equations (19.20), can reliably be evaluated by using, e.g., density-functional perturbation theory (Debernardi, 1998). In

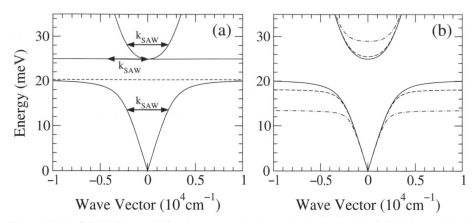

FIG. 19.7. SAW-driven TO-phonon polaritons in bulk CuCl. (a) Schematic of the polariton states resonantly coupled through one-SAW-phonon transition. The TO-phonon polariton dispersion branches and the LO-phonon dispersion are shown by the solid lines. The energies of the TO- and LO-phonons are given by $\hbar\Omega_{\text{TO}} = 20.28\,\text{meV}$ and $\hbar\Omega_{\text{LO}} = 24.95\,\text{meV}$, respectively. The SAW velocity is given by $v_{\text{saw}} = 1.1 \times 10^5\,\text{cm/s}$ (Takeuchi and Yamashita, 1981). (b) The SAW-induced change of the restrahlen band: $\langle I_{\text{saw}}\rangle = 1\,\text{kW/cm}^2$ (dashed line) and $10\,\text{kW/cm}^2$ (dash-dotted line). The surface acoustic wave is characterized by $\nu_{\text{saw}} = 1\,\text{GHz}$ and $k_{\text{saw}} = 5.71 \times 10^4\,\text{cm}^{-1}$, so that the SAW-coupled states of the lower polariton branch are outside the shown k-band ($\Delta k = 2 \times 10^4\,\text{cm}^{-1} < k_{\text{saw}}$).

order to have a well-developed resonant acousto-optic effect for TO-phonon polaritons, one needs large anharmonicities for scattering of a long-wavelength TO-phonon by a long-wavelength acoustic phonon. Recently, the strong anharmonicity of bulk CuCl (Göbel *et al.*, 1997; Ulrich *et al.*, 1999), GaP (Widulle *et al.*, 1999; Ves *et al.*, 2001) and ZnS (Tallman *et al.*, 2004) have been investigated by M. Cardona and co-workers. In these studies, the values of $K_0^{(3)} = K^{(3)}(\mathbf{q}, -\mathbf{q})$ relevant to the decay of a long-wavelength phonon in two outgoing short-wavelength LA-(TA-) phonons have been inferred.

If a bulk acoustic wave is used to realize the resonant acousto-optic effect for TO-phonon polaritons, $K^{(3)}(0,0)$ in equations (19.19) and (19.20) can be evaluated as $K^{(3)}(0,0) \simeq (ka_{\text{latt}})K_0^{(3)}$, where a_{latt} is the lattice constant. The smallness parameter $(ka_{\text{latt}}) \ll 1$ is due to the long-wavelength acoustic pump wave. Not much is known about the third-order anharmonicity of the process "TO-phonon + SAW phonon $\longleftrightarrow$ TO-phonon". An estimate of $K^{(3)}(0,0)$ for bulk TO-phonon polaritons driven by a surface acoustic wave is given by

$$K^{(3)}(0,0) \simeq (k_{\text{saw}}a_{\text{latt}})\ln\left(\frac{1}{k_{\text{saw}}a_{\text{latt}}}\right)K_0^{(3)}. \qquad (19.21)$$

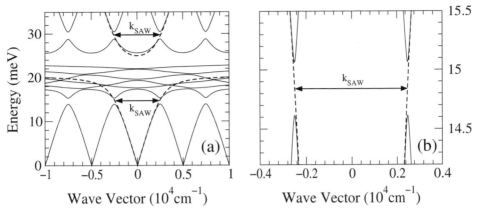

FIG. 19.8. The quasi-energy spectrum (solid lines) of TO-phonon polaritons in bulk CuCl driven by a surface acoustic wave of $\nu_{\text{saw}} = 87.5\,\text{MHz}$ and $k_{\text{saw}} = 0.5 \times 10^4\,\text{cm}^{-1}$. The SAW-unperturbed polariton spectrum is shown by the dashed lines. (a) $\langle I_{\text{saw}} \rangle = 10\,\text{kW/cm}^2$ and (b) $\langle I_{\text{saw}} \rangle = 1\,\text{kW/cm}^2$ (only the vicinity of the main SAW-induced spectral gap in the lower TO-phonon polariton dispersion branch is shown). In the calculations of the quasi-energy spectrum the TO-phonon damping rate is put equal to zero, $\gamma_{\text{TO}} = 0$.

The logarithmic function on the right-hand side of equation (19.21) arises from the evanescent envelope of the SAW.

In Figs. 19.7(b) and 19.8(a),(b) we plot the SAW-induced quasi-energy spectrum calculated for TO-phonon polaritons in bulk CuCl. In the evaluations, $K^{(3)}(0,0)$ is estimated with equation (19.21) by using the values of $K_0^{(3)}$ obtained in the experiments (Göbel *et al.*, 1997; Ulrich *et al.*, 1999). In Fig. 19.7(b), the SAW of frequency $\nu_{\text{saw}} = 1\,\text{GHz}$ resonantly couples two optically dark TO-phonon polariton states $p \simeq \pm 2.855 \times 10^4\,\text{cm}^{-1}$ of the lower dispersion branch, which are widely separated from the crossover points of the photon and TO-phonon dispersions. However, in this case the resonant acousto-optic effect still strongly influences the restrahlen band in the optically-bright part of the polariton spectrum, as clearly shown in Fig. 19.7(b). The quasi-energy spectrum becomes more complicated, if a low-frequency SAW, $\nu_{\text{saw}} \leq 100\,\text{MHz}$, is used (see Fig. 19.8). In this case the SAW-induced spectral gaps, due to the n-SAW-phonon assisted transitions, can optically be visualized in the TO-phonon polariton spectrum, as illustrated in Fig. 19.8 for $\nu_{\text{saw}} = 87.5\,\text{MHz}$.

The resonant acousto-optic effect for TO-phonon polaritons is promising for THz spectroscopy. In particular, the SAW-induced dynamical change of the restrahlen band can be used for the frequency resolved detection of THz emission.

19.7   Conclusions

The pioneering works by Léon Brillouin and Leonid V. Keldysh gave rise to the fascinating possibility to manipulate in a controlled way the light and polarization fields by using a coherent acoustic wave. In this brief review the author has tried to show how by using strong, coherent interaction between the light and polarization fields, i.e., the polariton effect, to enhance resonantly the acousto-optical nonlinearities. The resonant acousto-optics is very promising for device physics and technological applications.

Acknowledgments

The author thanks K. Cho, P. B. Littlewood, N. I. Nikolaev, and K. Okumoto for valuable discussions. Support of this work by EU RTN Project HPRN-2002-00298 is gratefully acknowledged.

References

Agranovich, V. M. and Ginzburg, V. L. (1984). *Crystal Optics with Spatial Dispersion and Excitons*. Springer, Berlin.

Berry, M. V. (1966). *The Diffraction of Light by Ultrasound*. Academic Press, London.

Blanch, G. (1965). Mathieu Functions, in: *Handbook of Mathematical Functions*, ed. by M. Abramowitz and I. A. Stegun. Dover Publications, New York. Chapter 20.

Born, M. and Wolf, E. (1970). *Principles of Optics* (4th edn). Pergamon Press, New York. Chapter 12.

Brillouin, L. (1922). *Ann. Phys. (Paris)* **17**, 88.

Brillouin, L. (1933). *Actual. Sci. Ind.* **59**, 1.

Campbell, C. (1989). *Surface Acoustic Wave Devices and Their Signal Processing Applications*. Academic Press, Boston. Chapter 2.

Cho, K. (2003). *Optical Response of Nanostructures: Microscopic Nonlocal Theory*. Springer Verlag, Heidelberg. Section 3.4.

Cho, K., Okumoto, K., Nikolaev, N. I. and Ivanov, A. L. (2005). *Phys. Rev. Lett.* **94**, 226406.

Chu, R. S. and Tamir, T. (1976). *J. Opt. Soc. Am.* **66**, 220.

Chu, R. S. and Kong, J. A. (1980). *J. Opt. Soc. Am.* **70**, 1.

Cunningham, J., Talyanskii, V. I., Shilton, J. M., Pepper, M., Kristensen, A. and Lindelof, P. E. (2000). *Phys. Rev. B* **62**, 1564.

Debernardi, A. (1998). *Phys. Rev. B* **57**, 12847.

de Lima, M. M., Jr., Hey, R., Santos, P. V. and Cantarero, A. (2005). *Phys. Rev. Lett.* **94**, 126805.

de Lima, M. M., Jr., and Santos, P. V. (2005). *Rep. Prog. Phys.* **68**, 1639.

de Lima, M. M., Jr., van der Poel, M., Santos, P. V. and Hvam, J. M. (2006). *Phys. Rev. Lett.* **97**, 045501.

Ebbecke, J., Bastian, G., Blöcker, M., Pierz, K. and Ahlers, F. J. (2000). *Appl. Phys. Lett.* **77**, 2601.

Farnell, G. W. (1978). Types and Properties of Surface Waves, in: *Acoustic Surface Waves* (Topics in Applied Physics, Vol. 24), ed. by A. A. Oliner. Springer-Verlag, Berlin. Chapter 2.

Göbel, A., Ruf, T., Lin, C.-T., Cardona, M., Merle, J.-C. and Joucla, M. (1997). *Phys. Rev. B* **56**, 210.

Gurevich, V. L. (1980). *Kinetics of Phonon Systems* (in Russian). Nauka, Moscow. Chapter 1.

Hess, P. (2002). *Physics Today* **55**, 42.

Ivanov, A. L. and Littlewood P. B. (2001). *Phys. Rev. Lett.* **87**, 136403.

Ivanov, A. L. and Littlewood P. B. (2003). *Semicond. Sci. Technol.* **18**, S428.

Ivanov, A. L. (2005). *Phys. stat. sol. (a)* **202**, 2657.

Ivanov, A. L., Nikolaev, N. I. and Cho, K. (2006). *IEE Proc.-Optoelectron.* **153**, 326.

Juhasz, T. and Bron, W. E. (1989). *Phys. Rev. Lett.* **63**, 2385.

Keldysh, L. V. (1962). *Fiz. Tverd. Tela* **4**, 2265 [*Sov. Phys. Solid State* **4**, 1658].

Klein W. R. and Cook, B. D. (1967). *IEEE Trans.* SU-**14**, 123.

Korpel, A. (1997). *Acousto-Optics* (2nd edn). Marcel Dekker, Inc., New York. Chapters 2 and 3.

Kukushkin, I. V., Smet, J. H., Höppel, L., Waizmann, U., Riek, M., Wegscheider, W. and von Klitzing, K. (2004). *Appl. Phys. Lett.* **85**, 4526.

Nägel, J. S., Stabenau, B., Böhne, G., Dreher, S., Ulbrich, R. G., Manzke, G. and Henneberger K. (2001). *Phys. Rev. B* **63**, 235202.

Östling, D. and Engan, H. E. (1995). *Opt. Lett.* **20**, 1247.

Sapriel, J. (1979). *Acousto-Optics*. Wiley & Sons, New York.

Shilton, J. M., Talyanskii, V. I., Pepper, M., Ritchie, D. A., Frost, J. E. F., Ford, C. J. B., Smith, C. G. and Jones, G. A. C. (1996). *J. Phys: Condens. Matter* **8**, L531.

Simon, S. H. (1996). *Phys. Rev. B* **54**, 13878.

Stepanov, A. G., Hebling, J. and Kuhl, J. (2001). *Phys. Rev B* **63**, 104304.

Takagaki, Y., Hesjedal, T., Brandt, O., and Ploog, K. H. (2004). *Semicond. Sci. Technol.* **19**, 256.

Takeuchi, H. and Yamashita, K. (1981). *J. Appl. Phys.* **52**, 7448.

Tallman, R. E., Ritter, T. M., Weinstein, B. A., Serrano, J., Lauck, R. and Cardona, M. (2004). *Phys. stat. sol. (b)* **241**, 491.

Ulrich, C., Göbel, A., Syassen, K. and Cardona, M. (1999). *Phys. Rev. Lett.* **82**, 351.

Ves, S., Loa, I., Syassen, K., Widulle, F. and Cardona, M. (2001). *Phys. stat. sol. (b)* **223**, 241.

Widulle, F., Ruf, T., Göbel, A., Schönherr, E. and Cardona, M. (1999). *Phys. Rev. Lett.* **82**, 5281.

Wilson, J. and Hawkes, J. F. B. (1983). *Optoelectronics*. Prentice-Hall International, New Jersey. Pages 111-116.

Zeldovich, Ya. B. (1966). *Zh. Eksp. Teor. Phys.* **51**, 1492 [*Sov. Phys. JETP* **24**, 1006].

20

INELASTIC TUNNELING SPECTROSCOPY OF SINGLE SURFACE ADSORBED MOLECULES

S. G. Tikhodeev[a] and H. Ueba[b]
[a] A. M. Prokhorov General Physics Institute, RAS, Vavilova 38, Moscow, 119991 Russia
[b] Department of Electronics, Toyama University, Gofuku, Toyama, Japan

Abstract
A theory of inelastic electron tunneling spectroscopy of a single molecule with a scanning tunneling microscope is given by using the Keldysh–Green's nonequilibrium function method for an adsorbate-induced resonance coupled to the molecular vibration.

20.1 Introduction

Scanning tunneling microscopy and spectroscopy are the ultimate techniques to study the structural and electronic properties of surfaces as well as adsorbates at atomic and molecular level spatial resolution. Soon after the invention of a scanning tunneling microscope (STM) Binnig, Garcia and Rohrer (1985) proposed a strategy to explore the vibrational properties of a single molecule based on a technique called inelastic electron tunneling spectroscopy (IETS) using tunneling electrons from/to a tip of the STM as an excitation source. The STM tip is placed on top of a target molecule and the bias voltage is ramped. When the energy of a tunneling electron matches the vibrational energy, the change in conductance is observed due to the opening of inelastic tunneling channel associated with electron-vibration scattering. Figure 20.1 shows a schematic illustration of the emergence of inelastic tunneling at the threshold for vibrational excitation. The change in the tunneling current due to vibrational excitation (inelastic tunneling current) is too small to be measured from the $I - V$ curve or the differential conductance dI/dV. The vibrational feature is more clearly extracted from d^2I/dV^2 (the vibrational spectrum). It may be a peak, a dip, or a derivative-like structure, indicating a change in the tunneling current upon opening of the inelastic channel.

In spite of this rather simple working principle of STM-IETS, it was 1998 when the first successful observation of a vibrational mode of a single molecule using an STM was reported by Ho and coworkers for the C-H stretch of an isolated acetylene on the Cu(100) surface (Stipe *et al.*, 1998) [see also (Ho, 2002) for a comprehensive review on single molecule chemistry]. Another important achievement of this experiment was that it is possible to perform

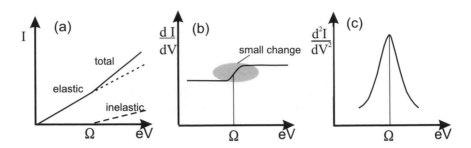

FIG. 20.1. Schematics of opening of inelastic tunneling (dashed line) at the threshold for vibrational excitation (with energy Ω). Gradual increase in $I-V$ (a) and dI/dV (b) curves and broadening in d^2I/dV^2 (c) around $eV = \Omega$ are due to a finite sample temperature, vibrational damping and modulation voltage. A situation is shown when the elastic component of tunneling current does not change with the inelastic channel opening (dotted curve in panel a). This case corresponds to a peak in the IETS spectra at positive bias voltage (panel c). However, as discussed below, the IETS spectrum feature may be also a dip, or a derivative-like structure, depending on the reaction of the elastic current upon the opening of the inelastic tunneling.

vibrational microscopy by mapping out the inelastic electron tunneling (IET) signal in real space with atomic-scale resolution. Using these new capabilities of the STM, the adsorbate atomic structure could be determined in real space with chemical specificity. Since then STM-IETS continues to provide rich information of single molecule chemistry at surfaces. Vibrational spectroscopies, such as infrared reflection absorption, electron energy loss and sum-frequency generation, are also a well established powerful tool to explore the vibrational properties of adsorbates on surfaces. Knowledge of the vibrational mode and the energies of a molecule allows us to identify its adsorption site, bonding to the surface atom and orientation. Vibrational spectrum thus serves as a "fingerprint" of adsorbates and can be used to identify chemical reaction products on surfaces (Moresco *et al.*, 1999; Pascual *et al.*, 2001; Kim *et al.*, 2002; Ho, 2002; Kushmerick *et al.*, 2004; Wang *et al.*, 2004).

Keldysh's (1964, 2003) real-time nonequilibrium Green's functions (NEGF) proved to be very well suited for the theoretical description of STM-IETS. Keldysh–Green's functions were first employed for macroscopic tunneling contacts by Caroli and coworkers (1972) to describe various features of electron current accompanied with the excitation of localized vibrations. However, the full self-consistent potential of the NEGF method becomes indispensable in IETS with microscopic tunneling contacts, due to the increased role of the relaxation of localized states and its influence on the stationary electron distribution functions (Arseyev and Maslova, 1992; Arseyev and Maslova, 1998). The Keldysh–Green's functions method has been applied to STM-IETS by

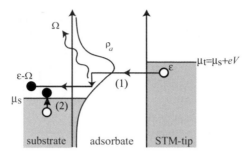

FIG. 20.2. Electron tunneling between an STM tip and a metal surface via adsorbate induced resonance state ρ_a near the Fermi level. Inelastic tunneling due to an opening of new channel associated with vibrational excitation (1), which decays through electron–hole pair excitation (2).

Tikhodeev *et al.* (2001), Mii *et al.* (2002), Mii *et al.* (2003) to generalize the pioneering work by Persson and Baratoff (1987) who calculated the inelastic tunneling fraction using the perturbation method for the adsorbate-induced resonance mode explained in Fig. 20.2. They found that the elastic contribution to the total current is always negative and may outweigh the positive contribution of the inelastic current for a particular vibrational mode. Such a negative feature (dip in the d^2I/dV^2-spectrum) has indeed been observed in single-molecule vibrational spectroscopy of an oxygen molecule chemisorbed on Ag (100) (Hahn *et al.*, 2000).

In addition to STM-IETS, the effect of electron–phonon coupling on the quantum transport in metallic nanowires including atomic contacts and chains and in molecular links sandwiched by two electrodes has also been intensively studied during the last decade. In particular, the conductance in Au atomic chains has been found to exhibit a very pronounced drop from a quantum conductance $G_0(= 2e^2/h)$ at voltages corresponding to the excitation of the vibrational modes of a finite suspended chain (Agraït *et al.*, 2002). A similar conductance drop has also been observed in the Pt-H-Pt system (Smit *et al.*, 2002). It is by now well established by the Keldysh–Green's functions theory that the sign of the conductance correction, which is negative for a large transmission, evolves into a positive correction for a low transmission. This behavior can be understood as a competition between elastic and inelastic processes (Paulsson *et al.*, 2005; Viljas *et al.*, 2005; de la Vega *et al.*, 2006; Ueba *et al.*, submitted).

Keldysh–Green's functions serve also as a theoretical basis for the density functional calculation of IETS as well as STM-imaging for acetylene on copper (Lorente and Persson, 2000).

Inelastic tunneling through a single adsorbed molecule opens the way to its manipulation by STM. Since the pioneering demonstration of a single atomic switch by Eigler *et al.* (1991) in which high- and low-current states are realized via controlling Xe atom transfer between a Ni substrate and a tip of STM, there

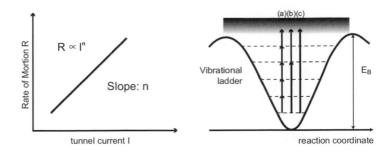

FIG. 20.3. Schematic log-log plot of the rate of motion $R \propto I^n$, and the vibrational ladders in the potential well along the reaction coordinate. (a) Incoherent multiple single-stepwise ladder climbing; (b) coherent superposition of multistep excitation by a single electron; (c) single electron excitation for $eV > E_B$. n stands for the number of tunneling electrons in each case of ladder climbing; in the case of incoherent ladder climbing process it corresponds also to the number of steps on vibrational ladder, from the ground vibrational state up to the dissociation barrier.

have been very exciting developments in the unique ability of STM to induce the motions and reactions of single adsorbates at surfaces. Examples of such novel experiments are dissociation of decaborane on Si(111) (Dujardin *et al.*, 1992), desorption of hydrogen from hydrogen-terminated Si(100) (Shen *et al.*, 1995), a step-by-step control of chemical reaction to form a biphenyl molecule from a iodobenzene on Cu(111) (Hla *et al.*, 2000), rotation and dissociation of an oxygen molecule on Pt(111) (Stipe *et al.*, 1997), rotation of acetylene molecule on Cu (100) (Stipe *et al.*, 1998), hopping of CO on Pd(110) (Komeda *et al.*, 2002) and hopping and desorption of ammonia on Cu(100) (Pascual *et al.*, 2003).

The theoretical understanding of the physical mechanisms behind these motions and reactions of single adsorbates has made slow but steady progress toward a full understanding of how tunneling electrons couple to nuclear motion of an adsorbate to overcome the potential barrier along the relevant reaction coordinate (Ueba, 2003). The power-law dependence of the atom transfer rate as a function of applied voltage or tunneling current observed for the Eigler switch has been modeled as a potential barrier crossing between the potential wells formed by the interaction of the adatom with the tip and the substrate. The atom overcomes the potential barrier through the stepwise vibrational ladder climbing by inelastic tunneling electrons [see the right-hand side panel in Fig. 20.3, process labelled (a)]. The so-called vibrational heating due to inelastic tunneling electrons has been proposed by Gao *et al.* (1992), Walkup *et al.* (1993) and Brandbyge and Hedegard (1994) [see also a detailed discussion in (Gao *et al.*, 1997)]. Arrhenius-like expression for the rate of motion is characterized by an effective temperature T_{eff} in the presence of inelas-

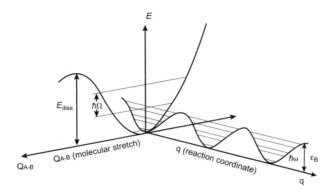

FIG. 20.4. Schematic of a two-dimensional potential well along the molecular stretch Q_{AB} and along the reaction coordinate q. It is implicitly assumed that the barrier for the AB bond-breaking (dissociation) E_{diss} and the vibrational energy Ω are much higher than, respectively, the barrier ϵ_B for the motion associated with the reaction coordinate mode and its energy ω.

tic tunneling current, otherwise a vibrational mode is in thermal equilibrium with a substrate temperature T. In comparison with the earlier papers (Gao *et al.*, 1992; Brandbyge and Hedegard, 1994), Tikhodeev and Ueba (2004), Tikhodeev and Ueba (2005) take into account self-consistently the effects of the finite vibration lifetime on the IET spectra as well as on the vibration generation rate and vibrational heating.

Another possibility of coherent multistep vibrational excitation by a single electron (Fig. 20.3c) has been proposed by Salam *et al.* (1994) in analogy to a mechanism of desorption induced by electronic transition developed by Gadzuk (1991). The coherent multiple excitation mechanism was shown to be important at low tunneling currents, where the average time between successive electrons is longer than the vibrational lifetime. The dissociation rate of single O_2 on Pt(111), where the vibrational relaxation rate of the O-O stretch mode due to electron–hole pair excitations in the substrate is much larger than the tunneling current (Persson, 1987), has been satisfactory described by this coherent multiple excitation process (Stipe *et al.*, 1997). However, one should bear in mind that since both coherent and incoherent multiple excitation mechanisms show a power-law dependence over a limited range of the bias voltage or the tunneling current, they cannot be simply distinguished.

Anharmonic coupling between different vibrational modes, e.g., between molecular stretch and reaction coordinate, opens a way to an indirect excitation of single molecule motions and reactions (Komeda *et al.*, 2002; Ueba, 2003; Ueba and Persson, 2004), see a schematic explanation in Fig. 20.4 [from (Ueba, 2003)]. According to this mechanism, the tunneling electrons excite inelastically, e.g., a molecular stretch mode Q_{AB}, instead of a direct excitation of the reaction coordinate mode q. The former transfers the excitation into reaction coordinate mode via anharmonic coupling. An interesting example of such anharmonic coupling has been observed recently (Backus *et al.*, 2005) for the most elementary reaction taking place on a surface: the lateral motion of a molecule (carbon monoxide) on the platinum surface, see also a discussion in (Ueba and Wolf, 2005). In the latter example the initial driving source is not the tunneling but the heating of the Pt electrons by ultrafast laser pulses. On the next stage the energy transfer is indirect, like that shown in Fig. 20.4, with the replacement of Q_{A-B} to frustrated rotation: the hot electrons first excite the frustrated rotation mode of the molecule (molecular rocking), and the energy is then passed to the translational mode (lateral hopping) via anharmonic coupling.

In what follows, we give a detailed introduction to the nonequilibrium Keldysh–Green's functions method in application to single-molecule spectroscopy and manipulation via inelastic electron tunneling.

20.2   Keldysh–Green's functions and inelastic tunneling

Keldysh's technique (Keldysh, 1964; Keldysh, 2003) has been used to describe the inelastic tunneling in many recent publications. Usually the formalism developed originally for macroscopic tunneling contacts (Caroli *et al.*, 1971*b*; Caroli *et al.*, 1971*a*; Combescot, 1971; Caroli *et al.*, 1972) is used also for a tunneling through a single molecule (Lorente *et al.*, 2001; Asai, 2004; Lorente, 2004; Lorente *et al.*, 2005) or a molecular bridge (Troisi *et al.*, 2003; Galperin *et al.*, 2004*a*; Galperin *et al.*, 2004*b*). Namely, the tunneling current is calculated via the renormalized (because of the elastic and inelastic tunneling) electron density of states of the adsorbate assuming equilibrium vibrational distribution function. However for tunneling through a single molecule the generation of vibrational excitations and overheating effects are known to be important (Gao *et al.*, 1992; Gao, 1994; Gao *et al.*, 1997; Tikhodeev and Ueba, 2004; Arseyev and Maslova, 2006). The self-consistent influence of this effect on the tunneling current was taken into account in (Mii *et al.*, 2002; Mii *et al.*, 2003; Arseyev and Maslova, 2005).

This effect is of crucial importance for IET overheating and thus for single molecule manipulation. The physics under this is that although the electron–phonon interaction is weak and the population changes are slow, it may cause eventually a strong renormalization of the stationary vibrational occupation number – the system has a long time to establish this stationary number. For example, the corrections to phonon and adsorbate distribution functions start from the second order over electron–phonon coupling constant χ, but the correction to the vibrational occupation number appears already in the zeroth order over χ (Tikhodeev *et al.*, 2001; Mii *et al.*, 2002; Tikhodeev and Ueba, 2004), see the discussion below.

The theory of inelastic tunneling electron spectroscopy gives a very instructive example of Keldysh's technique in a fully self-consistent manner. For tutorial purposes, we give here the milestones of the technique in the so-called $\pm$ and triangular representations which are usually left behind the screen in most papers.

We start from the formulation of the problem of electron tunneling between the STM tip and substrate through an adsorbed molecule.

20.2.1 *Hamiltonian and formulation of the inelastic tunneling problem*

The Newns–Anderson type Hamiltonian (Newns, 1969) with electron–phonon part is usually used to describe the joint influence of electron tunneling through the adsorbate and possibility of its vibrational excitation (Persson and Baratoff, 1987; Gao *et al.*, 1997; Tikhodeev *et al.*, 2001; Mii *et al.*, 2003):

$$\hat{H} = \sum_k \varepsilon_k c_k^\dagger c_k + \sum_p \varepsilon_p c_p^\dagger c_p + \varepsilon_a c_a^\dagger c_a + \Omega(b^\dagger b + 1/2)$$
$$+ \sum_k \left(U_{ka} c_k^\dagger c_a + \text{h.c.} \right) + \sum_p \left(U'_{pa} c_p^\dagger c_a + \text{h.c.} \right)$$
$$+ \chi c_a^\dagger c_a (b^\dagger + b) \,. \tag{20.1}$$

Here the energies and annihilation operators of a substrate, a tip, an adsorbate orbital and a vibrational mode are denoted by $\varepsilon_p, \varepsilon_k, \varepsilon_a, \Omega$ and c_p, c_k, c_a, b, respectively. The tip and substrate systems are assumed to be in thermal equilibrium at the same temperature T and to have independent chemical potentials μ_t and μ_s, respectively, whose difference corresponds to the bias voltage $eV = \mu_s - \mu_t$. Electronic tunneling matrix elements U_{ka} (tip–adsorbate) and U'_{pa} (substrate–adsorbate) give rise to stationary current between the tip and substrate through the adsorbate orbital. The distribution function for electrons of the tip and substrate systems, n_t and n_s respectively, are given by the Fermi distribution function. χ is the vibration–adsorbate orbital coupling constant.

Only one adsorbate orbital is taken into account in this simplest approach. The case of two adsorbate levels was recently discussed in (Arseyev and Maslova, 2005; Arseyev and Maslova, 2006).

Our goal is to calculate the tunneling current $I(V)$ between the STM tip and substrate through the adsorbate as a function of bias voltage in the stationary case, long after switching on of the bias voltage. It can be calculated as a net arrival flow of electrons to the substrate (we fix the chemical potential, not the number of electrons),

$$\frac{I}{e} = \sum_k \left\langle \frac{d\hat{N}_k(t)}{dt} \right\rangle = \sum_k \left\langle \left[\hat{H}, \hat{N}_k(t) \right]_- \right\rangle = -iU \sum_k \left\langle c_k^\dagger(t) c_a(t) - c_a^\dagger(t) c_k(t) \right\rangle \,. \tag{20.2}$$

Here all the operators are in the Heisenberg representation, and averaging $\langle (\cdots) \rangle = \text{Tr}\hat{\rho}(\cdots)$ is carried out over the stationary density matrix $\hat{\rho}$ established after switching on of the bias voltage at $t = -\infty$. Using Keldysh's

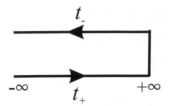

FIG. 20.5. Keldysh's closed time-ordering contour C. The time-ordering T_c is causal T-ordering on the lower branch t_+ of C, anticausal $\tilde{T}$ on the upper branch t_-. All times t_- are in absolute future compared with the times t_+, so that $T_c \hat{A}(t_+)\hat{B}(t'_-) = \eta \hat{B}(t')A(t)$ for any t, t' ($\eta = \pm 1$ for fermionic/bosonic operators).

closed C-contour of time ordering (Fig. 20.5) and introducing the real-time Keldysh–Green's function

$$G_{ak}^{+-} = -i\langle T_c\langle c_a(t, +)c_k^\dagger(t', -)\rangle \equiv i\langle T_c\langle c_k^\dagger(t')c_a(t)\rangle\,, \qquad (20.3)$$

the total tunneling current becomes

$$\frac{I}{e} = U \sum_k \left(G_{ka}^{+-}(t, t) - G_{ak}^{+-}(t, t)\right)\,. \qquad (20.4)$$

Thus we see that the real-time nonequilibrium Keldysh–Green's functions arise naturally in the formulation of the inelastic tunneling problem.

20.2.2 *Keldysh–Green's functions*

Equation (20.4) is exact for the Hamiltonian given by equation (20.1); for its evaluation we have to introduce the full set of Keldysh–Green's functions which are (2×2) matrices over Keldysh's $\pm$ indices $\alpha, \alpha' = \pm$ on the closed C-contour Fig. 20.5:

$$G_k(t, t') = -i\langle T_c\langle c_k(t, \alpha)c_k^\dagger(t', \alpha')\rangle = \begin{pmatrix} G_k^{++} & G_k^{+-} \\ G_k^{-+} & G_k^{--} \end{pmatrix}$$

$$\equiv \begin{pmatrix} G_k^c & G_k^+ \\ G_k^- & \tilde{G}_k^c \end{pmatrix} = \begin{pmatrix} -i\langle T\langle c_k(t)c_k^\dagger(t')\rangle & i\langle c_k^\dagger(t')c_k(t)\rangle \\ -i\langle c_k(t)c_k^\dagger(t')\rangle & -i\langle \tilde{T}\langle c_k(t)c_k^\dagger(t')\rangle \end{pmatrix}, (20.5)$$

the full Keldysh–Green's function of substrate electrons in $\pm$ representation, the same for adsorbate and tip electrons (substituting indices $k \to a \to p$), and

$$D(t, t') = -i\langle T_c\langle b(t, \alpha)b^\dagger(t', \alpha')\rangle = \begin{pmatrix} D^{++} & D^{+-} \\ D^{-+} & D^{--} \end{pmatrix}$$

$$\equiv \begin{pmatrix} D^c & D^+ \\ D^- & \tilde{D}^c \end{pmatrix} = \begin{pmatrix} -i\langle T\langle b(t)b^\dagger(t')\rangle & -i\langle b^\dagger(t')b(t)\rangle \\ -i\langle b(t)b^\dagger(t')\rangle & -i\langle \tilde{T}\langle b(t)b^\dagger(t')\rangle \end{pmatrix}, \qquad (20.6)$$

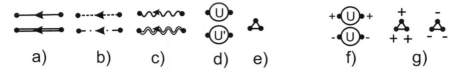

FIG. 20.6. Keldysh–Green's functions of the adsorbate electrons $G_{0,a}$ (a, top) and G_a (a, bottom), of the substrate and tip electrons G_k and G_p (b), and of the vibrational mode D_0 and D (c). Vertices of the tunneling between substrate and adsorbate $U = U_{ka}$ and tip and adsorbate $U' = U'_{pa}$ (d), and of the adsorbate electron-vibrational mode interaction (e). Only two types of vertices are nonzero in the $\pm$ representation, with all ends being pluses or minuses: iU (f, top) and $-iU$ (f, top); $\sqrt{i}\chi$ (g, left) and $-\sqrt{i}\chi$ (g, right).

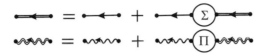

FIG. 20.7. Keldysh–Dyson equations for the adsorbate (top) and vibrational mode (bottom).

the full Keldysh–Green's function of vibrational excitation of the adsorbate (phonon). As in equation (20.3), all the operators here are in the Heisenberg representation, and averaging is made over the full density matrix $\hat{\rho}$.

The perturbation theory over powers of tunneling and electron–phonon interaction constants U, U', χ allows us to calculate equations (20.3), (20.5) and (20.6) as sums of diagrams containing free propagators

$$G_0 = -i\mathrm{Tr}\,\hat{\rho}_{-\infty}T_c\,c_{k,0}(t,\alpha)c_{k,0}^{\dagger}(t',\alpha')\,, \tag{20.7}$$

$$D_0 = -i\mathrm{Tr}\,\hat{\rho}_{-\infty}T_c\,b_0(t,\alpha)b_0^{\dagger}(t',\alpha') \tag{20.8}$$

for operators in the interaction representation averaged over the initial equilibrium density matrix $\hat{\rho}_{-\infty}$, coupled together via tunneling and electron–phonon vertices (see in Fig. 20.6). For example, $G_{a,0}^{+-} = in_{a,0}\exp{-i\varepsilon_a(t-t')}$ and $D_0^{+-} = in_{\mathrm{ph},0}\exp{-i\Omega(t-t')}$, where $n_{a,0}$ and $n_{\mathrm{ph},0}$ are the initial occupation numbers of the adsorbate and vibrational mode.

Some words about the notations. We follow the notations of the original paper by Keldysh (1964). Since then quite a number of different notations have been widely used. For example, in the famous textbook (Lifshitz and Pitaevskii, 1981) the notation of the Keldysh's time contour $\pm$ indices is reversed. The causal T-ordered Green's function which is $G^c = G^{++}$ in Keldysh's notations becomes G^{--}, etc. Another quite widely used set of notations [see, e.g., (Datta, 1995)] goes back to the Kadanoff–Baym (1962) extension of analytical continuation method by Schwinger (1961). Then the functions $G^>$ and $G^<$ remain for Keldysh's functions G^{+-} and G^{-+}, respectively.

Keldysh diagram technique for nonequilibrium system is a natural extension of Feynman's diagram technique for equilibrium system in the ground state (Keldysh, 2003). The price to be paid for the possibility to describe the nonequilibrium situations is that all the propagators and vertices become matrices over Keldysh's time indices, and the number of graphs grows significantly.[22] However the main advantage of the Feynman's graphs method, the independence of any part of any term of the perturbation theory on the perturbation order of this term, is valid. This allows us to introduce graphs, and to make partial summations. For example, Dyson's equations have the standard Feynman's form,

$$G = G_0 + G_0 \Sigma G , \quad D = D_0 + D_0 \Pi D , \qquad (20.9)$$

where Σ and Π are, respectively, the electron self-energy and the phonon polarization operators, now (2×2) matrices; see examples of the graphic representation of equations (20.9) in Fig. 20.7.

20.2.3 *Triangular and $\pm$ representations*

Not all components of the Keldysh–Green's functions are independent (Keldysh, 1964). For example,

$$G^{++}(t, t') = \theta(t - t')G^{-+}(t, t') + \theta(t' - t)G^{+-}(t, t'). \qquad (20.10)$$

This allows (Keldysh, 1964) to introduce the so-called triangular representation, after rotation

$$\overline{G} = RGR^{-1}, \overline{D} = RDR^{-1}, R = \frac{1}{\sqrt{2}} \begin{pmatrix} 1 & -1 \\ 1 & 1 \end{pmatrix}, R^{-1} = \frac{1}{\sqrt{2}} \begin{pmatrix} 1 & 1 \\ -1 & 1 \end{pmatrix}. \qquad (20.11)$$

The Keldysh–Green's functions in the triangular representation are very transparent physically, because they couple together the retarded, advanced and statistical (Keldysh's) functions G^r, G^a and F (D^r, D^a and S in case of phonons):

$$\overline{G} = \begin{pmatrix} 0 & G^a \\ G^r & F \end{pmatrix} = \begin{pmatrix} 0 & i\theta(t' - t)\langle[c(t), c^\dagger(t')]_+\rangle \\ -i\theta(t - t')\langle[c(t), c^\dagger(t')]_+\rangle & -i\langle[c(t), c^\dagger(t')]_-\rangle \end{pmatrix},$$
$$(20.12)$$

$$\overline{D} = \begin{pmatrix} 0 & D^a \\ D^r & S \end{pmatrix} = \begin{pmatrix} 0 & i\theta(t' - t)\langle[b(t), b^\dagger(t')]_-\rangle \\ i\theta(t - t')\langle[b(t), b^\dagger(t')]_-\rangle & -i\langle[b(t), b^\dagger(t')]_+\rangle \end{pmatrix},$$
$$(20.13)$$

where $[\cdots]_-$ ($[\cdots]_+$) stands for commutator (anticommutator). As seen from the definitions in equations (20.12)–(20.13), the retarded/advanced functions

---

[22]In a more general case when higher order correlations are important, the number of graphs grows even more: the correlation functions have to be included (Hall, 1975*a*; Hall, 1975*b*; Kukharenko and Tikhodeev, 1982; Fanchenko, 1983).

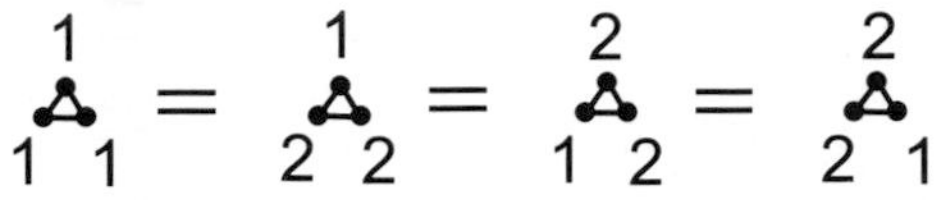

$$\underset{1\ \ 1}{\overset{1}{\triangle}} = \underset{2\ \ 2}{\overset{1}{\triangle}} = \underset{1\ \ 2}{\overset{2}{\triangle}} = \underset{2\ \ 1}{\overset{2}{\triangle}}$$

FIG. 20.8. Only four electron–phonon vertices are nonzero in triangular representation, all equal to $\sqrt{i/2}\,\chi$. The general rule is that there should be an odd number of type 1 ends.

are responsible for the dynamic response of the system, and statistical functions describe the population statistics (Keldysh, 1964).

In the Wigner representation, making the Fourier transform over $\tau = t - t'$, we deal with the Fourier transformed Keldysh–Green's functions

$$G(T, \varepsilon) = \int_{-\infty}^{\infty} d\tau\, G(T, \tau) e^{-i\varepsilon\tau}\,, \tag{20.14}$$

and the dependence on time $T = (t + t')/2$ disappears in the stationary state of interest.

The retarded Greens functions are connected with the density of states: of substrate electrons

$$\rho_s(\varepsilon) = -\frac{1}{\pi} \sum_k \operatorname{Im} G_k^r(\varepsilon)\,, \tag{20.15}$$

adsorbate,

$$\rho_a(\varepsilon) = -\frac{1}{\pi} \operatorname{Im} G_a^r(\varepsilon)\,, \tag{20.16}$$

etc..

As to the tunneling vertices in the triangular representation, the diagonal components vanish, $\overline{U}_{11} = \overline{U}_{22} = 0$, and $\overline{U}_{12} = \overline{U}_{21} = iU$. There are four nonzero components of the electron–phonon vertex, see Fig. 20.8.

Equations analogous to equation (20.10) are valid for the self-energy matrices. For example,

$$\Sigma^{++}(t, t') = -\theta(t - t')\Sigma^{-+}(t, t') - \theta(t' - t)\Sigma^{+-}(t, t')\,. \tag{20.17}$$

Generally, it can be proved (Tikhodeev, 1982) that any Keldysh diagram does not depend on the time index of the end point with the maximum time [do not forget to change the sign if the end point belongs to the interaction vertex, as in case of self-energy matrix, equation (20.17)]. Thus, the self-energies are triangular matrices too:

$$\overline{\Sigma} = \begin{pmatrix} \Omega & \Sigma^r \\ \Sigma^a & 0 \end{pmatrix}, \quad \overline{\Pi} = \begin{pmatrix} \Delta & \Pi^r \\ \Pi^a & 0 \end{pmatrix}. \tag{20.18}$$

As follows from the triangular shape of all the matrices equations (20.12), (20.13) and (20.18), the Keldysh–Dyson equations for dynamical functions decouple. For example,

$$G^r = G_0^r + G_0^r \Sigma^r G^r \,, \tag{20.19}$$

where $G_0^r \Sigma^r G^r = \int_{-\infty}^{\infty} dt_1 dt_2 G_0^r(t, t_1) \Sigma^r(t_1, t_2) G^r(t_2, t')$ in the time representation or just a simple product $G_0^r(\varepsilon) \Sigma^r(\varepsilon) G^r(\varepsilon)$ in the Wigner representation (in the stationary case). Then, for example, $G_{a,0}^{-1} G_{a,0}^r \equiv (\varepsilon - \varepsilon_a) G_{a,0}^r = 1$, $D_0^{-1} D_0^r \equiv (\omega - \Omega) D_0^r(\omega) = 1$ and solutions of equation (20.19) can be easily written as

$$G_a^r(\varepsilon) = \frac{1}{\varepsilon - \varepsilon_a - \Sigma^r(\varepsilon)} \,, \quad D^r(\omega) = \frac{1}{\omega - \Omega - \Pi^r(\omega)} \,. \tag{20.20}$$

Thus, the real part of Σ^r, Π^r describes the shift, and the imaginary part describes the broadening of the energy levels of the interacting nonequilibrium system in a stationary state.[23]

The solutions of the Dyson equations for statistical functions,

$$F = F_0(1 + \Sigma^a G^a) + G_0^r(\Sigma^r F + \Omega G^a) \,, \tag{20.21}$$

in the stationary case can be found (Keldysh, 1964) using the conditions $G_0^{-1} F_0 = 0$,

$$F(\varepsilon) = G^r(\varepsilon)\Omega(\varepsilon)G^a(\varepsilon) \,, \quad S(\omega) = D^r(\omega)\Delta(\omega)D^a(\omega) \,. \tag{20.22}$$

Although the triangular representation is more convenient for the retarded/advanced functions, the kinetic equation is simpler in the $\pm$ representation (Keldysh, 1964). For example, for the adsorbate occupation number we have

$$\frac{\partial N_a}{\partial t} = \int \frac{d\varepsilon}{2\pi} \left[\Sigma_a^{+-}(\varepsilon)G_a^{-+}(\varepsilon) - \Sigma_a^{-+}(\varepsilon)G_a^{+-}(\varepsilon) \right] \,, \tag{20.23}$$

and $\partial N_a / \partial t = 0$ in the stationary case. The same for the vibrational population number:

$$\frac{\partial N_{\text{ph}}}{\partial t} = \int \frac{d\omega}{2\pi} \left[\Pi^{+-}(\varepsilon)D^{-+}(\varepsilon) - \Pi^{-+}(\varepsilon)D^{+-}(\varepsilon) \right] = 0 \,. \tag{20.24}$$

The solutions for the Keldysh–Green's functions $G^{\pm\mp}$, $D^{\pm\mp}$ can be written analogously to equation (20.22) as

$$G^{\pm\mp} = G^r \Sigma^{\pm\mp} G^a, \ D^{\pm\mp} = D^r \Pi^{\pm\mp} D^a \,. \tag{20.25}$$

We will use the so-called τ-approximation (Keldysh and Tikhodeev, 1986) of slowly varied self-energies and replace the solutions of equations (20.22) and (20.25) with

$$\sum_k F_k(\varepsilon) = -2i\pi \left[1 - 2n_s(\varepsilon) \right] \rho_s(\varepsilon) \,,$$

---

[23]In what follows, we use the single-pole description for the vibrational Green's functions and double the number of diagrams with each phonon line, $D_{\alpha,\alpha'}(\omega) \rightarrow D_{\alpha,\alpha'}(\omega) + D_{\alpha',\alpha}(-\omega)$ to take into account the fact that the phonon field is real.

Fig. 20.9. The diagrams for the adsorbate self-energy. There are two adsorbate-phonon diagrams with oppositely directed phonon propagators, because we use the single-pole approximation for phonons.

$$\sum_p F_p(\varepsilon) = -2i\pi \left[1 - 2n_t(\varepsilon)\right] \rho_t(\varepsilon)\,,$$

$$F_a(\varepsilon) = -2i\pi \left[1 - 2n_a(\varepsilon)\right] \rho_a(\varepsilon)\,,$$

$$\sum_k G_k^{+-}(\varepsilon) = 2i\pi n_s(\varepsilon)\rho_s(\varepsilon)\,,$$

$$\sum_p G_p^{-+}(\varepsilon) = -2i\pi \left[1 - n_t(\varepsilon)\right] \rho_t(\varepsilon)\,,$$

$$G_a^{+-}(\varepsilon) = 2i\pi n_a(\varepsilon)\rho_a(\varepsilon)\,,$$

$$G_a^{-+}(\varepsilon) = -2i\pi \left[1 - n_a(\varepsilon)\right] \rho_a(\varepsilon)\,,$$

$$S(\omega) = -2i\pi \left[1 + 2n_{\rm ph}(\omega)\right] \rho_{\rm ph}(\omega)\,,$$

$$D^{+-}(\omega) = -2i\pi n_{\rm ph}(\omega)\rho_{\rm ph}(\omega)\,,$$

$$D^{-+}(\omega) = -2i\pi \left[1 + n_{\rm ph}(\omega)\right] \rho_{\rm ph}(\omega)\,, \tag{20.26}$$

where the occupation functions n_i, $i = a, s, t, \mathrm{ph}$ are connected with the corresponding occupation numbers via $N_i = \int (d\varepsilon/2\pi)n_i(\varepsilon)\rho_i(\varepsilon)$.

20.3   Calculation of adsorbate and vibrational densities and tunneling current

Now we are well equipped to solve the inelastic tunneling problem. We will restrict ourselves to the second order over the adsorbate–phonon interaction, and suppose that the occupation densities in the substrate and tip are partially thermalized in the sense that both are Fermi distributions,

$$n_s(\varepsilon) = n_s^{\rm F}(\varepsilon) = \left[\exp(\varepsilon - \mu_s) + 1\right]^{-1}\,, \quad n_t(\varepsilon) = n_t^{\rm F}(\varepsilon) = \left[\exp(\varepsilon - \mu_t) + 1\right]^{-1}\,, \tag{20.27}$$

with difference $\mu_s - \mu_t = eV$ fixed by the external bias voltage between the tip and substrate. We will also work in a wide band approximation of slowly changing tip and substrate densities of states, replacing them by some constants.

20.3.1   *Adsorbate retarded Green's function*

We start with the adsorbate retarded Green's function, equation (20.20), and take into account the diagrams for self-energy Σ_a shown in Fig. 20.9. The first diagram (substrate–adsorbate tunneling) gives

$$\Sigma_{as}^r(\varepsilon) = U^2 \sum_k G_k^r(\varepsilon)\,, \tag{20.28}$$

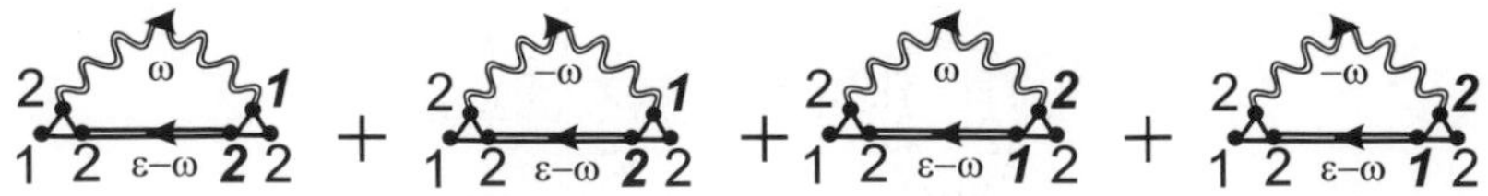

FIG. 20.10. Four electron–phonon diagrams for retarded adsorbate self-energy $\overline{\Sigma}_{12} \equiv \Sigma^r(\varepsilon)$. As follows from Fig. 20.8, the only freedom which is left for choosing the Keldysh indices of the triangular representation is replacing $1 \leftrightarrows 2$ (emphasized). Vertices with all "1" are absent because otherwise $\overline{G}_{11} = 0$ or $\overline{D}_{11} = 0$ appear in the diagrams.

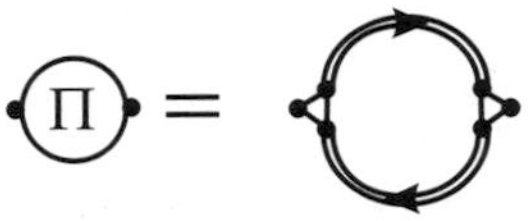

FIG. 20.11. One-loop diagrams for the vibrational mode self-energy (polarization operator) Π.

and neglecting the renormalization of the adsorbate energy due to tunneling,

$$\operatorname{Im} \Sigma_{as}^r = -\pi U^2 \rho_s(\varepsilon) \equiv -\Gamma_s . \tag{20.29}$$

Analogously, the tip–adsorbate tunneling (the second diagram in Fig. 20.9) gives

$$\operatorname{Im} \Sigma_{at}^r = -\pi U'^2 \rho_t(\varepsilon) \equiv -\Gamma_t . \tag{20.30}$$

In the wide band approximation both Γ_s, Γ_t are constants which we will treat as fitting parameters; $\Gamma_s \gg \Gamma_t$ due to the stronger coupling between the adsorbate and substrate.

As to the adsorbate–phonon diagrams in the second order of perturbation theory, only four terms are nonzero, as explained in Fig. 20.10. Using equation (20.26), we obtain for the adsorbate level shift and broadening:

$$\Delta_{a,\mathrm{ph}}(\varepsilon) \equiv \operatorname{Re} \Sigma_{a,\mathrm{ph}}^r = \frac{\chi^2}{2} \int d\omega \left\{ [P_{\mathrm{ph}}(\omega) + P_{\mathrm{ph}}(-\omega)] \left[1 - 2n_a(\varepsilon - \omega)\right] \rho_a(\varepsilon - \omega) \right.$$
$$\left. + \rho_{\mathrm{ph}}(\omega) \left[1 + 2n_{\mathrm{ph}}(\omega)\right] [P_a(\varepsilon + \omega) + P_a(\varepsilon - \omega)] \right\} , \tag{20.31}$$

$$\Gamma_{a,\mathrm{ph}}(\varepsilon) \equiv -\operatorname{Im} \Sigma_{a,\mathrm{ph}}^r = \pi \chi^2 \int d\omega \rho_{\mathrm{ph}}(\omega) \left\{ n_{\mathrm{ph}}(\omega) \left[\rho_a(\varepsilon - \omega) + \rho_a(\varepsilon + \omega)\right] \right.$$
$$\left. + \left[1 - n_a(\varepsilon - \omega)\right] \rho_a(\varepsilon - \omega) + n_a(\varepsilon + \omega)\rho_a(\varepsilon + \omega) \right\} , \tag{20.32}$$

where $P_{\mathrm{ph}} = \operatorname{Re} D^r, \rho_{\mathrm{ph}} = -1/\pi \operatorname{Im} D^r$, $P_a = \operatorname{Re} G_a^r$, $\rho_a = -1/\pi \operatorname{Im} G_a^r$.

20.3.2   *Vibrational retarded Green's function*

The one loop diagrams for vibrational polarization operator are shown in Fig. 20.11, in the same second order over χ; the nonzero graphs are explained

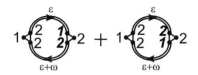

FIG. 20.12. Two diagrams for retarded vibrational mode self-energy $\overline{\Pi}_{12} \equiv \Pi^r(\varepsilon)$. As follows from Fig. 20.8, the only freedom which is left for choosing the Keldysh indices of the triangular representation is replacing $1 \leftrightarrows 2$ (emphasized). Vertices with all "1" are absent because otherwise $\overline{G}_{11} = 0$ appear in the diagrams.

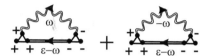

FIG. 20.13. Only two diagrams for the adsorbate self-energy $\Sigma^{+-}_{a,\text{ph}}$ are nonzero, as follows from Fig. 20.6(g). However, these diagrams, together with the corresponding term $\Sigma^{-+}_{a,\text{ph}}$ produce a zero impact into collision integral in kinetic equation (20.23), $\Sigma^{+-}_{a,\text{ph}} G_a^{-+} - \Sigma^{-+}_{a,\text{ph}} G_a^{+-} = 0$.

in Fig. 20.12. Then the vibration energy shift and broadening due to excitation of virtual electron–hole pairs are

$$\Delta_{eh}(\omega) \equiv \operatorname{Re} \Pi^r_{\text{ph}} = -\frac{\chi^2}{2} \int d\varepsilon \, P_a(\varepsilon)$$
$$\times \{[(1 - 2n_a(\varepsilon - \omega)) \, \rho_a(\varepsilon - \omega) + (1 - 2n_a(\varepsilon + \omega)) \, \rho_a(\varepsilon + \omega)]\} , \qquad (20.33)$$
$$\gamma_{eh}(\omega) \equiv -\operatorname{Im} \Pi^r_{\text{ph}} = \pi\chi^2 \int d\varepsilon \rho_a(\varepsilon) \rho_a(\varepsilon + \omega) \left[n_a(\varepsilon) - n_a(\varepsilon + \omega)\right] . \qquad (20.34)$$

20.3.3 *Adsorbate occupation function*

It can be easily shown that for the Hamiltonian, equation (20.1), only tunneling terms enter the kinetic equation (20.23) for the adsorbate occupation number (two left diagrams in Fig. 20.9), and the impact of electron-vibrational terms vanishes (see in Fig. 20.13). Then,

$$0 = \frac{\partial N_a}{\partial t} = 2 \int d\varepsilon \left[(\Gamma_s n^{\text{F}}_s(\varepsilon) + \Gamma_t n^{\text{F}}_t(\varepsilon)) - (\Gamma_s + \Gamma_t) \, n_a(\varepsilon) \right] \rho_a(\varepsilon). \qquad (20.35)$$

The adsorbate occupation function then takes a simple balance form:

$$n_a(\varepsilon) = \frac{\Gamma_s n^{\text{F}}_s(\varepsilon) + \Gamma_t n^{\text{F}}_t(\varepsilon)}{\Gamma_s + \Gamma_t} . \qquad (20.36)$$

20.3.4 *Vibrational occupation function*

Accounting for one loop diagrams of the type shown in Fig. 20.14, the kinetic eqn (20.24) for the vibrational occupation number takes the form:

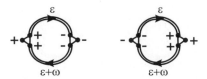

FIG. 20.14. The only nonzero one-loop diagrams for the vibrational mode self-energies $\Pi^{+-}(\varepsilon)$ (left) and $\Pi^{-+}(\varepsilon)$ (right), as follows from Fig. 20.6(g).

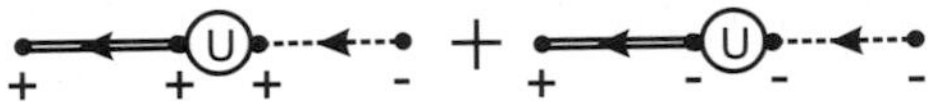

FIG. 20.15. Two diagrams for the substrate–adsorbate correlation function G_{ka}^{+-}.

$$\frac{\partial N_{\mathrm{ph}}}{\partial t} = 2\pi\chi^2 \int d\varepsilon d\omega \rho_a(\varepsilon)\rho_a(\varepsilon+\omega)\rho_{\mathrm{ph}}(\omega)\left[1-n_a(\varepsilon)\right]n_a(\varepsilon+\omega)$$
$$- \int d\omega \gamma_{eh}(\omega)n_{\mathrm{ph}}(\omega)\rho_{\mathrm{ph}}(\omega)\,, \tag{20.37}$$

with damping rate γ_{eh} due to electron–hole pair excitation, equation (20.34). Adding and subtracting integral term $\int d\omega \gamma_{eh}(\omega)n_T(\omega)\rho_{\mathrm{ph}}(\omega)$ to the kinetic equation (20.37), it can be transferred to a more transparent form, showing that in the absence of the applied voltage the stationary phonon distribution function is the equilibrium one, $n_T(\omega)=(e^{\omega/k_{\mathrm{B}}T}-1)^{-1}$,

$$0 = \frac{\partial N_{\mathrm{ph}}}{\partial t} = \int d\omega \rho_{\mathrm{ph}}(\omega)\left\{\Gamma_{\mathrm{in}}(\omega)-\gamma_{eh}(\omega)\left[n_{\mathrm{ph}}(\omega)-n_T(\omega)\right]\right\}\,, \tag{20.38}$$

or

$$n_{\mathrm{ph}}(\omega) = n_T(\omega) + \frac{\Gamma_{\mathrm{in}}(\omega)}{\gamma_{eh}}\,. \tag{20.39}$$

Here

$$\Gamma_{\mathrm{in}}(\omega) = 2\pi\chi^2\frac{\Gamma_s\Gamma_t}{(\Gamma_s+\Gamma_t)^2}\int d\varepsilon \rho_a(\varepsilon)\rho_a(\varepsilon+\omega)$$
$$\times \left[n_s^{\mathrm{F}}(\varepsilon+\omega)-n_t^{\mathrm{F}}(\varepsilon+\omega)\right]\left[n_s^{\mathrm{F}}(\varepsilon)-n_t^{\mathrm{F}}(\varepsilon)\right]\,, \tag{20.40}$$

is the vibrational generation rate function.

20.3.5 *Tunneling current*

Now we have everything to calculate the tunneling current, equation (20.4). In the same approximation as before, there are only two terms for the substrate–adsorbate correlation function G_{ka}^{+-} shown in Fig. 20.15 (and corresponding terms for G_{ka}^{-+}). Thus, in the time representation,

$$G_{ka}^{+-}(t,t) = U\sum_k \int_{-\infty}^{\infty} dt'\left[G_k^{++}(t,t')G_a^{+-}(t',t)-G_k^{+-}(t,t')G_a^{--}(t',t)\right]$$

$$= U \sum_k \int_{-\infty}^t dt' \left[G_k^{-+}(t,t') G_a^{+-}(t',t) - G_k^{+-}(t,t') G_a^{-+}(t',t) \right] ; \qquad (20.41)$$

the latter transformation is due to equation (20.10).
Calculating the corresponding terms for G_{ka}^{+-} and making the Wigner transformation equation (20.14), we get instead of equation (20.4)

$$\frac{I}{e} = U^2 \sum_k \int \frac{d\varepsilon}{2\pi} \left[G_k^{-+}(\varepsilon) G_a^{+-}(\varepsilon) - G_k^{+-}(\varepsilon) G_a^{-+}(\varepsilon) \right] . \qquad (20.42)$$

Substituting the solutions given above, we arrive finally at a very transparent formula for the tunneling current

$$\frac{I}{e} = 2 \int d\varepsilon \, \frac{\Gamma_s \Gamma_t}{(\Gamma_s + \Gamma_t)^2} \left[n_t^F(\varepsilon) - n_s^F(\varepsilon) \right] \rho_a(\varepsilon) . \qquad (20.43)$$

20.4   Discussion and some numerical results

The total tunneling current given by equation (20.43) is controlled by the bias voltage mostly through $n_t^F(\varepsilon) - n_s^F(\varepsilon)$, its magnitude depends mostly on the substrate–adsorbate and tip–adsorbate tunneling coupling constants $\Gamma_{s(t)} = \pi \sum_{k(p)} |U_{k(p)a}|^2 \, \delta(\varepsilon - \varepsilon_{k(p)})$ (see equations 20.16, 20.29 and 20.30). Inelastic effects are manifesting themselves through the adsorbate density of states $\rho_a = -(1/\pi) \text{Im} \, G_a^r$, and the adsorbate retarded Green's function G_a^r can be written as

$$G_a^r(\varepsilon) = \frac{1}{\varepsilon - \varepsilon_a - i\Gamma_{st} - \Sigma_{a,\text{ph}}^r(\varepsilon)} , \qquad (20.44)$$

$\Gamma_{st} = \Gamma_s + \Gamma_t$. The real and the imaginary part of the self-energy $\Sigma_{a,\text{ph}}^r(\varepsilon)$ are calculated to the second-order of χ, see equation (20.31), with the replacement (to start the iterative calculations) $\rho_a \to \rho_a^{(0)}$, where the unperturbed adsorbate density of states is

$$\rho_a^{(0)}(\varepsilon) = \frac{1}{\pi} \frac{\Gamma_{st}}{(\varepsilon - \varepsilon_a)^2 + \Gamma_{st}^2} . \qquad (20.45)$$

The vibrational density of states is

$$\rho_{\text{ph}}(\omega) = \frac{1}{\pi} \frac{\gamma_{eh}(\omega)}{(\omega - \Omega - \Delta_{eh}(\omega))^2 + \gamma_{eh}^2(\omega)} \qquad (20.46)$$

with the vibrational energy shift Δ_{eh}, equation (20.33), and damping rate γ_{eh}, equation (20.34), due to electron–hole excitation

$$\gamma_{eh}(\omega) = \pi \chi^2 \int d\varepsilon \left[n_a(\varepsilon) - n_a(\varepsilon + \omega) \right] \rho_a^{(0)}(\varepsilon) \rho_a^{(0)}(\varepsilon + \omega)$$

$$\approx 2\pi\chi^2 \frac{\Gamma_s \left[\rho_a^{(0)}(\mu_s)\right]^2 + \Gamma_t \left[\rho_a^{(0)}(\mu_t)\right]^2}{\Gamma_{st}} \, \omega \,. \tag{20.47}$$

The latter approximate equation (20.47) is valid in the limit of low-temperature, $k_B T \ll \Omega$ and slowly-varying $\rho_a^{(0)}(\varepsilon)$ over eV. Furthermore, the effective occupation function for adsorbate electrons n_a is given by equation (20.36), and the effective occupation function for phonons is

$$n_{\mathrm{ph}}(\omega) = \frac{1}{e^{\omega/k_B T} - 1} + \frac{\pi\chi^2}{\gamma_{eh}} \frac{\Gamma_s \Gamma_t}{\Gamma_{st}^2} \int d\varepsilon \rho_a^{(0)}(\varepsilon + \omega)\rho_a^{(0)}(\varepsilon)$$
$$\times [n_s(\varepsilon + \omega) - n_t(\varepsilon + \omega)][n_s(\varepsilon) - n_t(\varepsilon)] \,, \tag{20.48}$$

see equations (20.39)–(20.40). It includes the vibrational heating by electron–vibration coupling (Mii *et al.*, 2002; Tikhodeev and Ueba, 2004), the second term in equation (20.48). The important thing about this term is that both the denominator and nominator are proportional to χ^2, and, as a result, the deviation of vibrational population from the equilibrium one, $n_T(\omega)$, is proportional to χ^0,

$$n_{\mathrm{ph}}(\omega) = n_T(\omega) + \frac{\Gamma_s \Gamma_t}{\Gamma_{st}}$$
$$\times \frac{\int d\varepsilon \rho_a^{(0)}(\varepsilon + \omega)\rho_a^{(0)}(\varepsilon)[n_s(\varepsilon + \omega) - n_t(\varepsilon + \omega)][n_s(\varepsilon) - n_t(\varepsilon)]}{\int d\varepsilon \rho_a^{(0)}(\varepsilon + \omega)\rho_a^{(0)}(\varepsilon)[\Gamma_s (n_s(\varepsilon) - n_s(\varepsilon + \omega)) + \Gamma_t (n_t(\varepsilon) - n_t(\varepsilon + \omega))]}$$
$$\approx n_T(\omega) + \frac{\Gamma_s \Gamma_t}{\Gamma_{st}^2} \frac{|eV| - \Omega}{\Omega} \theta(|eV| - \Omega) \,. \tag{20.49}$$

The latter approximate equation is written in the low-temperature limit $k_B T \ll \Omega$ and neglecting the vibrational broadening. The effective vibrational temperature T_{eff} can be estimated from the condition $n_{\mathrm{ph}}(\Omega) = [\exp(\Omega/k_B T_{\mathrm{eff}}) - 1]^{-1}$. The overheating develops in the zeroth perturbation order over χ and does not depend on the vibrational properties at all. The overheating effects may become strong, what is very important for understanding the mechanisms of single molecule manipulation by STM [see the detailed discussion in (Tikhodeev and Ueba, 2004)].

Neglecting temporarily the adsorbate energy renormalization term, and since $\Gamma_{a,\mathrm{ph}}(\varepsilon)$ is proportional to χ^2 and small compared to Γ_{st}, the total tunneling current equation (20.43) can be decomposed into terms of the zero and the second order of χ (Ueba *et al.*, submitted):

$$I_{\mathrm{el}}^{(0)} = 2e \frac{\Gamma_s \Gamma_t}{\Gamma_{st}} \int d\varepsilon \rho_a^{(0)}(\varepsilon)[n_s^{\mathrm{F}}(\varepsilon) - n_t^{\mathrm{F}}(\varepsilon)] \,, \tag{20.50}$$

and

$$I^{(2)} = I_{\mathrm{el}}^{(2)} + I_{\mathrm{in}}^{(2)} \,, \tag{20.51}$$

where

$$I_{el}^{(2)} = (-2\pi)2e\frac{\Gamma_s\Gamma_t}{\Gamma_{st}}\int d\varepsilon[\rho_a^{(0)}(\varepsilon)]^2\Gamma_{a,ph}(\varepsilon)([n_s^F(\varepsilon) - n_t^F(\varepsilon)] \qquad (20.52)$$

and

$$I_{in}^{(2)} = \frac{2e}{\hbar}\frac{\Gamma_s\Gamma_t}{\Gamma_{st}^2}\int d\varepsilon\rho_a^{(0)}(\varepsilon)\Gamma_{a,ph}(\varepsilon)[n_s^F(\varepsilon) - n_t^F(\varepsilon)] \qquad (20.53)$$

describe the elastic and inelastic corrections to $I_{el}^{(0)}$ due to the opening of the electron–vibration scattering, respectively. Within this decomposition, the elastic contribution due to the back scattering induces a transition into an intermediate virtual state resulting in a renormalization of the transmission (Caroli *et al.*, 1972), while the inelastic contribution corresponds to the real emission or absorption of a vibrational mode. The inelastic contribution is positive, while the elastic one always gives a reduction of the total current. As a result of this competition the sign of the correction,

$$I^{(2)} = 2e\frac{\Gamma_s\Gamma_t}{\Gamma_{st}^2}\int d\varepsilon\rho_a^{(0)}(\varepsilon)\mathcal{R}(\varepsilon)\Gamma_{a,ph}(\varepsilon)[n_s^F(\varepsilon) - n_t^F(\varepsilon)], \qquad (20.54)$$

is determined by

$$\mathcal{R}(\varepsilon) = 1 - 2\pi\Gamma_{st}\rho_a^{(0)}(\varepsilon) = 1 - \frac{\Gamma_{st}^2}{2\Gamma_s\Gamma_t}\mathcal{T}(\varepsilon), \qquad (20.55)$$

where

$$\mathcal{T}(\varepsilon) = 4\pi\frac{\Gamma_s\Gamma_t}{\Gamma_{st}}\rho_a^{(0)}(\varepsilon) \qquad (20.56)$$

is the transmission probability (Hyldgaard *et al.*, 1994).

For ε_a close to the Fermi levels, i.e., $|\mu_{t/s} - \varepsilon_a| \ll \Gamma_{st}$ and $\mathcal{R}(\varepsilon) \simeq -1$, the negative elastic correction $I_{el}^{(2)}$ becomes two times larger than the positive inelastic correction $I_{in}^{(2)}$ (Persson and Baratoff, 1987). In this case we obtain $I^{(2)} \simeq -I_{in}^{(2)} \simeq I_{el}^{(2)}/2$. In the opposite limit $|\mu_{t/s} - \varepsilon_a| \gg \Gamma_{st}$ and $\mathcal{R}(\varepsilon) \simeq 1$, the inelastic currents dominate over the elastic correction; $I_{in}^{(2)} \gg |I_{el}^{(2)}|$ so that $I^{(2)} \simeq I_{in}^{(2)}$. It is also noted here that in the case of atomic- or molecular-wire junctions with symmetric contacts to the electrodes with $\Gamma_s = \Gamma_t$, equation (20.55) reduces to $\mathcal{R}(\varepsilon) = 1 - 2\mathcal{T}(\varepsilon_F)$ in agreement with that derived by Paulsson *et al.* (2005) and de la Vega *et al.* (2006) for the universal features of electron–phonon coupling in atomic wires. They demonstrated that the conductance increases due to phonon scattering for low transmission ($\mathcal{T} < 1/2$), while it decreases for high transmission ($\mathcal{T} > 1/2$). For an atomic gold wire with $\mathcal{T} = 1$, any scattering will decrease the conductance since it backscatters electrons. De la Vega *et al.* (2006) showed that if $\mathcal{T} \to 1$ ($\mathcal{T} \to 0$), the inelastic transitions correspond mainly to back scattering (forward scattering).

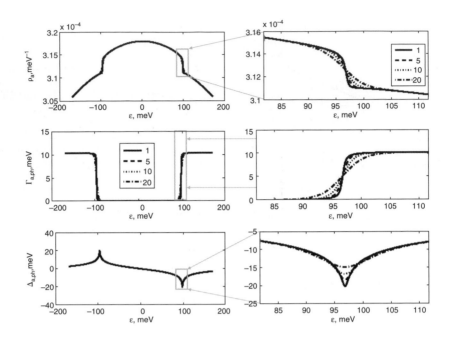

FIG. 20.16. Calculated energy dependencies of the adsorbate density of states (top panels), electron-vibrational damping (middle panels) and energy shift (bottom panels) for the adsorbate level in a resonance with the substrate Fermi level $\varepsilon_a = \mu_s$ (see values of other parameters in the text). The right-hand side panels show the magnified region around opening of the inelastic tunneling at positive bias voltage. Different curves (explained in the legend) demonstrate the role of increasing temperature $T = 1, 5, 10, 20$ K.

The evolution of the conductance as a function of $\mathcal{T}$ shows a crossover from a decrease to an increase at $\mathcal{T} = 1/2$.

For low temperatures, we obtain from equation (20.54),

$$\frac{dI^{(2)}}{dV} = \frac{2e^2}{\hbar} \frac{\Gamma_s \Gamma_t}{\Gamma_{st}^2} \rho_a^{(0)}(\mu_s) \mathcal{R}(\mu_s) \Gamma_{a,\mathrm{ph}}(\mu_s) , \qquad (20.57)$$

where, from equation (20.32) for $eV > 0$

$$\Gamma_{a,\mathrm{ph}}(\mu_s) = \pi \chi^2 \frac{\Gamma_t}{\Gamma_{st}} \int_0^{eV} d\omega \rho_{\mathrm{ph}}(\omega) \rho_a^{(0)}(\mu_s - \omega) . \qquad (20.58)$$

Since $\rho_{\mathrm{ph}}(\omega)$ is peaked at $\omega = \Omega$ (neglecting the vibrational energy renormalization equation 20.33) and $\rho_a^{(0)}(\mu_s - \omega)$ is a slowly varying function over $0 \leq \omega \leq eV$, we obtain finally

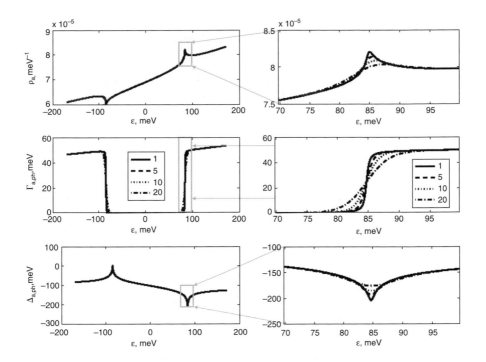

FIG. 20.17. Same as in Fig. 20.16, calculated for intermediate detuning between the adsorbate level and a substrate Fermi level $\varepsilon_a - \mu_s = 1$ eV (see values of other parameters in the text).

$$\frac{d^2 I^{(2)}}{dV^2} = \frac{2\pi e^3}{\hbar} \frac{\Gamma_s \Gamma_t^2}{\Gamma_{st}^3} \chi^2 \rho_a^{(0)}(\mu_s) \rho_a^{(0)}(\mu_s + eV) \mathcal{R}(\mu_s) \rho_{\mathrm{ph}}(eV) . \qquad (20.59)$$

This reveals that the overall $d^2 I/dV^2$-lineshape is determined by the vibrational density of states $\rho_{\mathrm{ph}}(eV)$. However, as was shown in (Mii *et al.*, 2003), the self-energy shift $\Delta_{a,\mathrm{ph}}(\varepsilon)$, equation (20.31), also plays an important role in the spectral features including the skewed, asymmetric or derivative lineshape directly calculated from equation (20.43); see the detailed analysis in (Ueba *et al.*, submitted) and the numerical illustration below.

In what follows we illustrate numerically the results obtained above for the tunneling current, for different temperatures. We compare here the situations when the adsorbate level is close to the substrate Fermi level, $\varepsilon_a = \mu_s$ (Fig. 20.16) and has an intermediate detuning $\varepsilon_a - \mu_s = 1$ eV (Fig. 20.17). The other parameters in Fig. 20.16 are $\Gamma_s = 1.0$ eV, $\Gamma_t = 1$ meV, $\Omega = 100$ meV, $\chi = 100$ meV. In Fig. 20.17, Γ_s, Γ_t, Ω are the same and $\chi = 500$ meV. These sets of parameters give the vibrational lifetime of about $\tau_{eh} = 1/\gamma_{eh} = 1$ ps. The temperatures are $T = 1, 5, 10, 20$ K. Figures 20.16 and 20.17 show

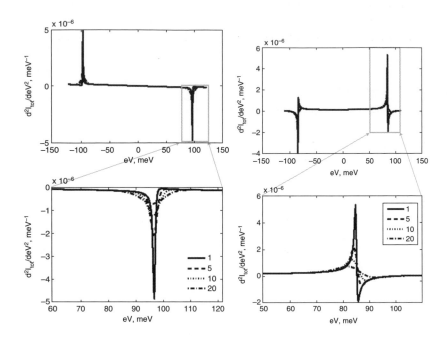

FIG. 20.18. Second derivatives d^2I/dV^2 of the total tunneling current equation (20.43) for parameters from Fig. 20.16 (left-hand side panels) and Fig. 20.17 (right-hand side panels).

the energy dependencies of the calculated adsorbate densities of states (upper panels), adsorbate-vibrational dampings (middle panels) and energy shifts (bottom panels). The corresponding energy spectra of the second derivatives d^2I/dV^2 are compared in Fig. 20.18. The left-hand side panels show the case $\varepsilon_a = \mu_s$, and the right-hand side panels are for $\varepsilon_a - \mu_s = 1$ eV.

In agreement with the analysis in the text, in the resonance case the overwhelming elastic contribution causes a decrease of the adsorbate electron density of states with the opening of the inelastic channel (top panel in Fig. 20.16), which is reflected in the conductance decrease and a negative dip in the IETS spectra (left-hand side panels in Fig. 20.18). With the increase of the adsorbate level detuning the inelastic term plays an increasing role, and the IETS spectrum modifies through a derivative-like one at intermediate detuning $\varepsilon - \mu_s \sim \Gamma_{st}$ (right panels in Fig. 20.18), to positive peaks at very large detuning $\varepsilon - \mu_s \gg \Gamma_{st}$ (not shown).

To conclude, the nonequilibrium Keldysh–Green's function method provides a theoretical basis for inelastic electron tunneling spectroscopy and manipulation of single adsorbed molecules. It also describes the nonequilibrium electron transport in nanodevices based on atomic or molecular wires. In particular, it becomes possible to understand how the conductance, the effective temperature and reaction coordinate of a single adsorbate or a molecular wire change upon opening of the inelastic channels, via the effect of the electron–phonon (vibration) interaction.

Acknowledgments

The authors are grateful to Professor Leonid Veniaminovich Keldysh for many enlightening discussions. His works have influenced and inspired the authors' whole scientific lives. It is a great honor for us to contribute to a volume dedicated to his 75th birthday.

References

Agraït, N., Untiedt, C., Rubio-Bollinger, G. and Vieira, S. (2002). *Phys. Rev. Lett.* **88**, 216803.

Arseyev, P. I. and Maslova, N. S. (1992). *Zh. Eksp. Teor. Fiz.* **102**, 1056 [*Sov. Phys. JETP* **75**, 575].

Arseyev, P. I. and Maslova, N. S. (1998). *Solid State Commun.* **108**, 717.

Arseyev, P. I. and Maslova, N. S. (2005). *Pisma Zh. Eksp. Teor. Fiz.* **82**, 331 [*JETP Lett.* **82**, 297].

Arseyev, P. I. and Maslova, N. S. (2006). *Pisma Zh. Eksp. Teor. Fiz.* **84**, 99 [*JETP Lett.* **84**, 93].

Asai, Y. (2004). *Phys. Rev. Lett.* **93**, 246102 [Erratum: (2005) *Phys. Rev. Lett.* **94**, 099901].

Backus, E. H. G., Eichler, A., Kleyn, A. W. and Bonn, M. (2005). *Science* **310**, 1790.

Binnig, G., Garcia, N. and Rohrer, H. (1985). *Phys. Rev. B* **32**, 1336.

Brandbyge, M. and Hedegard, P. (1994). *Phys. Rev. Lett.* **72**,2919.

Caroli, C., Combescot, R., Lederer, D., Nizieres, P. and Saint-James, D. (1971*a*). *J. Phys.* C **4**, 2598.

Caroli, C., Combescot, R., Nozieres, P. and Saint-James, D. (1971*b*). *J. Phys.* C **4**, 916.

Caroli, C., Combescot, R., Nozieres, P. and Saint-James, D. (1972). *J. Phys.* C **5**, 21.

Combescot, R. (1971). *J. Phys.* C **4**, 2611.

Datta, S. (1995). *Electronic Transport in Mesoscopic Systems*. Cambridge University Press, Cambridge.

de la Vega, L., Martín-Rodero, A., Agraït, N. and Levy Yeyati, A. (2006). *Phys. Rev. B* **73**, 075428.

Dujardin, G., Walkup, R. E. and Avouris, Ph. (1992). *Science* **255**, 1232.

Eigler, D. M., Lutz, C. P. and Rudge, W. E. (1991). *Nature* **352**, 600.

Fanchenko, S. S. (1983). *Teor. Mat. Fiz.* **55**, 137 [*Theor. Math. Phys.* **55**, 406].

Gadzuk, J. W. (1991). *Phys. Rev. B* **44**, 13466.

Galperin, M., Ratner, M. A. and Nitzan, A. (2004a). *J. Chem. Phys.* **121**, 11965.

Galperin, M., Ratner, M. A. and Nitzan, A. (2004b). *Nano. Lett.*. **4**, 1605.

Gao, S., Persson, M. and Lundqvist, B. I. (1992). *Solid State Commun.* **84**, 271.

Gao, S. W. (1994). *Surf. Sci.* **313**, 448.

Gao, S. W., Persson, M. and Lundqvist, B. I. (1997). *Phys. Rev.* B **55**, 4825.

Hahn, J. R., Lee, H. J. and Ho, W. (2000). *Phys. Rev. Lett.* **85**, 1914.

Hall, A. G. (1975a). *J. Phys.* A **8**, 214.

Hall, A. G. (1975b). *Physica* A **80**, 369.

Hla, S.-W., Bartels, L., Meyer, G. and Rieder, K.-H. (2000). *Phys. Rev. Lett.* **85**, 2777.

Ho, W. (2002). *J. Chem. Phys.* **117**, 11033.

Hyldgaard, P., Hershfield, S., Davies, J.H. and Wilkins, J.W. (1994). *Ann. of Physics* **236**, 1.

Kadanoff, L. P. and Baym, G. (1962). *Quantum Statistical Mechanics*. Benjamin, New York.

Keldysh, L. V. (1964). *Zh. Eksp. Teor. Fiz.* **47**, 1515 [(1965) *Sov.Phys. JETP* **20**, 1018].

Keldysh, L. V. and Tikhodeev, S. G. (1986). *Zh. Eksp. Teor. Fiz.* **90**, 1852 [*Sov.Phys. JETP* **63**, 1086].

Keldysh, L. V. (2003). Real-time nonequilibrium Green's functions. In *Progress in Noneequilibrium Green's Functions II* (ed. M. Bonitz and D. Semkat). World Scientific, Singapore.

Kim, Y., Komeda, T. and Kawai, M. (2002). *Phys. Rev. Lett.* **89**, 126104.

Komeda, T., Kim, Y., Kawai, M., Persson, B. N. J. and Ueba, H. (2002). *Science* **295**, 2055.

Kukharenko, Yu. A. and Tikhodeev, S. G. (1982). *Zh. Eksp. Teor. Fiz.* **83**, 1444 [*Sov. Phys. JETP* **56**, 831].

Kushmerick, J. G., Lazorcik, J., Patterson, C. H. Shashidhar, R., Seferos, D. S. and Bazan, G. C. (2004). *Nano. Lett.*. **4**, 639.

Lifshitz, E. M. and Pitaevskii, L. P. (1981). *Physical kinetics*. Pergamon Press, Oxford, New York.

Lorente, N. and Persson, M. (2000). *Phys. Rev. Lett.* **85**, 2997.

Lorente, N., Persson, M., Lauhon, L. J. and Ho, W. (2001). *Phys. Rev. Lett.* **86**, 2593.

Lorente, N. (2004). *Appl. Phys.* A **78**,799.

Lorente, N., Rurali, R. and Tang, H. (2005). *J. Phys.* C **17**, S1049.

Mii, T., Tikhodeev, S. and Ueba, H. (2002). *Surf. Sci.* **502**, 26.

Mii, T., Tikhodeev, S. G. and Ueba, H. (2003). *Phys. Rev.* B **68**, 205406.

Moresco, F., Meyer, G. and Rieder, K. H. (1999). *Mod. Phys. Lett.* B **13**, 709.

Newns, D. M. (1969). *Phys. Rev.* **178**, 1123.

Pascual, J. I., Jackiw, J. J., Song, Z., Weiss, P. S., Conrad, H. and Rust,

H.-P. (2001). *Phys. Rev. Lett.* **86**, 1050.

Pascual, J. I., Lorente, N., Song, Z., Conrad, H. and Rust, H.-P. (2003). *Nature* **423**, 525.

Paulsson, M., Frederiksen, T. and Brandbyge, M. (2005). *Phys. Rev. B* **72**, 201101(R).

Persson, B. N. J. (1987). *Chem. Phys. Lett.* **139**, 457.

Persson, B. N. J. and Baratoff, A. (1987). *Phys. Rev. Lett.* **59**, 339.

Salam, G. P., Persson, M. and Palmer, R. E. (1994). *Phys. Rev. B* **49**, 10655.

Schwinger, J. (1961). *J. Math. Phys.* **2**, 407.

Shen, T. C., Wang, C., Abeln, G. C., Tucker, J. R., Lyding, J. W., Avouris, P. and Walkup, R. E. (1995). *Science* **268**, 1590.

Smit, R. H. M., Noat, Y., Untiedt, C., Lang, N. D., van Hemert, M. C. and van Ruitenbeek, J. M. (2002). *Nature* **419**, 906.

Stipe, B. C., Rezaei, M. A., Ho, W., Gao, S., Persson, M. and Lundqvist, B. I. (1997). *Phys. Rev. Lett.* **78**, 4410.

Stipe, B. C., Rezaei, M. A. and Ho, W. (1998). *Science* **280**, 1732.

Tikhodeev, S. G. (1982). *Dokl. Akad. Nauk SSSR* **264**, 1378 [*Sov. Phys. Dokl.* **27**, 492].

Tikhodeev, S., Natario, M., Makoshi, K., Mii, T. and Ueba, H. (2001). *Surf. Sci.* **493**, 63.

Tikhodeev, S. G. and Ueba, H. (2004). *Phys. Rev. B* **70**, 125414.

Tikhodeev, S. G. and Ueba, H. (2005). *Surf. Sci.* **587**, 25.

Troisi, A., Ratner, M. A. and Nitzan, A. (2003). *J. Chem. Phys.* **118**, 6072.

Ueba, H. (2003). *Surf. Rev. Lett.* **10**, 771.

Ueba, H. and Persson, B. N. J. (2004). *Surf. Sci.* **566**, 1.

Ueba, H. and Wolf, M. (2005). *Science* **310**, 1774.

Ueba, H., Mii, T. and Tikhodeev, S. G. (2006). *Surf. Sci.*, submitted.

Viljas, J. K., Cuevas, J. C., Pauly, F. and Häfner, M. (2005). *Phys. Rev. B* **72**, 245415.

Walkup, R. E., Newns, D. M. and Avouris, P. (1993). *Phys. Rev. B* **48**, 1858.

Wang, W., Lee, T., Kretzschmar, I. and Reed, M. A. (2004). *Nano. Lett..* **4**, 643.

INDEX